Sustainable Development Goals Series

The **Sustainable Development Goals Series** is Springer Nature's inaugural cross-imprint book series that addresses and supports the United Nations' seventeen Sustainable Development Goals. The series fosters comprehensive research focused on these global targets and endeavours to address some of society's greatest grand challenges. The SDGs are inherently multidisciplinary, and they bring people working across different fields together and working towards a common goal. In this spirit, the Sustainable Development Goals series is the first at Springer Nature to publish books under both the Springer and Palgrave Macmillan imprints, bringing the strengths of our imprints together.

The Sustainable Development Goals Series is organized into eighteen subseries: one subseries based around each of the seventeen respective Sustainable Development Goals, and an eighteenth subseries, "Connecting the Goals", which serves as a home for volumes addressing multiple goals or studying the SDGs as a whole. Each subseries is guided by an expert Subseries Advisor with years or decades of experience studying and addressing core components of their respective Goal.

The SDG Series has a remit as broad as the SDGs themselves, and contributions are welcome from scientists, academics, policymakers, and researchers working in fields related to any of the seventeen goals. If you are interested in contributing a monograph or curated volume to the series, please contact the Publishers: Zachary Romano [Springer; zachary.romano@springer.com] and Rachael Ballard [Palgrave Macmillan; rachael.ballard@palgrave.com].

Ibrahim Nandom Yakubu
Editor

Strategic Approaches to Banking Business and Sustainable Development Goals

Springer

Editor
Ibrahim Nandom Yakubu
School of Business
University for Development Studies
Tamale, Ghana

ISSN 2523-3084 ISSN 2523-3092 (electronic)
Sustainable Development Goals Series
ISBN 978-3-031-80743-5 ISBN 978-3-031-80744-2 (eBook)
https://doi.org/10.1007/978-3-031-80744-2

This Springer imprint is published by the registered company Springer Nature Switzerland AG
The registered company address is: Gewerbestrasse 11, 6330 Cham, Switzerland

Contents

Part I

Sustainable Banking Practices and Green Finance

Banking Business Models and Sustainable Development: Challenges and Opportunities

1

Richard Stephens Cromwell
and James Atta Peprah

1.1 Introduction

Sustainable development is a critical focus for policymakers and scholars because it fosters growth that preserves the environment. Essentially, sustainable development integrates advancements across social, environmental, and economic dimensions. To achieve sustainable development, the United Nations (UN) established the Sustainable Development Goals (SDGs), aiming for completion by 2030. While there are other policies like Africa's Agenda 2063, the European Green Deal, and among others, the SDGs are widely recognized as a key standard for driving social, economic, and environmental progress necessary for social recovery and prosperity. The programme stands out because governments have explicitly committed to engaging countries in a transformative journey towards enhanced social and economic stability. The SDGs represent a paradigm shift that has been embraced by many academics and non-governmental organizations (NGOs), who have examined their principles, advantages, and disadvantages (Fukuda-Parr, 2023). Furthermore, it holds the potential to prioritize human and ecosystem well-being in international relations (Fukuda-Parr, 2023). After the adoption of SDGs in 2015, the UN also implemented an additional policy, that is, Integrated National Financing Frameworks (INFF) in 2018 that aims to mobilize finance to fund the SDGs. These policies have caused various institutions including banks to modify their business model.

A business model outlines how a company creates, delivers, and captures value. It includes the core aspects of the business, such as its purpose, target customers, value proposition, operational processes, revenue streams, and cost structures (Ensign, 2022). Initially, the banking business model (BBM) was the core functions of banks such as accepting deposits, providing loans, and managing risks (Abor, 2023). However, these core functions have been revised to emphasize environmental, social, and economic protection, as well as activities aimed at combating climate change (Wachira et al., 2023). Previously, banks primarily concentrated on extending credit to clients, such as fossil fuel companies, to maximize profits. Now, there is a gradual shift towards financing cleaner energy alternatives (Valls Martínez et al., 2020). Additionally, banks have been recognized as significant contributors to climate finance within the private sector (Buchner et al., 2023). Since the Paris Agreement, financial institutions increased deforestation-linked investments by 40%, provid-

R. S. Cromwell (✉)
Institute of Oil and Gas, University of Cape Coast, Cape Coast, Ghana
e-mail: richard.cromwell@stu.ucc.edu.gh

J. A. Peprah
School of Economics, University of Cape Coast, Cape Coast, Ghana
e-mail: jpeprah@ucc.edu.gh

I. N. Yakubu (ed.), *Strategic Approaches to Banking Business and Sustainable Development Goals*, Sustainable Development Goals Series, https://doi.org/10.1007/978-3-031-80744-2_1

Table 1.1 Banking Business Models

Type of BBMs	Purpose/Activities
Traditional BBM	This model relies on deposits and loans, generating revenue from the interest margin between them
Investment BBM	Investment banks specialize in underwriting, M&A, market making, and trading derivatives, fixed income, FX, commodities, and equities
Universal BBM	Universal banks integrate retail and investment banking, offering comprehensive financial services as a one-stop shop
Private Banking and Wealth Management	This model serves high-net-worth individuals (HNWIs) and ultra-high-net-worth individuals (UHNWIs), offering personalized financial services and investment advice
Digital BBM	Digital banks operate online, leveraging technology to provide efficient, cost-effective banking services without branches
Islamic BBM	Islamic banks operate in compliance with *Sharia* (i.e., Islamic law), which prohibits interest (i.e., *Riba*) and promotes risk-sharing
Sustainable Banking	Emphasizes environmentally friendly practices and supports projects that contribute to sustainability
Ethical Banking	Operates based on ethical values such as transparency, fairness, and social justice

Source: Authors
Note: In reality, most banks do not stick to one business model but may adopt two or more business models based on their objectives

ing US$153.9 billion for forest-risk commodities.[1] This has caused a shift from the core functions of banks to other functions which has given rise to various banking business models (BBMs). The BBMs' direction has prompted discussions on their role and potential contributions to sustainable development. The chapter aims to explore the challenges and opportunities associated with BBMs in promoting sustainable development.

The rest of the chapter is structured as follows: Section 1.2 captures the overview of BBMs whereas Sect. 1.3 discusses the meaning and policies of sustainable development. The link between BBMs and sustainable development is discussed in Sect. 1.4. While Sect. 1.5 discusses the challenges associated with the link between BBMs and sustainable development, Section 1.6 discusses the challenges. The chapter ends with concluding remarks in Sect. 1.7.

1.2 Overview of the Banking Business Models

Unlike some business models which are standardized, BBMs are very difficult to define. For example, the European Banking Authority (2022) described BBMs as fundamentally involving the acceptance of deposits and provision of credit. However, while banks may have shared the same elements, they are increasingly distinct in practice. Ayadi and De Groen (2014) also explain BBMs as mechanisms for banks to achieve their objectives. These mechanisms can vary depending on whether the bank prioritizes stakeholder or shareholder value. According to Hryckiewicz and Kozłowski (2017), BBM is heterogeneous and closely linked to both capital and the financial system. Roengpitya et al. (2014) also indicated that BBMs have three main business models, which include a retail-funded commercial bank, a wholesale-funded commercial bank, and a capital market-oriented bank.

From the above concepts and definitions from various authors, we conclude that the BBMs have evolved (see Table 1.1), with each bank selecting a business model that best aligns with its objectives. BBMs started with the traditional model or the fundamental model which mostly focused on allowing credit and accepting deposits (Abor, 2023; Ensign, 2022). The traditional model relies heavily on physical branches to deliver services. While it serves all customer types (i.e., individuals, businesses, and institutions), it tends to prioritize less risky customers (Valls Martínez et al., 2020). However, various situations and conditions have transformed the

[1] https://www.whistleblowers.org/the-role-of-banking-in--climate-change/. Accessed July 30, 2024.

traditional model into alternative models. For instance, the need to support and facilitate large-scale financial transactions and capital raising, which were crucial for industrial and economic development, gave rise to investment banking. Investment banks evolved from nineteenth-century merchant bankers who handled foreign exchange and project financing.[2] They financed major infrastructure like the Erie Canal and early railroads, inspired by European predecessors. During the post-Civil War, modern investment banking emerged with firms like J.P. Morgan and Goldman Sachs, focusing on public securities and underwriting[4]. The twentieth century saw massive growth, notable deals like U.S. Steel, and regulatory changes, including the Glass-Steagall Act of 1933[4]. Ordinary, investment banking has three main practices such as advisory services,[3] capital market participation,[4] and trading and sales[5](Davis, 2003; Smith, 1986). Some banks began to use both the traditional and investment banking models simultaneously, leading to the development of the universal banking model.

Universal banking began to take shape in the late nineteenth and early twentieth centuries, but its modern form was significantly influenced by regulatory and financial reforms in the 1980s and 1990s (Tilly, 1998). One notable milestone was the Glass-Steagall Act in the United States, enacted in 1933, which initially separated commercial banking from investment banking. However, this separation was effectively dismantled with the Gramm-Leach-Bliley Act of 1999, which allowed banks to offer a broad range of financial services under one roof, marking a key development in the evolution of universal banking. In Europe, universal banking models were more common historically, with institutions like Deutsche Bank and Credit Suisse offering a wide range of services long before these regulatory changes. While universal banking was on the rise, the focus increasingly shifted to its investment aspects. The repeal of Glass-Steagall in 1999 also spurred a surge in investment banking. However, the 2008 financial crisis transformed the industry, leading to a greater emphasis on private banking and wealth management and heightened competition from financial technology (Fintech) firms.

In 2000, 2% of the world's adult population held over 50% of global wealth, with the top one percent holding 40%. In the United States and United Kingdom, neoliberal policies and market booms increased the number of super-rich. High-Net-Worth Individuals (HNWIs) with assets over US$1 million numbered 10.1 million in 2008, with a wealth of US$40.7 trillion (Beaverstock, 2011). Banks seized the opportunity to address this gap by offering a business model known as private banking and wealth management. While private banking is personalized banking and financial services provided to HNWIs and UHNWIs by private banks or specialized financial institutions, wealth management is comprehensive financial planning and investment management services aimed at sustaining and growing a client's wealth over time (Beaverstock, 2011; Dechtman, 2024). During these periods, financial technology (Fintech) started to gain attention with some innovations such as mobile money,[6] Bitcoin,[7] and some digital wallets.[8] Many customers were satisfied with some of these innovations, as they could access financial services from the comfort of their homes (Navaretti et al., 2018). This motivated the banking sector to incorporate these Fintech innovations into their operations to retain existing

[2] https://vault.com/industries/investment-banking/background. Accessed July 31, 2024.

[3] Advisory services include the provision of strategic advice for M&A, restructuring, and other corporate finance activities.

[4] Assisting clients in raising capital through equity and debt issuance.

[5] Engaging in market-making activities, proprietary trading, and brokerage services.

[6] Mobile money allows users to store, send, and receive money via mobile phones, providing financial services in underbanked regions.

[7] Bitcoin is a decentralized digital currency that operates without a central authority or intermediaries, using blockchain technology for secure transactions.

[8] Digital wallets, or e-wallets (e.g., Google Wallet and Apple Pay), are electronic systems that allow users to store payment information, make electronic transactions, and manage financial activities online.

customers and attract new ones (Iddrisu, 2024). Some countries such as the United States, United Kingdom, China, and others even established digital banks[9] that have few or no branches. In Ghana, nearly all banks now offer mobile money services, and Fintech services have garnered significant global attention.

Although Islamic banking has roots dating back to the 1960s, its growth accelerated significantly as institutions[10] expanded into non-Muslim-majority countries and developed sophisticated financial products like *sukuk* (i.e., Islamic bonds) and *takaful* (i.e., Islamic insurance) (Abdullah Saif Al Nasser et al., 2013; Dwi Sari, 2016). Islamic banking principles are based on the teachings of the Quran and *Hadith*, which emphasize justice, fairness, and the prohibition of interest (i.e., *riba*) (Abdullah Saif Al Nasser et al., 2013; Dwi Sari, 2016). Traditional banking often prioritized highly profitable and lower-risk customers, thereby neglecting marginalized sectors such as Small and Medium Enterprises (SMEs), women, and low-income individuals. As a result, advocacy from religious groups and development partners led to the concept of ethical banking, where banks finance projects aligned with ethical principles and serve marginalized sectors. This business model gained attention after the establishment of some early ethical banks.[11]

The effort to combat climate change and promote sustainable development has led the banking sector to adopt an additional business model. This business model is sustainable banking, which includes climate and green banking practices. Sustainable banking refers to financial practices and business models that incorporate environmental, social, and governance (ESG) considerations into banking operations and decision-making (Abor, 2023; Wachira et al., 2023). This business model has led to the establishment of green banks.[12] It also incorporates other business models, such as digital and ethical banking, into its operations. Currently, this form of business model is gaining traction, with increasing interest from customers and businesses, especially in developed countries, in how banks can operate sustainably. Central banks and financial regulators are also implementing policies to encourage sustainability in banking activities. Additionally, development partners prefer to collaborate with banks that incorporate sustainability into their operations to support national development and execute projects effectively. These reasons have made some banks redefine their business models.

In contrast to traditional banking where revenue is mainly obtained from interest income (i.e., spread between interest on loan and deposit), other BBMs obtained revenue from other sources like grants, bonds, rewards, other fees and fines, inter alia.[13] Examining the various discussed forms suggests that BBMs emerge in response to community issues and needs. Therefore, with the growing focus on climate change and sustainable development, it is crucial to identify how BBMs can contribute to achieving these sustainable development.

1.3 Sustainable Development: Meaning and Policies

1.3.1 Meaning of Sustainable Development

Sustainable development is defined in various ways by different authors, but there are common themes including development, the environment, equity, and society. Scholars of sustainable development differ in their emphasis on what should be sustained, what should be developed, how to link environmental and developmental goals, and

[9] Some of these banks are Chime, N26, Revolut, Monzo, Starling Bank, Ally Bank and Simple.

[10] Islamic banks in non-Islamic countries are Al Rayan Bank (UK), Bank of London and the Middle East (UK), Dubai Islamic Bank (USA), HSBC's Islamic Banking Division (USA), KT Bank AG (Germany), Baraka Bank (operate in France), Rothschild & Co (Australia).

[11] Early ethical banks are ShoreBank in 1973 (US), Triodos Bank in 1980 (Netherlands), Cooperative Bank in 1997 (England), Banca Popolare in 1995 (Italy) and JAK Medlemsbank in 1997 (Sweden)

[12] Some examples include the Montgomery County Green Bank, New York Green Bank, ING Bank, and IBank

[13] https://www.thebanker.com/Sustainable-Banking-Revenues-Ranking-2023-1696233648. Accessed August 1, 2024.

the duration over which sustainable development should be achieved. The well-known definition by the World Commission on Environment and Development (WCED) (1987) describes sustainable development as the kind of development that satisfies the needs of the current generation without jeopardizing the ability of future generations to meet their own needs. Sustainable Footprint also explains that central themes in sustainable development are ecological responsibility, economic efficiency, and social well-being. The potential goals of sustainable development include enhancing economic efficiency, protecting and restoring ecological systems, and improving human well-being, either individually or in combination.[14] The Commission on Sustainable Development (CSD) of the United Nations defines sustainable development as the capacity to fulfil human needs for food, water, and health, while also providing opportunities for education, employment, and participation.

The concept of sustainable development dates back to 1713, when the emphasis was placed on forestry sustainability by harvesting only regrown timber to preserve soil fertility (Mannan, 2012). To promote sustainable development, the U.S. government implemented the National Environmental Policy Act (NEPA) in 1969 (Rome, 2003). This legislation was influenced by advocacy aimed at protecting the environment and conserving resources (see, e.g., Carson, 1962). NEPA sought to ensure that federal agencies carefully evaluated the environmental impacts of their actions and considered environmentally friendly alternatives. The focus moved from national to global concerns with the 1972 Stockholm Declaration on the Human Environment, which advanced sustainable development by highlighting the importance of global environmental protection and human welfare. As one of the first international agreements to acknowledge the link between environmental health and human development, the Declaration set the stage for incorporating environmental considerations into economic and social planning. It advocated for policies that harmonize environmental conservation with developmental needs and provided a foundation for future international environmental agreements and conventions. The concept of sustainable development was not explicitly defined in the Declaration until 15 years later when the WCED report fully articulated the idea of sustainable development.

Following the WCED report, the concept of sustainable development became more comprehensive. The 1992 United Nations Conference on Environment and Development (UNCED) Earth Summit in Rio de Janeiro, known as the Earth Summit, resulted in Agenda 21 (i.e., a global action plan for sustainable development) as well as the Statement of Principles for Sustainable Management of Forests and the Rio Declaration on Environment and Development.[15] Over 178 governments participated, focusing on post-Cold War development issues. Mechanisms established to support this included the Commission on Sustainable Development, the Inter-agency Committee on Sustainable Development, and the High-level Advisory Board on Sustainable Development. The Rio Declaration outlined 27 principles for achieving sustainable development. Two years later, the 1994 Global Conference on the Sustainable Development of Small Island Developing States in Barbados produced the Barbados Programme of Action (BPOA), addressing environmental and social vulnerabilities faced by islands.[16] In 2000, the United Nations Millennium Declaration (UNMD) committed global leaders to combating poverty, hunger, disease, illiteracy, environmental degradation, and gender discrimination.[17] The Millennium Development Goals (MDGs) emerged from this, encompassing eight goals. However, the MDGs had limitations, particularly in governance, transparency, participation, and human rights. These issues were addressed in the SDGs, which incorporate the full scope of sus-

[14] https://sustainablefootprint.org/about-this-project/sustainable-footprint/. Accessed August 2, 2024.

[15] https://publications.gc.ca/Collection-R/LoPBdP/BP/bp317-e.htm#:~:text=From%203%2D14%20June%201992,environment%20in%20a%20political%20context. Accessed August 3, 2024.

[16] https://www.un.org/en/conferences/small-islands/bridgetown1994. Accessed August 4, 2024.

[17] https://www.ohchr.org/en/instruments-mechanisms/instruments/united-nations-millennium-declaration. Accessed August 4, 2024.

Table 1.2 Major Regional Sustainable Development Policies

Institution	Policies	Brief Description
African Union	Agenda 2063	A strategic framework for Africa's socio-economic transformation by 2063. It emphasises inclusive growth, sustainable development, and continental integration, with aspirations including economic prosperity, good governance, and environmental sustainability.
Arctic Council	Sustainable Development Strategy for the Arctic Region	A strategy to promote sustainable development in the Arctic region, addressing challenges such as climate change, environmental protection, and the sustainable use of resources while respecting indigenous peoples' rights.
Association of Southeast Asian Nations (ASEAN)	ASEAN Socio-Cultural Community (ASCC) Blueprint 2025	A blueprint outlining goals for fostering a cohesive, inclusive, and sustainable ASEAN community by enhancing regional cooperation in areas such as social welfare, environmental sustainability, and cultural preservation.
Caribbean Community (CARICOM)	Regional Framework for Achieving Development Resilient to Climate Change	A framework designed to enhance the resilience of Caribbean nations to climate change through regional cooperation, disaster risk management, and sustainable development initiatives
Central American Integration System (SICA)	Central American Strategy for Sustainable Development (ECAD)	A regional strategy aimed at promoting sustainable development in Central America through integrated policies on economic growth, environmental management, and social equity
European Union	European Green Deal	It aims for climate neutrality by 2050, focusing on reducing emissions, promoting sustainable industries, and enhancing environmental protection through clean energy and biodiversity initiatives.
North America	NAFTA Environmental Cooperation Agreement	An agreement focused on environmental cooperation and enforcement among the United States, Canada, and Mexico, aiming to ensure that trade and environmental policies support each other.
Nordic Council	Strategy on Sustainable Development for the Arctic	A strategy focusing on promoting sustainable development in the Arctic region, including environmental conservation, sustainable resource management, and support for indigenous communities.
Pacific Islands Forum	Pacific Plan for Strengthening Regional Cooperation and Integration	A framework aimed at enhancing regional integration and cooperation among Pacific Island countries, focusing on sustainable development, economic growth, and improved governance.
South America	Mercosur Green Protocol	A regional agreement within the Southern Common Market (Mercosur) to promote sustainable development practices, enhance environmental protection, and support green trade and production.
UNESC for Asia and the Pacific (ESCAP)	2030 Agenda for Sustainable Development	A regional strategy aimed at promoting sustainable development in Central America through integrated policies on economic growth, environmental management, and social equity

Source: Authors
Note: NAFTA is the North America Free Trade Agreement; UNESC is the United Nations Economic and Social Commission

tainable development.[18] Today, sustainable development is essential for nearly every region and country, with regional and national leaders actively working to implement policies that advance it. Many of these policies are built upon or have their foundation on the SDGs.

[18]https://www.sdgfund.org/mdgs-sdgs. Accessed August 4, 2024.

1.3.2 Policies of Sustainable Development

Various sustainable development policies have been revised to address different aspects of sustainability. However, for the purpose of this chapter, we discussed only global policy since the regional policies (see Table 1.2) either align with SDGs or as a way to achieve SDGs.

Sustainable Development Goals

The SDGs build on decades of global efforts, including the 1992 Earth Summit's Agenda 21, which aimed to improve lives and protect the environment. The 2000 Millennium Declaration led to eight MDGs targeting extreme poverty. The 2002 Johannesburg Declaration and Plan of Implementation reaffirmed commitments to poverty eradication and environmental protection. At the 2012 Rio + 20 Conference, "The Future We Want" led to the development of the SDGs, culminating in the adoption of the 2030 Agenda in September 2015. Key agreements in 2015 also included the Sendai Framework, the Addis Ababa Action Agenda, and the Paris Agreement. The UN's Division for SDGs supports the implementation and evaluation of the SDGs. The SDGs consist of 17 goals (see Table 1.3), supported by 169 targets and 232 indicators. These components are designed to monitor global, regional, and national progress, offering an evidence-based framework for tracking advancements over 15 years. Each year, the UN Secretary-General releases an annual SDG Progress Report, created in collaboration with the UN System. This report relies on the global indicator framework and data from national statistical systems and regional information. Additionally, the Global Sustainable Development Report, published every 4 years, contributes to the quadrennial SDG review discussions at the General Assembly. This report is prepared by an Independent Group of Scientists appointed by the Secretary-General.

The progress on SDGs indicates that the COVID-19 pandemic has some level of effect on its success. For instance, the pandemic and other factors increased extreme poverty by 23 million and hunger by 123 million in 2022 compared to

Table 1.3 Sustainable Development Goals

Goal	Definition
1. No Poverty	End poverty in all its forms everywhere.
2. Zero Hunger	End hunger, achieve food security and improved nutrition, and promote sustainable agriculture.
3. Good Health and Well-being	Ensure healthy lives and promote well-being for all at all ages.
4. Quality Education	Ensure inclusive and equitable quality education and promote lifelong learning opportunities for all.
5. Gender Equality	Achieve gender equality and empower all women and girls.
6. Clean Water and Sanitation	Ensure availability and sustainable management of water and sanitation for all.
7. Affordable and Clean Energy	Ensure access to affordable, reliable, sustainable, and modern energy for all.
8. Decent Work and Economic Growth	Promote sustained, inclusive, and sustainable economic growth, full and productive employment, and decent work for all.
9. Industry, Innovation, and Infrastructure	Build resilient infrastructure, promote inclusive and sustainable industrialization, and foster innovation.
10. Reduced Inequality	Reduce inequality within and among countries.
11. Sustainable Cities and Communities	Make cities and human settlements inclusive, safe, resilient, and sustainable.
12. Responsible Consumption and Production	Ensure sustainable consumption and production patterns.
13. Climate Action	Take urgent action to combat climate change and its impacts.
14. Life Below Water	Conserve and sustainably use the oceans, seas, and marine resources
15. Life on Land	Protect, restore, and promote sustainable use of terrestrial ecosystems, manage forests sustainably, combat desertification, and halt and reverse land degradation and halt biodiversity loss
16. Peace, Justice, and Strong Institutions	Promote peaceful and inclusive societies for sustainable development, provide access to justice for all, and build effective, accountable, and inclusive institutions at all levels.
17. Partnerships for the Goals	Strengthen the means of implementation and revitalize the Global Partnership for Sustainable Development.

Source: United Nations (2015)

2019, exacerbating income inequality between rich and poor countries.[19] The global working poverty rate decreased from 8.4% in 2015 to 6.9% in 2023, but 241 million remain in extreme poverty[21]. Global health progress has slowed, education remains concerning with only 58% of students meeting reading proficiency, and gender equality goals are still far from being achieved, with persistent violence against women and girls[21].

While progress on some goals has been slow, others have experienced significant advances. In 2023 for example, around 67% of the global population, or 5.4 billion people, were online[21]. This marks a 4.7% increase from 2022, which is higher than the 3.5% growth observed from 2021 to 2022.

1.4 Banking Business Models and Sustainable Development

The BBMs can promote sustainable development by integrating sustainability principles into its core strategies, operations, and financial practices. Here are several ways the BBMs can support and drive sustainable development.

1.4.1 Financing Green Projects

Some BBMs such as ethical and sustainable BBM prioritize green financing by offering loans and investments specifically for environmentally friendly projects, such as renewable energy, energy efficiency, and sustainable agriculture. For instance, the green bond market in China stands out globally due to the significant role played by commercial banks. In the years 2016, 2017, and 2018, commercial banks were responsible for issuing a substantial majority of green bonds in China, accounting for 77%, 62%, and 68% of all green bonds issued, respectively (Cao et al., 2021). Allocating financing to green projects contributes to the achievement of SDGs 1, 2, 7, 13, and 17. Additionally, it generates employment opportunities and fosters economic growth, as outlined in SDG 8.

1.4.2 Risk Assessment Using Environmental, Social, and Governance Criteria

Sustainable BBM integrates ESG criteria into the bank's risk assessment and decision-making processes which help to ensure that investments and loans support sustainable outcomes. This involves ESG screening[20] and Sustainable Investment Funds.[21] Banks can evaluate whether projects funded through loans or investments have a positive environmental impact. For instance, a bank may prioritize funding for renewable energy projects (solar, wind) over fossil fuel projects due to their lower environmental footprint. This help prevent greenwashing[22] and corruptible acts by the borrowers.

1.4.3 Innovation and Product Development

Most of the non-conventional BBM present opportunities for banks to develop innovative financial products that align with SDGs. Some of these innovative products are green mortgages,[23] green deposit,[24] sustainable savings accounts,[25]

[19] https://unstats.un.org/sdgs/files/report/2024/SG-SDG-Progress-Report-2024-advanced-unedited-version.pdf. Accessed August 4, 2024.

[20] Assessing potential investments for their environmental and social impact before making decisions.

[21] Creating and promoting investment funds that focus on companies with strong ESG performance.

[22] Greenwashing refers to the practice where companies, organizations, or financial institutions present an exaggerated or misleading impression of their environmental efforts or sustainability practices to appear more eco-friendly than they truly are.

[23] Green mortgages are specialized financial products designed to support the financing of energy-efficient and environmentally sustainable housing.

[24] Green deposits are a type of banking deposit that allows individuals and businesses to invest their funds in environmentally sustainable and socially responsible projects.

[25] Sustainable savings accounts are financial products designed to align with ESG principles, focusing on promoting sustainability through banking practices.

green credit cards[26] and others. Introducing innovative financial products opens new opportunities for private sector entities committed to environmental protection but seeking secure investment channels for their funds. Additionally, these products provide banks with alternative revenue streams, enhancing their profitability and increasing shareholder value. For instance, DBS Bank has launched the Green Deposit, a pioneering product designed to support SDGs (see Box 1.1).

Box 1.1: DBS Bank India Launched Its Green Deposit Programme

On March 15, 2022, DBS Bank India launched its Green Deposit Programme, aimed at corporate clients seeking to invest in environmentally sustainable projects. This initiative aligns with the bank's commitment to integrating Sustainable Development Goals (SDGs) into its financial offerings, complementing sustainable lending and trade credit solutions. The Green Deposit Programme supports green industries and initiatives, including green buildings, sustainable water management, renewable energy, and clean transportation. It combines the benefits of traditional fixed deposits with a focus on funding environmentally friendly projects, offering corporates a way to incorporate sustainability into their treasury operations. Divyesh Dalal, Managing Director and Head of Global Transaction Services at DBS Bank India, highlighted the program's role in advancing ESG-friendly practices and supporting SDGs. Green Deposits provide three main advantages: they showcase a commitment to environmental sustainability, allow investment in projects with positive environmental impacts, and offer stable returns without compromising deposit security. DBS Bank has a track record in green finance, evidenced by its 2021 partnership with CapitaLand to place its first green deposit and a total of INR 1546 Crores in outstanding green loans in India. The bank is also involved in scaling sustainable finance across Asia, including collaborating with Inditex to finance organic cotton procurement for India's cotton industry.

Source: DBS (2024)

Another example is the introduction of green fixed deposit by IndusInd Bank where these deposits will support sectors such as energy efficiency, renewable energy, green transport, and waste management (see Box 1.2).

Box 1.2: IndusInd Bank India Launched Its Green Fixed Deposit

On December 28, 2021, IndusInd Bank launched its "Green Fixed Deposits," where deposit proceeds are allocated to projects and firms supporting the United Nations Sustainable Development Goals (SDGs). This initiative integrates SDGs into a traditional fixed deposit product, available to both retail and corporate customers. The funds will support sectors such as energy efficiency, renewable energy, green transport, and waste management. The Green Fixed Deposits offer competitive interest rates, with an additional 50 basis points for senior citizens. Depositors receive a "Green" certificate and an 'Assurance' certificate confirming the use of funds. This launch reflects IndusInd Bank's dedication to sustainable economic growth and value creation for stakeholders.

Source: IndusInd Bank (2021)

[26]Green credit cards are financial products designed to support environmental sustainability and social responsibility through various features and benefits.

1.4.4 Promoting Inclusive Finance

One of the effective ways to reduce poverty is through financial inclusion (Demirguc-Kunt et al., 2018; Doku et al., 2023; World Bank, 2024). However, banks have promoted high level of financial inclusion through the various BBMs. Through BBMs like traditional, digital, and ethical banking, banks have established more bank branches including some villages; use digital transaction like mobile money, Automated Teller Machines (ATMs), and mobile wallets; and extend financial services to the marginalized areas. Promoting financial inclusion not only supports the development of business ideas through targeted financing but also encourages savings and job creation. By funding new ventures, banks can create employment opportunities, including positions for branch staff and security personnel who protect ATMs. Additionally, the emphasis on financial inclusion drives investments in infrastructure, such as ICT, reflecting the increasing reliance on technology in banking to enhance financial access and services. Banks have increase financial inclusion in Uganda through various BBMs (see Box 1.3).

Box 1.3: Banking Business Model Has Increased Financial Inclusion in Uganda

Uganda's financial sector, predominantly composed of banks which hold 83% of the sector's assets, has experienced significant transformation over the past two decades due to financial innovations. From 2000 to 2010, advancements included ATMs, mobile banking, internet banking, and payment cards. Between 2011 and 2020, the focus shifted to leveraging technology for new products and processes, enhancing market reach and risk management. Major payment service providers introduced self-service kiosks for bill payments, POS transactions, and e-money services across various sectors. Financial inclusion in Uganda has notably improved, with access to financial services rising from 58% in 2009 to 78% in 2018. Formal account ownership increased from 52% in 2013 to 66% in 2021, with a notable rise in female account ownership from 53% in 2017 to 65% in 2021, reducing the gender gap from 13% in 2014 to 2% in 2021.

Source: Bank of Uganda and Ministry of Finance (2023)

1.4.5 Partnerships and Collaboration

The uniqueness of BBMs positions banks as key partners in developmental projects, fostering collaboration with multilateral agencies and NGOs. Banks' ability to handle large-scale financial operations and their extensive networks make them crucial for implementing projects aimed at economic growth and social improvement. Multilateral agencies like the World Bank and the IMF utilize banks' infrastructure to efficiently channel funds for poverty alleviation, infrastructure, and health care. Similarly, NGOs benefit from banks' financial tools and channels, enhancing their ability to execute initiatives such as microfinance, education, and environmental conservation. This partnership maximizes the impact of developmental programmes and advances SDGs. For example, European Bank for Reconstruction and Development (EBRD) and Commercial International Bank (CIB) partnered to finance green investments and women-led businesses in Egypt (see Box 1.4).

Box 1.4: EBRD and Commercial International Bank Sign US$ 60 Million Financing Agreement

The European Bank for Reconstruction and Development (EBRD) and Commercial International Bank (CIB) have signed a US$60 million (€55.9 million) financing agreement to support green investments and women-led businesses in Egypt. This package includes a US$50 million Green

Economy Financing Facility (GEFF II), with US$7.5 million co-financed by the Green Climate Fund (GCF). The GEFF II will fund technologies and services for climate change mitigation and adaptation, benefiting private businesses across agribusiness, manufacturing, logistics, and ICT sectors. Additionally, up to €5.2 million in investment incentive grants from the European Union (EU) will support local micro, small, and medium-sized enterprises (MSMEs) in adopting greener technologies and developing resilient supply chains for green equipment. This is the third GEFF II credit line provided in Egypt to foster a greener, more inclusive economy. CIB will implement the Green Technology Selector, a tool from the EBRD to promote collaboration among technology providers, increasing the availability of climate technologies locally. The EBRD has already invested over US$240 million in Egypt through the GEFF program to support energy efficiency, climate adaptation, and circular economy initiatives. The second component of the financing package is a US$10 million EBRD Women in Business loan aimed at enhancing female participation in the economy. This initiative will offer financial, advisory, and mentoring support to women-led businesses with fewer than 250 employees and less than €50 million in annual turnover, covering various sectors. The agreements were signed at the EU Investment Conference in Cairo. EBRD's First Vice President emphasized the commitment to supporting Egypt's transition to a greener economy and enhancing women's business participation. CIB's Deputy CEO highlighted the bank's dedication to providing sustainable and inclusive financial solutions. Since 2012, the EBRD has invested over €12 billion in Egypt, focusing on sectors such as finance, agribusiness, and infrastructure.

Source: Zgheib (2024)

1.4.6 Capacity Building and Education

Education and capacity building are crucial elements in the BBMs (Abor, 2023; European Banking Authority, 2022; Nosratabadi et al., 2020). These activities enable banks to provide guidance on sustainable practices to clients, employees, and the wider community. By imparting knowledge about sustainability and its relevance to financial decisions, banks empower stakeholders to make informed choices that support sustainable development. This proactive approach helps integrate environmental and social considerations into financial strategies, fostering a culture of sustainability. Enhanced understanding leads to more responsible decision-making and drives positive change, aligning financial activities with broader goals of sustainable development and contributing to a more resilient and equitable economy. For example, Access Bank Ghana Plc train SMEs on the essence and use of digital finance products (see Box 1.5).

Box 1.5: Access Bank Ghana Plc Provide Education and Capacity Building for SMEs

Access Bank (Ghana) Plc organises its fifth SME engagement workshop in Takoradi, concentrating on digitalization for business growth in partnership with Open Labs. The Zonal Head for Business Banking emphasized the bank's commitment to SME development through various engagements, recognizing the significant impact SMEs have on Ghana's economy. He highlighted research indicating that 20% of SMEs fail within their first year and 50% within 5 years due to funding and technical knowledge gaps. To mitigate these issues, Access Bank has hosted capacity-building workshops in Accra and Kumasi, benefiting over 6000 SMEs by offering business advisory services and loan facilities to boost profitability. The Campus Head for Open Labs, who facilitated the workshop, emphasized the importance of digital technology for

(contined)

Box 1.5: (continued)

business growth, noting that many SMEs miss out on technological advantages due to a lack of education. SMEs were informed that smartphones, often used for basic activities, could greatly enhance business knowledge and innovation if utilized effectively. Additionally, SMEs were encouraged to adopt digital technology to remain competitive and successful. Participants were introduced to Access Bank's digital banking services, including AccessAfrica, Access Pay, Merchant Pay, POS, and Access Cards, designed to improve business operations. They also received guidance on leveraging targeted internet surfing for business growth through digitalization.

Source: Access Bank Ghana Plc (2024)

1.5 Challenges of Banking Business Models on Sustainable Development

The intersection of the BBMs and sustainable development presents both opportunities and challenges. The BBMs, which encompasses the strategies, operations, and financial structures banks use to generate revenue, must adapt to support sustainable development effectively. The following are some key challenges.

1.5.1 Short-Term Profit Focus

In practice, banks frequently utilize a mix of BBMs, combining both modern and traditional approaches. For instance, a bank may integrate cutting-edge digital banking solutions with traditional lending practices. While these banks can offer innovative products and services, they may also prioritize short-term profitability and shareholder returns, which can impact their investment strategies.[27] Sustainable development projects, such as renewable energy initiatives or green infrastructure, typically require extended time horizons to achieve financial returns and positive environmental impacts. Banks that focus primarily on immediate financial gains may overlook these long-term investments. For example, a bank with a strong focus on quarterly profits might favour financing high-yield projects or sectors with quicker returns, such as fossil fuel extraction, over funding sustainable projects like wind farms or energy-efficient housing. This short-term focus can undermine broader sustainability goals and limit the bank's contribution to long-term environmental and social benefits.

1.5.2 Gradual Transition

As previously mentioned, BBMs have evolved over time, and some banks are gradually transitioning from traditional banking to modern BBMs. However, many of these banks are still in the process of fully integrating ESG principles into their operations, despite their ongoing efforts. Consequently, some banks continue to invest heavily in fossil fuels rather than non-fossil fuel alternatives, or maintain business relationships with existing customers who may not prioritize sustainable development. This may lower investment towards sustainable development.

For instance, following the Paris Agreement in 2016, the world's 60 largest private banks have directed US$6.9 trillion towards fossil fuel financing, with almost half (US$3.3 trillion) allocated to expanding fossil fuel activities (see Box 1.6).

Box 1.6: Some Major Banks Still Invest in Fossil Fuels

The 15th annual Banking on Climate Chaos (BOCC) report employs a new and expanded dataset that credits each bank involved in a deal rather than only those in leading roles. This report, which aims to cut through greenwashing, examines the top 60 banks' lending and underwriting to over 4200 fossil fuel companies, as well as their financing of companies causing degradation in the Amazon

[27] https://zoetalentsolutions.com/balancing-short-term-profitability-with-long-term-sustainability-in-business-management/. Accessed August 5, 2024.

and Arctic. Since the Paris Agreement in 2016, the world's 60 largest private banks have financed fossil fuels with US$6.9 trillion, with nearly half (US$3.3 trillion) allocated towards fossil fuel expansion. In 2023 alone, banks provided US$705 billion in fossil fuel financing, with US$347 billion directed towards expansion. JPMorgan Chase emerged as the leading financier of fossil fuels in 2023, committing US$40.8 billion, and ranking first in fossil fuel expansion. Mizuho was the second largest financier, with US$37.0 billion in fossil fuel financing and US$18.8 billion for expansion. Citibank has been the worst funder of fossil fuel expansion since the Paris Agreement, providing US$204 billion since 2016. Some banks have recently increased their climate risk exposure by weakening their policies. Bank of America, which ranks third on the 2023 list of worst fossil fuel funders, exemplifies this trend by removing exclusions on Arctic drilling, thermal coal, and coal-fired power plants, lacking energy ratio disclosures and near-term emission targets, and abandoning the Equator Principles. This year's report methodology has been significantly adjusted to include more primary sources and to track bank participation in corporate finance deals comprehensively. Each bank's financial contributions to a deal are now exposed, and banks were contacted to confirm and review the deals attributed to them. The report highlights high bank financing for the most climate-damaging fossil fuel practices in 2023. CIBC, RBC, and Scotiabank each committed US$523 million to tar sands extraction, MUFG financed US$512 million in ultra-deepwater offshore drilling, JPMorgan Chase provided US$6 billion for fracking, and CITIC backed coal mining with US$7.6 billion. Furthermore, UniCredit committed US$265 million to Arctic drilling, and Bank of America funded US$162 million in oil and gas extraction in the Amazon biome.

Source: RainForest Action Network et al. (2024)

1.5.3 Geographic Location of Banks

When banks adopt BBMs such as digital banking, ethical banking and sustainable banking, it is anticipated that they will redirect funds towards sustainable development projects. However, the geographic location of these banks may limit their ability to reallocate such funds or fully implement these BBMs. For instance, banks in certain developing countries, such as those in Africa, may face challenges in effectively implementing a BBM like sustainable banking due to the underdeveloped financial sectors and weak governance and institutions. In contrast to their counterparts in Asia, where commercial banks actively issue green bonds to businesses (see Cao et al., 2021), only a few banks in Africa engage in issuing green bonds, and these are primarily directed towards governments rather than businesses. This limitation hinders their ability to trade in green bonds and fully adopt sustainable BBMs. Banks located in regions where clients lack understanding of sustainable development and its associated goals may struggle to be taken seriously by customers, particularly if these banks practice modern BBMs that have more impact on sustainable development.

1.5.4 Challenges in Aligning Products with Sustainability Goals

Developing innovative financial products that support sustainable development, such as green bonds, sustainability-linked loans, and impact investing, involves a significant investment in research and development. This process requires banks and financial institutions to conduct thorough research to understand the evolving needs of the market and the specific requirements of sustainable projects. For instance, creating green bonds necessitates assessing the environmental impact of potential investments and ensuring that the funds are allocated to projects with verified environmental benefits. Similarly, sustainability-linked loans require the development of metrics to link loan conditions with the borrower's performance on sustainability goals. Impact investing involves designing investment strategies that balance financial returns with measurable social or environmental impact, which requires comprehensive analysis and monitoring. The research and development (R&D) process not only involves understanding regulatory and market dynamics but also requires collaboration with experts, stakeholders, and technology providers to develop effective solutions. This commitment to R&D ensures that financial products not only meet regulatory standards but also align with broader sustainability goals, ultimately fostering a more responsible and impactful financial ecosystem.

1.5.5 Stakeholder Pressure

In countries that are aligned with sustainability principles, banks face increasing demands from a wide range of stakeholders, including customers, investors, and activists, to prioritize sustainability in their operations and financial offerings. Customers are progressively seeking banks that commit to environmentally and socially responsible practices, while investors are calling for enhanced transparency and a stronger focus on sustainable investments. Activists and advocacy groups further pressure banks to align their strategies with broader environmental and social objectives. Balancing these expectations with the need to maintain profitability and competitiveness is a challenging task. Banks must manage the delicate balance between advancing sustainability initiatives and ensuring strong financial performance, as failing to meet these demands can lead to serious reputational damage and loss of business. For example, a bank that disregards sustainability may encounter public criticism, loss of customer trust, and diminished investor confidence, potentially harming its market position and profitability. Thus, embedding sustainability into the core business strategy is essential not only for addressing stakeholder expectations but also for ensuring long-term success and retaining a competitive edge in a rapidly evolving financial landscape.

1.5.6 Misaligning Objectives with Partners

Sustainable development frequently necessitates collaboration among banks, governments, nonprofits, and other key stakeholders, each bringing unique perspectives and expertise to the table. However, forging effective partnerships can be challenging due to differences in objectives, priorities, and operational practices. Banks might focus on financial returns and risk management, while governments may prioritize policy implementation and regulatory frameworks, and NGOs often emphasize social and environmental impact. To overcome these challenges and achieve successful collaboration, it is crucial to establish clear communication channels and ensure that all parties align their goals and strategies. This involves setting common objectives, defining roles and responsibilities, and fostering a shared commitment to sustainability outcomes. Effective partnerships also require ongoing dialogue and flexibility to address and reconcile divergent interests, as well as mechanisms for evaluating progress and adapting strategies as needed. When these elements are in place, collaborative efforts can leverage the strengths of each partner to drive meaningful and sustainable development outcomes, creating a more cohesive approach to addressing complex environmental and social challenges.

1.6 Opportunity of Banking Business Models on Sustainable Development

Although challenges may hinder the ability of BBMs to promote sustainable development, several opportunities also arise.

1.6.1 Implementation of Policies

Globally, various regions and countries are actively advancing policies that foster sustainable development, which increasingly include incentives for banks to adopt sustainable banking practices. For example, the European Union has implemented the EU Sustainable Finance Action Plan, which includes regulations such as the EU Taxonomy and the Sustainable Finance Disclosure Regulation (SFDR). These regulations require banks to integrate ESG criteria into their investment decisions and disclose the sustainability impacts of their financial products. In Asia, countries like Japan and China have also made significant strides. Japan's Financial Services Agency introduced the Stewardship Code and Corporate Governance Code, encouraging banks and investors to focus on long-term sustainability and responsible investment.[28] China has incorporated sustainability into its broader economic plans, such as the 14th Five-Year Plan, which promotes green finance and supports banks in financing renewable energy projects and green bonds.[29]

Similarly, in the United States, the government in 2023 paid attention to the importance of sustainable finance through initiatives like the Green Bond Standard and the establishment of a Climate Finance Task Force.[30] These policies aim to guide financial institutions towards investments that support climate resilience and sustainable growth. These policies and initiatives collectively drive banks to align their operations with sustainability goals, offering various tools and frameworks for integrating ESG considerations into financial practices. By doing so, they not only contribute to global sustainability targets but also position themselves competitively in a market increasingly driven by environmental and social responsibility.

Box 1.7: National Development Banks

National Development Banks (NDBs) are government-established institutions focused on national economic development and policy goals. Unlike commercial banks, NDBs address market failures through targeted financial interventions, funding infrastructure, industrial development, social programs, and underfunded areas like renewable energy and affordable housing. They stabilize economies by investing in strategic sectors and mitigating economic downturns. NDBs align with national policies and often collaborate with the private sector to co-finance projects. Notable examples include the Development Bank of Latin America (CAF), the Asian Development Bank (ADB), and the European Investment Bank (EIB). National DFIs, known as NDBs, provide more funds than other development finance institutions (DFIs) and public climate finance sources. From 2012 to 2020, NDBs contributed about 44% of total public climate finance, peaking at 60% in 2017. Therefore, climate finance contributes a lot to the achievement of SDGs 13 and 14.

Sources: Buchner et al. (2023) and Giglio et al. (2021)

1.6.2 The Establishment of More National Development Banks

In the past, only a limited number of countries had established National Development Banks (NDBs) (see Box 1.7). However, there has been a notable increase in the number of countries now engaging in the creation of NDBs. These institutions play a crucial role in addressing the maturity mismatches that often exist between the short-term focus of commercial banks and the long-term nature of sustainable development projects. By specializing in this area, NDBs provide an essential bridge that facilitates investment in sustainable development. As more NDBs are established, they can support commercial banks that are typically risk-averse and focused on short-term profits and shareholder wealth. This support enables these banks to invest in sustainable development initiatives, thereby aligning financial practices with broader economic and environmental goals.

[28] https://www.fsa.go.jp/en/refer/councils/stewardship/20200324/01.pdf. Accessed August 5, 2024.

[29] http://english.eximbank.gov.cn/News/WhitePOGF/202301/P020230106613030856905.pdf. Accessed August 5, 2024.

[30] https://www.state.gov/progress-report-on-president-bidens-climate-finance-pledge/. Accessed August 5, 2024.

1.6.3 Increasing Awareness of Sustainable Development

Academics, multilateral development banks (MDBs), other development partners, some policymakers NGOs, and other stakeholders are constantly providing education and awareness on the need to pursue sustainable development. For example, some environmental NGOs are constantly encouraging banks to reduce investment in projects that do not lead to sustainable development (see Box 1.8).

Box 1.8: Environmental NGOs Urge Banks to Limit Unsustainable Investments

Bond financing is heavily relied upon by TotalEnergies, a French oil and gas company, to expand its oil and gas projects despite climate warnings. Concerns are raised by environmental NGOs regarding TotalEnergies' recent US$4.25 billion bond issuance, facilitated by banks such as BPCE/Natixis and Standard Chartered. Between 2016 and 2022, 45 bonds worth US$48.9 billion were issued by TotalEnergies, covering projects in 53 countries, including sensitive areas. Despite pledges made by French banks to cease direct financing for controversial projects, indirect support is still provided to TotalEnergies through bonds. Banks are called upon by NGOs to stop facilitating new bond issuances, and investors are urged to avoid funding oil and gas ventures. With 10 bonds maturing by 2025, appeals are made by NGOs to major financial institutions to discontinue their support.

Source: BankTrack (2024)

Box 1.9: Energy Consumption, Energy Efficiency, and Environmental Quality in Africa

Jinapor et al. (2023) analyse the impact of energy consumption and energy efficiency on environmental quality in Africa. Their findings reveal that while energy consumption tends to deteriorate environmental quality, energy efficiency has a positive effect by mitigating this impact. Specifically, even though increased energy consumption typically reduces environmental quality through carbon and nitrous emissions, the presence of energy efficiency measures can lessen these negative effects. As African nations aim to explore, utilize, and develop their energy resources to drive economic growth, they face the risk of harming the environment. However, by prioritizing energy efficiency, they can reduce greenhouse gas emissions, contributing to Sustainable Development Goal (SDG) 13. The study also reveals that while renewable energy promotes environmental quality, non-renewable energy minimizes environmental quality. This study highlights the significant levels of carbon and nitrous emissions in Africa and their implications for sustainable development. It underscores the importance of adopting energy-efficient strategies, such as designing open buildings and using energy-efficient appliances. To achieve SDG 7 and Aspiration 1.7 of Africa's Agenda 2063, the study recommends that policymakers in African countries incorporate renewable energy sources into their energy mix.

Source: Jinapor et al. (2023)

Also, Jinapor et al. (2023) create awareness on energy consumption, energy efficiency, and environmental quality in Africa highlighting the need to pursue investment in energy efficiency strategies and reduce reliance on non-renewable energy (see Box 1.9).

This education and awareness are gradually sinking into the mind of individuals, businesses and governments. Currently, some businesses and individuals are interested in conducting transactions with banks that have a green corporate image (Chang & Fong, 2010; Sharma & Choubey, 2022). In most countries especially the developed ones, banks can retain existing customers and attract potential ones by adopting business models that offer products supporting sustainable development (Stauropoulou et al., 2022).

1.6.4 Some MDBs Partner Domestic Banks for Project Execution

Some MDBs prefer to collaborate with local commercial banks or financial institutions to execute their projects.[31] This approach allows MDBs to leverage the local expertise and infrastructure of these banks, ensuring more effective implementation and management of development initiatives. By working with established financial institutions in the regions where projects are undertaken, MDBs can enhance their reach and impact, facilitate smoother operations, and mitigate risks associated with unfamiliar markets. For example, The African Development Bank (AfDB) collaborates with financial institutions that focus on supporting or delivering financial services to SMEs, providing tailored financing solutions for these businesses (see Box 1.10). Initiatives like this will encourage financial institutions to integrate sustainable BBM into their operations.

Box 1.10: African Development Bank Partner with Local Financial Institutions to Finance SMEs

The Africa SME Programme, initiated by the African Development Bank (AfDB), aims to support African local Financial Institutions (FIs) by providing long-term liquidity through Lines of Credit and offering technical assistance. This initiative is designed to help FIs offer relevant financing to small and medium-sized enterprises (SMEs) and develop robust SME loan portfolios. The Programme also aids SME clients in preparing better loan applications, such as developing business plans and financial literacy training. Eligible entities include banks, leasing companies, and other regulated Non-Bank Financing Institutions, including Micro-Financing Institutions with SME portfolios. The ultimate beneficiaries are local SMEs seeking reasonably priced financing for their business investments and activities. The AfDB does not directly finance SMEs but instead supports local FIs to recognize SMEs as viable market opportunities. The Programme addresses market supply and demand imbalances in financing between FIs and SMEs, facilitating easier access to funds for both parties. Initially set for 4 years, the Programme may be extended based on performance, aiming to support up to 25 FIs with a total envelope of USD 125 million. A portion of this funding is dedicated to institutions focusing on women entrepreneurs. FIs from any African country can participate, provided they meet specific criteria, including financial and commercial viability, compliance with local regulations, and adherence to high corporate governance and social and environmental standards. Lines of Credit range from USD one million to USD ten million, tailored to the size of the FI's balance sheet and SME funding needs, with loan terms of 4 to 7 years to address asset/liability mismatches. These loans are intended for on-lending to SMEs only and are competitively priced based on risk assessments and existing local market rates. Technical assistance is provided on a case-by-case basis to enhance the capabilities of borrowing FIs and improve SME loan applications. FIs are expected to co-finance a portion of this assistance to emphasize its importance. Additional benefits include standardized due diligence processes and ready-to-deploy technical support. Eligibility for the Programme is broad but subject to AfDB's discretion, prioritizing financial viability, risk profiles, and compliance with regulatory and governance standards.

Source: AfDB (2024)

[31] https://www.adb.org/sites/default/files/page/649056/mdb-brochure.pdf. Accessed August 5, 2024.

1.7 Conclusion

The emergence of global concerns such as climate change, inquality, poverty, coronavirus disease-2019 (COVID-19) and related issues have prompted countries and global development institutions to prioritize sustainable development in their decision-making processes. Effective implementation of sustainable development policies necessitates the involvement of both public and private institutions in providing funding, leading many to adopt suitable business models. The financial sector, including banks have also adopted various business models which is the BBMs. This chapter explores how BBMs can contribute to achieving sustainable development. The chapter begins with a discussion of BBMs and its various forms. While the traditional banking model remains prevalent, it has evolved to meet customer needs and adapt to changing circumstances. New models include investment banking, private banking and asset management, Islamic banking, digital banking, and sustainable and ethical banking.

The chapter continues by examining sustainable development and the policies aimed at achieving it. Sustainable development is defined as development that meets the needs of the current generation without compromising the ability of future generations to meet their own needs. Various policies support this goal, with the SDGs being a comprehensive global framework. Regional policies include Agenda 2063, the Sustainable Development Strategy for the Arctic Region, the ASCC Blueprint 2025, the Regional Framework for Achieving Development Resilient to Climate Change, the European Green Deal, the NAFTA Environmental Cooperation Agreement, the Strategy on Sustainable Development for the Arctic, the Mercosur Green Protocol, and the 2030 Agenda for Sustainable Development for Asia and the Pacific.

The chapter also identifies how BBMs can promote sustainability through financing green projects, incorporating ESG criteria in risk assessment, fostering innovation and product development, promoting inclusive finance, encouraging partnerships and collaboration, and supporting capacity building and education. However, challenges remain, such as short-term profit focus, gradual transition processes, geographic location of banks, difficulties in aligning products with sustainability goals, stakeholder pressure, and misalignment of objectives with partners. Finally, the chapter identifies factors that encourage banks to adopt BBMs that supports sustainable development. These opportunities include the implementation of supportive policies, the establishment of more NDBs, increasing awareness of sustainable development, and partnerships between MDBs and domestic banks for project execution.

References

Abdullah Saif Al Nasser, S., Datin, & Muhammed, J. (2013). Introduction to history of Islamic banking in Malaysia. *Humanomics, 29*(2), 80–87. https://doi.org/10.1108/08288661311319157

Abor, J. Y. (2023). *The changing role of national development banks in Africa*. Springer International Publishing. https://doi.org/10.1007/978-3-031-38639-8

Access Bank Ghana Plc. (2024). *Access Bank Builds SMEs Capacity on Digitalization*. Access News. https://www.ghana.accessbankplc.com/pages/Media/Access-News/ACCESS-BANK-BUILDS-SMEs-CAPACITY-ON-DIGITALIZATION.aspx

AfDB. (2024). *Access to finance for SMEs through FIs*. https://www.afdb.org/en/topics-and-sectors/initiatives-partnerships/access-to-finance-for-smes-through-fis#:~:text=The%20AfDB%20does%20not%20finance,real%20and%20successful%20market%20opportunity

Ayadi, R., & De Groen, W. P. (2014). *Banking Business Models Monitor 2014 Europe*.

Bank of Uganda, & Ministry of Finance, P. and E. D. (2023). *National Financial Inclusion Strategy 2023–2028*.

BankTrack. (2024, April 10). NGOS call on banks and investors to cut TotalEnergies' main source of finance: Some 60 NGOs from around the world signed an open letter calling on banks and investors to stop participating in bonds issued by TotalEnergies. BankTrack. https://www.banktrack.org/article/ngos_call_on_banks_and_investors_to_cut_totalenergies_main_source_of_finance

Beaverstock, J. V. (2011). The privileged world city: Private banking, wealth management and the bespoke servicing of the global super-rich. In *International handbook of globalization and world cities*. Edward Elgar Publishing. https://doi.org/10.4337/9781781001011.00047

Buchner, B., Naran, B., Padmanabhi, R., Stout, S., Strinati, C., Wignarajah, D., Miao, G., Connolly, J., & Marini, N. (2023). *Global landscape of climate finance 2023*.

Cao, X., Jin, C., & Ma, W. (2021). Motivation of Chinese commercial banks to issue green bonds: Financing costs or regulatory arbitrage? *China Economic Review, 66*, 101582. https://doi.org/10.1016/j.chieco.2020.101582

Carson, R. (1962). *Of man and the stream of time* (Vol. 36). Scripps College.

Chang, N., & Fong, C. (2010). Green product quality, green corporate image, green customer satisfaction, and green customer loyalty. *African Journal of Business Management, 4*(13), 2836–2844.

Davis, S. I. (2003). *Investment Banking*. Palgrave Macmillan UK. https://doi.org/10.1057/9780230001114

DBS. (2024). *DBS Bank India launches Green Deposits programme enabling institutions to support sustainable projects*. https://www.dbs.com/newsroom/DBS_Bank_India_launches_Green_Deposits_programme_enabling_institutions_to_support_sustainable_projects

Dechtman, S. (2024, February 14). *Private Banking Vs. Wealth Management*. Dechtman Wealth Management. https://dechtmanwealth.com/insights/blog/private-banking-vs-wealth-management/

Demirguc-Kunt, A., Klapper, L., Singer, D., & Ansar, S. (2018). *The Global Findex Database 2017: Measuring financial inclusion and the fintech revolution*. World Bank Publications.

Doku, J. N., Iddrisu, K., Bortey, D. N. A. S., & Ladime, J. (2023). Impact of digital financial technology on financial inclusion in Sub- Saharan Africa: The moderation role of institutional quality. *African Finance Journal, 25*(2).

Dwi Sari, M. (2016). History of Islamic Bank in Indonesia: Issues behind its establishment. *International Journal of Finance and Banking Research, 2*(5), 178. https://doi.org/10.11648/j.ijfbr.20160205.13

Ensign, P. C. (2022). Business models and sustainable development goals. *Sustainability, 14*(5), 2558. https://doi.org/10.3390/su14052558

European Banking Authority. (2022). *Banks, Business models and Beyond*. https://www.eba.europa.eu/sites/default/files/document_library/1038860/Eurofi%20 2022%20Business%20model.pdf

Fukuda-Parr, S. (2023). Sustainable development goals (SDGs) and the promise of a transformative agenda. In International organization and global governance (pp. 708-723). Routledge.

Giglio, S., Kelly, B., & Stroebel, J. (2021). Climate Finance. *Annual Review of Financial Economics, 13*(1), 15–36. https://doi.org/10.1146/annurev-financial-102620-103311

Hryckiewicz, A., & Kozłowski, Ł. (2017). Banking business models and the nature of financial crisis. *Journal of International Money and Finance, 71*, 1–24. https://doi.org/10.1016/j.jimonfin.2016.10.008

Iddrisu, K. (2024). Foreign bank presence and income inequality in Africa: What role does economic freedom play? *Future Business Journal*. https://doi.org/10.1186/s43093-024-00357-x

IndusInd Bank. (2021, December 28). *IndusInd Bank launches Green Fixed Deposits*. https://www.indusind.com/in/en/about-us/mediabrand/FY/2021-2022/December/indusInd-bank-launches-green-fixed-deposits.html

Jinapor, J. A., Suleman, S., & Cromwell, R. S. (2023). Energy consumption and environmental quality in Africa: Does energy efficiency make any difference? *Sustainability, 15*(3), 2375. https://doi.org/10.3390/su15032375

Mannan, M. S. (2012). *Lees' loss prevention in the process industries*. Elsevier. https://doi.org/10.1016/C2009-0-24104-3

Navaretti, G. B., Calzolari, G., Mansilla-Fernandez, J. M., & Pozzolo, A. F. (2018). Fintech and banking. Friends or foes?. *Friends or Foes*.

Nosratabadi, S., Pinter, G., Mosavi, A., & Semperger, S. (2020). Sustainable Banking; Evaluation of the European Business Models. *Sustainability, 12*(6), 2314. https://doi.org/10.3390/su12062314

RainForest Action Network, BankTrack, Indigenous Environmental Network, OilChange International, Reclaim Finance, Sierra Club, Urgewald, & Ceed. (2024). *Banking on Climate Chaos: Fossil Fuel Finance Report 2024*. https://www.bankingonclimatechaos.org/wp-content/uploads/2024/07/BOCC_2024_vF3.pdf

Roengpitya, R., Tarashev, N., & Tsatsaronis, K. (2014). Bank business models. *BIS Quarterly Review*.

Rome, A. (2003). "Give Earth a Chance": The environmental movement and the sixties. *Journal of American History, 90*(2), 525. https://doi.org/10.2307/3659443

Sharma, M., & Choubey, A. (2022). Green banking initiatives: A qualitative study on Indian banking sector. *Environment, Development and Sustainability, 24*(1), 293–319. https://doi.org/10.1007/s10668-021-01426-9

Smith, C. W. (1986). Investment banking and the capital acquisition process. *Journal of Financial Economics, 15*(1–2), 3–29. https://doi.org/10.1016/0304-405X(86)90048-6

Stauropoulou, A., Sardianou, E., Malindretos, G., Evangelinos, K., & Nikolaou, I. (2022). The role of customers' awareness towards the sustainable development goals (SDGs) of banks on their behavior. *Environmental Science and Pollution Research, 30*(5), 13495–13507. https://doi.org/10.1007/s11356-022-23111-8

Tilly, R. (1998). Universal banking in historical perspective. *Journal of Institutional and Theoretical Economics (JITE)/Zeitschrift Für Die Gesamte Staatswissenschaft, 154*(1), 7–32.

Valls Martínez, M. d. C., Cruz Rambaud, S., & Parra Oller, I. M. (2020). Sustainable and conventional banking in Europe. *PLoS One, 15*(2), e0229420. https://doi.org/10.1371/journal.pone.0229420

Wachira, M. M., Iddrisu, K., Abban, G., & Abor, J. Y. (2023). Sustainable financing in developing economies. In J. Y. Abor (Ed.), *Sustainable and Responsible Investment in Developing Markets* (pp. 159–171). Edward Elgar Publishing. https://doi.org/10.4337/9781803927060.00019

WCED. (1987). *Our Common Future*. Oxford University Press.

World Bank. (2024, April 17). *Financial Inclusion in Sub-Saharan Africa—Overview*. The Global Findex Database 2021. https://www.worldbank.org/en/publication/globalfindex/brief/financial-inclusion-in-sub-saharan-africa-overview

Zgheib, N. (2024, June 30). *EBRD, EU and GCF boost green and inclusive financing for Egyptian MSMEs*. https://www.ebrd.com/news/2024/ebrd-eu-and-gcf-boost-green-and-inclusive-financing-for-egyptian-msmes-.html#:~:text=The%20European%20Bank%20for%20Reconstruction,women-led%20businesses%20in%20Egypt

Banking and Insurance Sector Dynamics and Greening Considerations

2

Ibrahim Nandom Yakubu, Alhassan Musah, and Ibrahim Mohammed

2.1 Introduction

The concern over global warming has sparked debate worldwide over the years. It is commonly believed that excessive emissions of greenhouse gases (GHGs), particularly carbon dioxide (CO_2), are the root cause of this problem (Yakubu et al., 2021a). Several researchers (see Dogan & Seker, 2016a; Cosmas et al., 2019; Fan et al., 2019; Anwar et al., 2021; Yakubu et al., 2021b; Emir et al., 2023; Tukhtamurodov et al., 2024) have looked at the link between a number of factors and carbon dioxide (CO_2) emissions, with suggestions on how these emissions can be reduced. The pursuit of economic growth (ECG) and development in general often comes at a cost to the environment. For instance, agricultural production, industrialization, urbanization, and other economic activities are recognized as contributing factors to increasing levels of greenhouse gas emissions. In recent years, researchers contend that economic growth, financial development, and trade are crucial factors that could largely impact carbon dioxide emissions. Some argue that financial development increases environmental pollution through an aggravated energy consumption route. The intuition may well rest on the potential of financial sector development to enhance liquidity for borrowers in particular and the citizenry in general. For instance, Sadorsky (2010) postulated that with a well-developed financial sector, banks can advance loans for the purchase of high-energy consumption goods such as machinery and household electrical devices, which possibly leads to increased emissions. Along similar lines, Abokyi et al. (2019) reasoned that financial sector deepening leads to an increase in capital investments, which boosts productivity and consequently degrades the environment, especially in countries that rely heavily on fossil fuels. However, there are some scholars who maintain that financial sector development aids environmental sustainability by encouraging firms to adopt environmentally friendly technologies. According to Shahbaz (2009), a thriving financial sector decreases energy emissions through increasing proficiency in the energy zone. The growth of the financial sector also facilitates financing mechanisms for environmentally friendly activities at lower costs and also supports the influx of foreign capital that is linked to advanced technology

I. N. Yakubu (✉)
School of Business, University for Development Studies, Tamale, Ghana
e-mail: ibrahim.yakubu@uds.edu.gh

A. Musah
Department of Sustainable Development Studies, University for Development Studies, Tamale, Ghana
e-mail: alhassan.musah@uds.edu.gh

I. Mohammed
The Hong Kong Polytechnic University, Hung Hom, Kowloon, Hong Kong
e-mail: ibrah.mohammed@polyu.edu.hk

I. N. Yakubu (ed.), *Strategic Approaches to Banking Business and Sustainable Development Goals*, Sustainable Development Goals Series, https://doi.org/10.1007/978-3-031-80744-2_2

that contributes to environmental sustainability (Abbasi & Riaz, 2016).

Following from the foregoing, we argue that prior studies on the relationship between financial development and environmental degradation have produced diverse results and the evidence is far from being conclusive. Consequently, this study seeks to add to the ongoing debate by empirically examining the influence of financial development on environmental considerations in the case of West Africa where research on the financial development–environmental quality nexus is still scanty. Despite the fact that West African countries have experienced some significant levels of economic growth in recent years, the growth dynamics have decelerated due to deteriorating environmental quality (Musah et al., 2022). Notwithstanding the region's degrading environmental condition, studies that have explored the underlying causes of this problem are sparse.

The novelty of this study is that it accounts for the impact of the insurance sector on environmental pollution. The insurance sector, which is an integral aspect of the overall financial system, is generally not at the forefront of the discourse when measuring financial development. The insurance sector, like the banking sector and stock markets, plays a significant role in the economy and thus can potentially impact environmental quality. Insurance companies assist firms in risk management, which adds to the growth and survival of these firms and, as a result, promotes economic growth (Alhassan & Fiador, 2014). Hence, it is critical to consider the effect of the insurance sector when examining the impact of the financial system on environmental quality.

This study contributes significantly to the literature on environmental degradation in several ways. To begin with, we glean from the literature that carbon dioxide (CO_2) emission is commonly used as a metric of environmental degradation (see Cetin et al., 2018; Jiang & Ma, 2019; Acheampong et al., 2020; Aluko & Obalade, 2020; Kayani et al., 2020; Le & Ozturk, 2020; Bulus & Koc, 2021; Omri et al., 2021; Abid et al., 2022; Musah et al., 2022; Hassan et al., 2024). This proxy, however, has been criticized for not being a suitable metric because it ignores resources like oil, forest, mining, soil, and fishing (Ulucak & Apergis, 2018). As a result, we broaden the frontiers of the metrics in this study by employing ecological footprint as an indicator of environmental degradation. The ecological footprint measure considers the effect of anthropogenic actions with respect to carbon, fishing, forest, soil, air, and land. Hence, it serves as an all-inclusive measure of degradation, unlike CO_2 emissions. To the best of our knowledge, this study presents an initial attempt to investigate the financial development and environmental quality nexus in West Africa using ecological footprint as a proxy for environmental degradation.

Until now, existing studies have largely used single measures of financial development, which are either the banking sector or stock market-based measures. This study departs from the use of single-based indicators of financial development by employing a financial sector development index, which captures the development of the banking sector and the insurance sector. The researchers are not aware of any study that has employed a financial development index to investigate the financial development–environmental quality nexus in the case of West Africa. Hence, this study fills a significant gap in the literature.

For the rest of this study, Sect. 2.2 reviews the literature. Section 2.3 covers the methodology, and Sect. 2.4 presents the empirical results and discussions of the findings. Section 2.5 concludes this study with some recommendations.

2.2 Review of Literature on Financial Development and Environmental Pollution Nexus

In theory, the financial sector, according to Schumpeter (1911), boosts economic growth through the provision of the requisite financial services and resources. Several empirical studies have provided compelling evidence in support of this proposition (see Bist 2018; Asteriou & Spanos, 2019; Kumar & Paramanik, 2020; Lenka & Sharma, 2020; Paudel, 2020; Rahman et al., 2020; Afonso & Blanco-Arana, 2022). By enhancing overall economic growth, financial development has been noted to play a significant role in influencing environmental performance. The financial sector, for instance, is perceived to contribute to mitigating pollution by fostering technological innovation in energy-related sectors (Abbasi & Raiz, 2016). The financial sector also provides financing for research and development and investment in environmentally sustainable projects. A number of empirical investigations have demonstrated that environmental pollution is reduced by the development of the financial sector. For instance, Al-Mulali et al. (2015) examined the impact of financial development on environmental degradation using a sample of 129 countries. Applying the dynamic ordinary least squares (DOLS) technique, the authors found that financial development has a negative long- and short-term influence on CO_2 emissions. Using data from 58 Middle Eastern and African (MEA) countries and a sample of 41 countries in the European Union (EU), Abid (2017) documented that financial development abets environmental degradation. Employing the bootstrapping autoregressive distributed lag (ARDL) technique, Shahbaz et al. (2018) reported a negative effect of financial development on pollution in France. Dogan and Seker (2016b) discovered that the development of the financial sector reduces carbon emissions in the top renewable energy countries. According to Adom et al. (2018), the development of the financial sector decreases CO_2 emissions in Ghana. Omoke et al. (2020) showed that an increase in financial development significantly enhances environmental sustainability in Nigeria. Applying the fully modified OLS, Raghutla and Chittedi (2020) found a negative impact of financial development on environmental pollution in the Brazil, Russia, India, China, and South Africa (BRICS) economies. Naqvi et al. (2021) revealed that financial development has a negative and statistically significant impact on environmental deterioration in countries with high-income statuses.

On the contrary, some researchers suggest that financial sector development contributes to environmental pollution. For instance, using data from the top ten CO_2 emission countries, Kayani et al. (2020) established a positive effect of financial development on pollution. By adopting the ARDL and fully modified least squares (FMOLS) techniques, Kwakwa (2020) found that environmental pollution increases with the development of the financial sector in Tunisia. Ali et al. (2019) posited a negative effect of financial sector development on environmental degradation in Nigeria. Using the Westerlund and Edgerton Lagrange multiplier (LM) bootstrap cointegration test, Bayar and Maxim (2020) reported that environmental degradation in the long term is increased by financial development in the post-transition European Union countries. In Japan, Ahmed et al. (2021) found a positive effect of financial development on pollution levels. Yang et al. (2021a, b) showed that environmental sustainability in the Gulf Cooperation Council (GCC) is reduced as the financial sector develops. Employing the system generalized method of moments (GMM) technique, Ibrahim and Vo (2021) documented that financial development increases environmental degradation in some selected industrialized countries. Appiah-Otoo and Acheampong (2021) reported a positive influence of insurance sector development (INSD) on carbon dioxide emissions in the BRICS countries. Finally, Musah et al. (2022) found that financial development increases the emission of carbon dioxide in West Africa.

From the extensive review of the literature, research in the case of West Africa is scanty, thus necessitating more research efforts. With the notable exception of the work of Musah et al. (2022), there is no other study on the impact of financial development on environmental pollution in West Africa. Even with the work of Musah et al. (2022), the authors employed a single measure of financial development and used carbon dioxide emissions as a proxy for environmental degradation. We distinguish our work from that of Musah et al. (2022) by adopting an index of financial development while using ecological footprint as an indicator of environmental degradation.

Table 2.1 provides a summary of the empirical studies on the relationship between financial development and environmental pollution.

Table 2.1 Summary of the empirical literature on the financial development–pollution nexus

Author(s)	Country/region	Indicators of financial development	Technique	Impact of financial development on environmental pollution
Al-Mulali et al. (2015)	Lower-, upper-, and high-income countries	Domestic credit to the private sector	Dynamic OLS	Negative
Abbasi and Riaz (2016)	Pakistan	Total credit and private sector credit	ARDL	Positive
Dogan and Seker (2016b)	Top renewable energy countries	Domestic credit to the private sector	FMOLS and DOLS	Negative
Abid (2017)	MEA and EU countries	Domestic credit to the private sector	System GMM	Negative
Pata (2018)	Turkey	Domestic credit to the private sector	ARDL, FMOLS, and CCR	Positive
Shahbaz et al. (2018)	France	Domestic credit to the private sector	Bootstrapping ARDL	Negative
Ali et al. (2019)	Nigeria	Domestic credit to the private sector	ARDL	Positive
Zakaria and Bibi (2019)	South Africa	Domestic credit to the private sector	2SLS and GLS	Positive
Bayar and Maxim (2020)	Post-transition European Union countries	IMF financial development index	Westerlund and Edgerton LM bootstrap cointegration test	Positive
Kayani et al. (2020)	Top 10 carbon dioxide emission countries	Domestic credit to the private sector	FMOLS and VECM	Positive
Khan et al. (2020)	Pakistan	Stock market development and domestic credit	ARDL	Positive/negative
Kwakwa (2020)	Tunisia	Domestic credit by the financial sector	ARDL and FMOLS	Positive
Omoke et al. (2020)	Nigeria	Index of private credit by banks, domestic credit to the private sector, broad money, and financial sector credit to the private sector	Nonlinear ARDL	Negative
Raghutla and Chittedi (2020)	BRICS	Gross fixed capital formation	FMOLS	Negative
Ahmed et al. (2021)	Japan	Domestic credit to the private sector	ARDL	Positive
Appiah-Otoo and Acheampong (2021)	BRICS	Insurance sector development	IV-GMM	Positive
Ibrahim and Vo (2021)	Industrialized countries	Financial development index based on access, depth, and efficiency of financial markets and institutions	System GMM	Positive
Naqvi et al. (2021)	Income groups	Domestic credit by the financial sector	AMG	Negative
Yang et al. (2021a, b)	Gulf Cooperation Council	IMF financial development index	FMOLS and CCE-2SLS	Positive
Musah et al. (2022)	West Africa	Domestic credit to the private	CS-ARDL, CS-DL, and CAEC	Positive

Source: Authors

IMF International Monetary Fund; *CCR* Canonical cointegrating regression; *2SLS* Two-stage least squares; *GLS* Generalized least squares; *VECM* Vector error correction model; *CCE-2SLS* Common Correlated Effects - Two-Stage Least Squares; *CS-ARDL* Cross-sectional autoregressive distributed lag; *CS-DL* Cross-sectionally augmented distributed lag; *CAEC* Cross-sectional augmented error correction

2.3 Research Methodology

2.3.1 Data and Variables

To achieve the research aim, this study uses annual data covering from 2000 to 2017 from 11 West African countries. The sampled countries include Benin, Burkina Faso, Ivory Coast, Ghana, Guinea, Mali, Niger, Nigeria, Senegal, Sierra Leone, and Togo. The availability of data informed the choice of these countries and the timeframe.

Ecological footprint (in gha per person) is employed as an indicator of environmental degradation. The ecological footprint is calculated using six productive surface areas, which include "carbon demand on land, grazing land, built-up land, agriculture, fishing grounds, and wildlife habitat." Financial development is the main independent variable computed as an index of banking sector development (BSD) and insurance sector development. This index is constructed using the principal component analysis (PCA). The variables constituting the banking sector sub-index are domestic credit by the financial sector (% of gross domestic product [GDP]), bank assets, broad money, and bank deposits. The insurance sector development sub-index is based on life insurance premium volume (% of GDP) and non-life insurance premium volume (% of GDP). Stock market development is not considered in the computation of the financial development index due to data unavailability on stock market indicators for most of the countries in our sample.

In applying the principal component analysis (PCA), the index for the *jth* factor is specified as follows:

$$\text{FDEX}_j = W_{j1}X_1 + W_{j2}X_2 \ldots\ldots + W_{jP}X_P \quad (2.1)$$

where FDEX_j is the composite index for financial development. The weight of the factor score is denoted by W_j. The original figure of the specific components is denoted by *X*, and the number of parameters included in the equation is signified by the letter *P*. Thus, the FDEX is a function of banking sector development (BSD) and insurance sector development (INSD) as indicated below:

$$\text{FDEX} = f(\text{BSD,INSD}) \quad (2.2)$$

This study controls for the impact of foreign direct investment (FDI), economic growth (ECG), and trade openness (TOP). FDI is measured by the net inflows of FDI in terms of the percentage of GDP. Trade openness is the sum of exports and imports of goods divided by GDP, and economic growth is proxied by gross domestic product (GDP) per capita.

The data for ecological footprint are gleaned from the Global Footprint Network (GFN), while the data for the rest of the variables are sourced from the World Development Indicators (WDI).

2.3.2 Empirical Model

In specifying the baseline models for this study, the researcher first considers the impact of the financial development index on environmental degradation. Second, the effect of the various sub-indices (i.e., banking sector development and insurance sector development) on environmental degradation is examined, respectively. Hence, the basic empirical models are stated as follows:

$$\begin{aligned}\text{ECF}_{it} &= \alpha_0 + \beta_1\text{FDEX}_{it} + \beta_2\text{FDI}_{it} + \beta_3\text{ECG}_{it} \\ &+ \beta_4\text{TOP}_{it} + \varepsilon_{it}\end{aligned} \quad (2.3)$$

$$\begin{aligned}\text{ECF}_{it} &= \alpha_0 + \beta_1\text{BSD}_{it} + \beta_2\text{FDI}_{it} + \beta_3\text{ECG}_{it} \\ &+ \beta_4\text{TOP}_{it} + \varepsilon_{it}\end{aligned} \quad (2.4)$$

$$\begin{aligned}\text{ECF}_{it} &= \alpha_0 + \beta_1\text{INSD}_{it} + \beta_2\text{FDI}_{it} + \beta_3\text{ECG}_{it} \\ &+ \beta_4\text{TOP}_{it} + \varepsilon_{it}\end{aligned} \quad (2.5)$$

where ECF, FDEX, BDS, INSD, FDI, ECG, and TOP are ecological footprint, financial development index, banking sector development, insurance sector development, foreign direct investment, economic growth, and trade openness, respectively. β_1 to β_4 are the slope parameters of the independent factors. The constant and error term are denoted by α and ε, respectively. The subscript i signifies the cross-sectional dimension, and the subscript t represents the timeframe.

2.3.3 Analytical Approach

As a first stage, this study uses the CD test by Pesaran (2004) to examine cross-sectional dependence (CD) in the dataset. In the second stage, the cross-sectional augmented Dickey–Fuller (CADF) test for stationarity is employed to establish the order of integration of the variables. The CADF test is robust to residual cross-sectional dependence in the panel dataset (Musah et al., 2022). This study applies the Pedroni residual cointegration test and the Johansen Fisher panel cointegration test to determine the long-term relationship among the variables.

Following these preliminary analyses, the panel Fully Modified Least Squares (FMOLS) technique is employed to examine the impact of the explanatory factors on environmental degradation. The FMOLS approach is better suited for overcoming the bias, endogeneity, and serial correlation problems inherent in panel estimations (Yang et al., 2021a, b).

The FMOLS estimator for a panel dataset is specified as follows:

$$\hat{\alpha}^*_{\text{GFM}} = N^{-1}\sum_{n-1}^{N}\hat{\alpha}^*\text{FM},n \qquad (2.6)$$

From Eq. (2.6), the FMOLS estimator, which is applied to the individual countries in the panel, is connoted by $\hat{\alpha}^*_{\text{GFM}}$.

2.4 Empirical Results and Discussions

2.4.1 Cross-Sectional Dependence Test Results

In panel studies, it is imperative to examine the presence of cross-sectional dependence (CD). Failure to do so may lead to biased estimates (Zeraibi et al., 2021). The CD test by Pesaran (2004) is utilized in this study to ascertain whether there is CD among the sampled countries. Table 2.2 summarizes the results of the CD test. From the three test results (Breusch–Pagan LM, Pesaran scaled LM, and bias-corrected scaled LM), this study rejects the null hypothesis, which stipulates that there is no cross-sectional dependence, given the significance of the variables at 1%. The presence of cross-section dependence suggests that if a random shock happens in one of the sample nations, it will have a spillover effect on the other countries in the group.

2.4.2 Panel Unit Root Test

Since there is cross-sectional dependence in the dataset, using the traditional panel unit, root tests could produce biased results. Hence, the modern unit root testing approach within the Cross-sectional Augmented Dickey-Fuller (CADF) method is applied in this study. Table 2.3 presents the results of the CADF panel unit root test. The test results indicate that all the variables are stationary at the first difference. Following the integration

Table 2.2 Results of the cross-sectional dependence test

Series	Breusch–Pagan LM		Pesaran scaled LM		Bias-corrected scaled LM	
	Statistic	Probability	Statistic	Probability	Statistic	Probability
ECF	247.6510***	0.0000	18.3686***	0.0000	18.0450***	0.0000
FDEX	612.1436***	0.0000	53.1216***	0.0000	52.7980***	0.0000
BSD	668.6924***	0.0000	58.5133***	0.0000	58.1898***	0.0000
INSD	260.4391***	0.0000	19.5879***	0.0000	19.2643***	0.0000
FDI	125.0370***	0.0000	6.6778***	0.0000	6.35423***	0.0000
ECG	87.7208***	0.0033	3.1198***	0.0018	2.7963***	0.0052
TOP	199.4776***	0.0000	13.7754***	0.0000	13.4519***	0.0000

Note: *** indicates significance level at 1%
Source: Authors' Computation

Table 2.3 CADF unit root test results

Series	Level	First difference
	Z(t-bar)	Z(t-bar)
ECF	1.180	−3.449***
FDEX	0.235	−3.371***
BSD	−0.757	−5.151***
INSD	−0.389	−4.467***
FDI	0.770	−2.222**
ECG	0.227	−7.468***
TOP	0.932	−3.978***

Note: *** and ** indicate significance levels at 1% and 5%, respectively
Source: Authors' Computation

of the variables at order I(1), the null hypothesis of no unit root among the residual terms cannot be rejected.

2.4.3 Panel Cointegration Test

In this study, Pedroni's panel cointegration test and the standard Johansen Fisher panel cointegration test are employed. Hence, both parametric and non-parametric tests are combined to examine the long-term relationship among the variables. Table 2.4 displays the results of the cointegration test. From the results, it can be concluded that the series are cointegrated as evidenced by some of the tests. Thus, the null hypothesis of no long-run relationship among the variables is rejected. As a result, there is enough justification for the estimation of the models.

2.4.4 Regression Results

The FMOLS regression estimates are presented in Table 2.5. From the table, Model 1 represents the results of the impact of the financial development index on ecological footprint. While Model 2 examines the impact of banking sector development on environmental degradation, the estimates of Model 3 depict the results of the nexus between insurance sector development and environmental degradation.

From Model 1, the results show that the financial development index has a positive significant impact on the ecological footprint. This result holds for banking sector development and insurance sector development in Model 2 and Model 3, respectively. Thus, the finding suggests that increased financial development in West Africa does not contribute to environmental quality. There are several plausible explanations for this finding as financial development degrades environmental quality through various mechanisms. In terms of banks' activities, a well-developed banking sector makes financial resources more accessible, notably low-interest bank credit. This allows firms to purchase energy-intensive equipment and expand production, resulting in increased energy consumption. Likewise, with a boom in bank credit due to a well-developed banking sector, individuals tend to increase their spending on household goods and automobiles that rely heavily on non-renewable energy sources, thus increasing environmental pollution. Being largely an agricultural-based economic zone, we also surmise that a booming financial sector may expand agricultural credit, resulting in increased investments in the sector, including large-scale adoption and application of chemical fertilizers that have been proven to have long-term negative consequences on soil and environmental quality. For the positive effect of insurance sector development on ecological footprint, it can be argued that a well-developed insurance sector allows economic units such as businesses to access funds generated through insurance premiums to undertake investments (Appiah-Otoo & Acheampong, 2021). This boosts economic activity, which by some means upsurges energy consumption and, as a result, increases pollution. Overall, the positive effect of the composite index of financial development on environmental degradation is congruent with most prior empirical studies (Ali et al., 2019; Bayar & Maxim, 2020; Kwakwa, 2020; Ibrahim & Vo, 2021; Yang et al., 2021a, b; Musah et al., 2022).

In all the model estimates, foreign direct investment exerts a positive significant effect on ecological footprint. This suggests that FDI, in the form of technology and other transfers, con-

Table 2.4 Panel cointegration test results

1. Pedroni's cointegration test				
Test	*Statistic*	*Probability*	*Weighted statistic*	*Probability*
Common autoregressive coefficients (within dimension)				
Panel v-statistic	−0.2689	0.6060	−0.3751	0.6462
Panel rho-statistic	1.0412	0.8511	0.4297	0.6663
Panel PP-statistic	−4.9068	0.0000	−5.0355	0.0000
Panel ADF-statistic	−2.4800	0.0066	−1.4727	0.0704
Individual autoregressive coefficients (between dimensions)				
Group rho-statistic	1.5933	0.9445		
Group PP-statistic	−6.2272	0.0000		
Group ADF-statistic	−1.2696	0.1021		
2. Johansen Fisher's panel cointegration test				
Null hypothesis	*Fisher's statistic[a] (trace test)*	*Probability*	*Fisher's statistic[a] (maximum eigenvalue)*	*Probability*
r = 0	387.8	0.0000	295.4	0.0000
r ≤ 1	164.2	0.0000	129.9	0.0000
r ≤ 2	62.07	0.0000	43.99	0.0036
r ≤ 3	35.94	0.0308	31.83	0.0803
r ≤ 4	35.02	0.0386	35.02	0.0386

Source: Authors' Computation
[a]Probabilities are computed using an asymptotic chi-square distribution

Table 2.5 FMOLS regression results

Variables	Model 1	Model 2	Model 3
FDEX	0.0473***		
	(0.0004)		
BSD		0.0205***	
		(0.0026)	
INSD			0.0595***
			(0.0016)
FDI	0.0072***	0.0075***	0.0073***
	(0.0042)	(0.0033)	(0.0040)
ECG	0.0071***	0.0078***	0.0074***
	(0.0013)	(0.0006)	(0.0009)
TOP	−0.0005	−0.0006	0.0001
	(0.5218)	(0.4279)	(0.9039)
R-squared	0.8210	0.8188	0.8101
Adjusted R-squared	0.8065	0.8041	0.7946
Standard error of regression	0.0896	0.0901	0.0923
Long-run variance	0.0116	0.0121	0.0117

Note: *** indicates significance level at 1%
Source: Authors' Computation

tributes to increased pollution in West Africa. This result supports the pollution haven hypothesis (PHH), which argues that emission reduction in many developed countries is partly due to the shifting of polluting activities to developing countries, which receive the largest shares of FDI inflows and global emissions (Demena & Afesorgbor, 2020). The direct impact of FDI on environmental degradation as documented in this study syncs with the findings of Bulut et al. (2021), Nawaz et al. (2021), and Musah et al. (2022).

The results further reveal that economic growth and ecological footprint are positively and significantly related. This indicates that an increase in economic activity upsurges pollution levels in West Africa. An obvious inference is that as the economy advances, so do manufacturing and industrialization activities. This increases energy consumption and, as a corollary, environmental pollution. In all scenarios, the findings show that trade openness has an insignificant effect on environmental degradation.

2.5 Conclusions and Policy Recommendations

Environmental degradation has emerged as a critical threat and has remained a major hurdle on the road to sustainable development, owing to the plethora of environmental challenges it poses, including climate change. The Paris Climate Conference and the Kyoto Protocol mandate governments to help combat global warming. In addition, many of the globally agreed Sustainable Development Goals (SDGs) invite actions from nation-states to promote ecological and environmental justice. In particular, SDGs 7, 12, 13, and 15 address affordable and clean energy, responsible consumption and production, climate action, and life on land, respectively. In the quest to enhance environmental quality, several studies have looked at how different factors impact pollution. However, research on the factors that influence environmental sustainability in West Africa is generally scanty. As a result, this study investigated the influence of financial development on environmental degradation in West Africa, adding to the limited research on pollution drivers in the region. We also extended this research genre by closely examining and highlighting the impact of a significant yet often neglected subsector—the insurance subsector—on greening considerations in the region. The researchers employ data from 11 West African countries covering the period 2000–2017. Prior to the empirical estimation, this study performs different preliminary analyses, including the cross-sectional dependency, panel unit root, and panel cointegration tests to ensure accurate findings. Based on the results of these preliminary analyses, this study applies the Fully Modified Least Squares (FMOLS) technique in the empirical estimation. The researchers first explore the impact of the financial development index on environmental degradation and then the effect of the various sub-indices (i.e., banking sector development and insurance sector development) on environmental degradation. From the model estimates, the findings reveal that financial development has a positive and statistically significant impact on environmental degradation, implying that the development of both the banking sector and the insurance sector impedes environmental quality in West Africa. The results establish that foreign direct investment increases pollution, confirming the pollution haven hypothesis (PHH) in the region. Also, economic growth is evidenced to positively affect environmental degradation, whereas the effect of trade openness is insignificant.

The results offer significant policy implications for policymakers in the region. First, to promote green growth and reduce greenhouse gas emissions, financial institutions in West Africa must focus on allocating financial resources to environmentally friendly projects. Green bonds should be considered as a way to further incentivize investment in environmentally sustainable projects in the region. Given the positive impact of the insurance sector on environmental degradation, insurers in West Africa should step up efforts to support and invest in projects that seek to mitigate climate change. As FDI reduces environmental quality, direct foreign investments in clean energy sources such as solar and wind should be encouraged in the region to reduce the over-reliance on non-renewable energy sources. Likewise, the respective governments should consider embracing measures to restrict the transfer of pollution generation equipment into the region. To further enhance environmental sustainability in the region and discourage what may be considered as outsourcing the sins of pollution to West Africa, foreign firms engaged in low-carbon-intensive operations may be given lower taxes, while levying higher taxes on firms with high pollution status.

This study is confined to West African nations. Therefore, it would be inapt to generalize the results to all African countries. The researchers recommend that future research undertake an analysis of the entire African continent to provide more extensive insights into how both the banking sector and the insurance sector development influence environmental quality.

References

Abbasi, F., & Riaz, K. (2016). CO2 emissions and financial development in an emerging economy: An augmented VAR approach. *Energy Policy, 90*, 102–114.

Abid, M. (2017). Does economic, financial and institutional developments matter for environmental quality? A comparative analysis of EU and MEA countries. *Journal of Environmental Management, 188*, 183–194.

Abid, A., Mehmood, U., Tariq, S., & Haq, Z. U. (2022). The effect of technological innovation, FDI, and financial development on CO2 emission: Evidence from the G8 countries. *Environmental Science and Pollution Research, 29*(8), 11654–11662.

Abokyi, E., Appiah-Konadu, P., Abokyi, F., & Oteng-Abayie, E. F. (2019). Industrial growth and emissions of CO2 in Ghana: The role of financial development and fossil fuel consumption. *Energy Reports, 5*, 1339–1353.

Acheampong, A. O., Amponsah, M., & Boateng, E. (2020). Does financial development mitigate carbon emissions? Evidence from heterogeneous financial economies. *Energy Economics, 88*, 104768.

Adom, P. K., Kwakwa, P. A., & Amankwaa, A. (2018). The long-run effects of economic, demographic, and political indices on actual and potential CO2 emissions. *Journal of Environmental Management, 218*, 516–526.

Afonso, A., & Blanco-Arana, M. C. (2022). Financial and economic development in the context of the global 2008-09 financial crisis. *International Economics, 169*, 30–42.

Ahmed, Z., Zhang, B., & Cary, M. (2021). Linking economic globalization, economic growth, financial development, and ecological footprint: Evidence from symmetric and asymmetric ARDL. *Ecological Indicators, 121*, 107060.

Alhassan, A. L., & Fiador, V. (2014). Insurance-growth nexus in Ghana: An autoregressive distributed lag bounds cointegration approach. *Review of Development Finance, 4*(2), 83–96.

Ali, H. S., Law, S. H., Lin, W. L., Yusop, Z., Chin, L., & Bare, U. A. A. (2019). Financial development and carbon dioxide emissions in Nigeria: Evidence from the ARDL bounds approach. *GeoJournal, 84*(3), 641–655.

Al-Mulali, U., Tang, C. F., & Ozturk, I. (2015). Does financial development reduce environmental degradation? Evidence from a panel study of 129 countries. *Environmental Science and Pollution Research, 22*(19), 14891–14900.

Aluko, O. A., & Obalade, A. A. (2020). Financial development and environmental quality in sub-Saharan Africa: Is there a technology effect? *Science of the Total Environment, 747*, 141515.

Anwar, A., Chaudhary, A. R., Malik, S., & Bassim, M. (2021). Modelling the macroeconomic determinants of carbon dioxide emissions in the G-7 countries: The roles of technological innovation and institutional quality improvement. *Global Business Review*, 09721509211039392.

Appiah-Otoo, I., & Acheampong, A. O. (2021). Does insurance sector development improve environmental quality? Evidence from BRICS. *Environmental Science and Pollution Research, 28*(23), 29432–29444.

Asteriou, D., & Spanos, K. (2019). The relationship between financial development and economic growth during the recent crisis: Evidence from the EU. *Finance Research Letters, 28*, 238–245.

Bayar, Y., & Maxim, A. (2020). Financial development and CO2 emissions in post-transition European Union countries. *Sustainability, 12*(7), 1–15.

Bist, J. P. (2018). Financial development and economic growth: Evidence from a panel of 16 African and non-African low-income countries. *Cogent Economics & Finance, 6*(1), 1449780.

Bulus, G. C., & Koc, S. (2021). The effects of FDI and government expenditures on environmental pollution in Korea: The pollution haven hypothesis revisited. *Environmental Science and Pollution Research, 28*(28), 38238–38253.

Bulut, U., Ucler, G., & Inglesi-Lotz, R. (2021). Does the pollution haven hypothesis prevail in Turkey? Empirical evidence from nonlinear smooth transition models. *Environmental Science and Pollution Research, 28*(29), 38563–38572.

Cetin, M., Ecevit, E., & Yucel, A. G. (2018). The impact of economic growth, energy consumption, trade openness, and financial development on carbon emissions: Empirical evidence from Turkey. *Environmental Science and Pollution Research, 25*(36), 36589–36603.

Cosmas, N. C., Chitedze, I., & Mourad, K. A. (2019). An econometric analysis of the macroeconomic determinants of carbon dioxide emissions in Nigeria. *Science of the Total Environment, 675*, 313–324.

Demena, B. A., & Afesorgbor, S. K. (2020). The effect of FDI on environmental emissions: Evidence from a meta-analysis. *Energy Policy, 138*, 111192.

Dogan, E., & Seker, F. (2016a). An investigation on the determinants of carbon emissions for OECD countries: Empirical evidence from panel models robust to heterogeneity and cross-sectional dependence. *Environmental Science and Pollution Research, 23*(14), 14646–14655.

Dogan, E., & Seker, F. (2016b). The influence of real output, renewable and non-renewable energy, trade and financial development on carbon emissions in the top renewable energy countries. *Renewable and Sustainable Energy Reviews, 60*, 1074–1085.

Emir, F., Udemba, E. N., & Philip, L. D. (2023). Determinants of carbon emissions: Nexus among carbon emissions, coal, agriculture, trade and innovations. *Environment, Development and Sustainability*, 1–15.

Fan, J. L., Da, Y. B., Wan, S. L., Zhang, M., Cao, Z., Wang, Y., & Zhang, X. (2019). Determinants of carbon emissions in 'belt and road initiative' countries: A production technology perspective. *Applied Energy, 239*, 268–279.

Hassan, A. Y., Mohamed, M. A., Ahmed, M. A., Mohamed, M. A., & Osman, B. M. (2024). Toward Environmental Sustainability: The Nexus between Agriculture, Renewable Energy, and Economic Growth in Mitigating CO2 Emissions in Somalia. *International Journal of Energy Economics and Policy, 14*(6), 127–135.

Ibrahim, M., & Vo, X. V. (2021). Exploring the relationships among innovation, financial sector development and environmental pollution in selected industrialized countries. *Journal of Environmental Management, 284*, 112057.

Jiang, C., & Ma, X. (2019). The impact of financial development on carbon emissions: A global perspective. *Sustainability, 11*(19), 1–22.

Kayani, G. M., Ashfaq, S., & Siddique, A. (2020). Assessment of financial development on environmental effect: Implications for sustainable development. *Journal of Cleaner Production, 261*, 120984.

Khan, M. I., Teng, J. Z., & Khan, M. K. (2020). The impact of macroeconomic and financial development on carbon dioxide emissions in Pakistan: Evidence with a novel dynamic simulated ARDL approach. *Environmental Science and Pollution Research, 27*(31), 39560–39571.

Kumar, K., & Paramanik, R. N. (2020). Nexus between Indian economic growth and financial development: A non-linear ARDL approach. *The Journal of Asian Finance, Economics and Business, 7*(6), 109–116.

Kwakwa, P. A. (2020). The long-run effects of energy use, urbanization and financial development on carbon dioxide emissions. *International Journal of Energy Sector Management, 14*(6), 1405–1424.

Le, H. P., & Ozturk, I. (2020). The impacts of globalization, financial development, government expenditures, and institutional quality on CO2 emissions in the presence of environmental Kuznets curve. *Environmental Science and Pollution Research, 27*(18), 22680–22697.

Lenka, S. K., & Sharma, R. (2020). Re-examining the effect of financial development on economic growth in India: Does the measurement of financial development matter? *Journal of Asia-Pacific Business, 21*(2), 124–142.

Musah, M., Owusu-Akomeah, M., Nyeadi, J. D., Alfred, M., & Mensah, I. A. (2022). Financial development and environmental sustainability in West Africa: Evidence from heterogeneous and cross-sectionally correlated models. *Environmental Science and Pollution Research, 29*(8), 12313–12335.

Naqvi, S. A. A., Shah, S. A. R., Anwar, S., & Raza, H. (2021). Renewable energy, economic development, and ecological footprint nexus: Fresh evidence of renewable energy environment Kuznets curve (RKC) from income groups. *Environmental Science and Pollution Research, 28*(2), 2031–2051.

Nawaz, S. M. N., Alvi, S., & Akmal, T. (2021). The impasse of energy consumption coupling with pollution haven hypothesis and environmental Kuznets curve: A case study of south Asian economies. *Environmental Science and Pollution Research, 28*(35), 48799–48807.

Omoke, P. C., Nwani, C., Effiong, E. L., Evbuomwan, O. O., & Emenekwe, C. C. (2020). The impact of financial development on carbon, non-carbon, and total ecological footprint in Nigeria: New evidence from asymmetric dynamic analysis. *Environmental Science and Pollution Research, 27*(17), 21628–21646.

Omri, A., Kahia, M., & Kahouli, B. (2021). Does good governance moderate the financial development-CO2 emissions relationship? *Environmental Science and Pollution Research, 28*(34), 47503–47516.

Pata, U. K. (2018). Renewable energy consumption, urbanization, financial development, income and CO2 emissions in Turkey: Testing EKC hypothesis with structural breaks. *Journal of Cleaner Production, 187*, 770–779.

Paudel, R. C. (2020). The role of financial development in economic growth of Nepal: ARDL approach of cointegration with structural break analysis. *Journal of Economics and Business, 3*(4), 1372–1386.

Pesaran, M. H. (2004). *General diagnostic tests for cross section dependence in panels*, IZA Discussion Paper (No. 1240). The Institute for the Study of Labor, Bonn.

Raghutla, C., & Chittedi, K. R. (2020). Financial development, energy consumption, technology, urbanization, economic output and carbon emissions nexus in BRICS countries: An empirical analysis. *Management of Environmental Quality: An International Journal, 32*(2), 290–307.

Rahman, A., Khan, M. A., & Charfeddine, L. (2020). Financial development–economic growth nexus in Pakistan: New evidence from the Markov switching model. *Cogent Economics & Finance, 8*(1), 1716446.

Sadorsky, P. (2010). The impact of financial development on energy consumption in emerging economies. *Energy Policy, 38*(5), 2528–2535.

Shahbaz, M. (2009). A reassessment of finance-growth nexus for Pakistan: Under the investigation of FMOLS and DOLS techniques. *IUP Journal of Applied Economics, 8*(1), 65–80.

Shahbaz, M., Nasir, M. A., & Roubaud, D. (2018). Environmental degradation in France: The effects of FDI, financial development, and energy innovations. *Energy Economics, 74*, 843–857.

Tukhtamurodov, A., Sobirov, Y., Toshalieva, S., Ibrayimova, D., & Feruz, M. (2024). Determinants of CO2 emissions in the BRICS. A dynamic panel ARDL approach. In *BIO web of conferences* (Vol. 82, p. 06002). EDP Sciences.

Ulucak, R., & Apergis, N. (2018). Does convergence really matter for the environment? An application based on club convergence and on the ecological footprint concept for the EU countries. *Environmental Science & Policy, 80*, 21–27.

Yakubu, I. N., Kapusuzoglu, A., & Ceylan, N. B. (2021a). Environmental impact of economic globalization and renewable energy consumption in the OPEC countries. In *Strategic approaches to energy management* (pp. 155–171). Springer.

Yakubu, I. N., Kapusuzoglu, A., & Ceylan, N. B. (2021b). Urbanization, fossil fuel consumption and carbon dioxide emission in Ghana: The STIRPAT model approach. In S. Yuksel & H. Dincer (Eds.), *Strategic approaches to energy management* (Contributions to management science) (pp. 201–216). Springer.

Yang, B., Jahanger, A., & Ali, M. (2021a). Remittance inflows affect the ecological footprint in BICS countries: Do technological innovation and financial development matter? *Environmental Science and Pollution Research, 28*(18), 23482–23500.

Yang, B., Jahanger, A., Usman, M., & Khan, M. A. (2021b). The dynamic linkage between globalization, financial development, energy utilization, and environmental sustainability in GCC countries. *Environmental Science and Pollution Research, 28*(13), 16568–16588.

Zakaria, M., & Bibi, S. (2019). Financial development and environment in South Asia: The role of institutional quality. *Environmental Science and Pollution Research, 26*(8), 7926–7937.

Zeraibi, A., Balsalobre-Lorente, D., & Murshed, M. (2021). The influences of renewable electricity generation, technological innovation, financial development, and economic growth on ecological footprints in ASEAN-5 countries. *Environmental Science and Pollution Research, 28*(37), 51003–51021.

Green Finance Initiatives in Banking Institutions

3

Khadijah Iddrisu, Ibrahim Nandom Yakubu, and Joshua Yindenaba Abor

3.1 Introduction

While the banking sector itself may not be a direct source of environmental harm, its financial activities and products can significantly influence environmental outcomes. Through the provision of loans, investments and other financial services, banks can either contribute to environmental degradation or promote sustainability, depending on the nature of the projects they finance. For instance, leading banks, including the Royal Bank of Canada (RBC), JPMorgan Chase, Citibank, Bank of America, Scotiabank, Mitsubishi UFJ Financial Group (MUFG) and Mizuho, continue to direct substantial funding into fossil fuel expansion (Oil Change International, 2023). These banks collectively funnel billions of dollars into the sector. In 2022 alone, the top 60 global banks provided US$150 billion in financing to the 100 largest companies driving the growth of the oil, gas and coal industries (Oil Change International, 2023). If the banking sector fails to proactively incorporate environmental sustainability into its products and services, it risks contributing to environmental degradation amidst rising environmental concerns and the urgent need for sustainable development (Wachira et al., 2023).

The financial sector has made an effort to foster environmental sustainability through the introduction of various initiatives designed to support and promote eco-friendly practices. These initiatives, collectively known as green finance initiatives, aim to channel financial resources towards projects and activities that have positive environmental impacts (Abor et al., 2019; Wachira et al., 2023). Examples of these initiatives include financing projects that have a low-carbon footprint; supporting endeavours that positively impact the environment; creating products that raise awareness about environmental sustainability and promoting offerings that protect the environment, among others. These initiatives have been adopted across various segments of the financial industry, including the banking sector. Thus, the primary aim of this chapter is to explore how these initiatives have been integrated into banking practices. As banks adopt these initiatives, they can reduce their ecological footprint and promote positive environmental change through the financial activities they facilitate.

K. Iddrisu (✉)
School of Business, Simon Diedong Dombo University of Business and Integrated Development Studies, Wa, Ghana
e-mail: khadijahiddris@iug.edu.gh

I. N. Yakubu
School of Business, University for Development Studies, Tamale, Ghana
e-mail: ibrahim.yakubu@uds.edu.gh

J. Y. Abor
Department of Finance, University of Ghana Business School, Legon, Ghana
e-mail: joshabor@ug.edu.gh

I. N. Yakubu (ed.), *Strategic Approaches to Banking Business and Sustainable Development Goals*, Sustainable Development Goals Series, https://doi.org/10.1007/978-3-031-80744-2_3

The rest of the chapter is as follows: Section 3.2 presents the discussion of green finance initiatives, whereas the motivation for green finance initiatives adoption and the potential metric for sustainable development are discussed in Sect. 3.3. We discuss the policies and principles that underpin the essence of integrating green finance initiatives in banking (GFIB) institutions in Sect. 3.4. Section 3.5 presents the challenges, while future trends and prospects are revealed in Sect. 3.6. We conclude this chapter in Sect. 3.7.

3.2 Green Finance and Its Initiatives

This section explores the concept of green finance, trends within the field, various green finance initiatives and the significance of integrating green finance principles.

3.2.1 What Is Green Finance

While various institutions and authors have proposed definitions, there is no universally accepted definition of green finance. According to Wachira et al. (2023), green finance is a subset of sustainable finance supporting an energy transition and the fight against climate change, with initiatives such as afforestation, flood prevention and wildfire mitigation to protect communities and wildlife. Ozili (2022) explained green finance as securing and allocating funds for activities focused on environmental preservation while also ensuring fair returns for investors or lenders. While the Organisation for Economic Cooperation and Development (OECD) (2020) defined green finance as financial support from both public and private sources aimed at fostering sustainability in a country's economy, the International Finance Corporation (IFC) (2017) views green finance as the financing of investments that contribute to environmental well-being. These definitions typically emphasise two key elements: the allocation of funds and the promotion of environmental sustainability. Therefore, any financial allocation directed towards environmental protection qualifies as green finance. For instance, when a government issues green bonds to fund solar electricity generation, such funding falls under the category of green finance.

3.2.2 Evolution of Green Finance

The evolution of green finance has been presented in Table 3.1. As shown in Table 3.1, the development of green finance began with the National Environmental Policy Act (NEPA) of 1969 in the USA, which raised global awareness about environmental protection. Since then, it has expanded through various policies, institutions and frameworks such as Climate Investment Funds (CIFs), Sustainable Development Goals (SDGs), Climate Action 100+, Equator Principles (EPs), Global Environment Facility (GEF), green bond principles, Establishment of Partnership for Action on Green Economy (PAGE), Establishment of Sustainable Banking Network, now Sustainable Banking and Finance Network (SBFN) and inter alia. These policies, institutions and frameworks have now placed green finance at the centre of most economies.

Apart from what's detailed in Table 3.1, additional regional and national policies allocate funds for environmental conservation, often inspired by or adapted from those in Table 3.1. For instance, the Bank of China issued Asia's first offshore Blue Bond to finance marine-related green projects, promoting a sustainable blue economy.[1] Furthermore, the European Union (EU) Sustainable Finance Taxonomy was established to assist in responsible investment by assessing project sustainability for potential investors. Green finance was further propelled in 2021 when New Zealand mandated that financial institutions disclose climate change impacts, setting a global precedent. The European Commission also introduced the Sustainable

[1] https://www.scmp.com/business/banking-finance/article/3253518/ocean-friendly-projects-funded-blue-bonds-present-bankability-challenge-climate-business-forum-hears

Table 3.1 Evolution of green finance

Year	Development/policies	Aim/objectives
1969	National Environmental Policy Act (NEPA) by the USA	Federal agencies conducted comprehensive evaluations of the environmental impacts of their actions and considered eco-friendly alternatives
1972	The Stockholm Declaration on Human Environment by the United Nations (UN)	Focuses on the global nature of environmental issues and the need for collective responsibility
1980	Launch of the green grant by Triodos Bank on the Amsterdam Stock Exchange	It was used to minimise pollution and fund environmentally friendly projects
1987	Brundtland Report by World Commission on Environment and Development (WCED)	Advocate for meeting present needs without compromising future generations' needs
1988	Establishment of Intergovernmental Panel on Climate Change	Assessing climate change
1989	France proposed to allocate additional funds (900) to the World Bank for environmental projects	Promoting investment in green projects
1990	Pilot Global Environment Facility (GEF) by 27 countries	To address global environmental challenges by providing financial assistance for environmental projects in developing countries
1990	Finland introduced the carbon tax	To reduce emissions
1991	Official operation of GEF	Provide finance to minimise environmental challenges
1992	Establishment of the United Nations Environment Programme Finance Initiative (UNEP FI)	To promote sustainable finance and environmental conservation
1992	Establishment of the United Nations Framework Convention on Climate Change (UNFCCC)	To address the growing global concern over climate change
!997	The Kyoto Protocol was adopted	It aimed to strengthen and operationalise the commitments made under the UNFCCC
2003	Equator Principles (EPs)	Assessing and managing the environmental and social risks associated with project financing
2005	Kyoto Protocol enters into force	Establishing binding emission reduction targets
2008	Launch of Climate Investment Funds	To promote low-carbon and climate-resilient development
2009	Proposed Green Bank Act in the USA	Promote clean energy and efficient tech with financing to leverage private sector investments in the USA
2010	Green Climate Fund under COP 16 and started operation in 2011	A global fund to support developing countries in their efforts to combat climate change and adapt to its effects
2012	Establishment of Sustainable Banking Network, now Sustainable Banking and Finance Network (SBFN)	Supportive regulations are crucial for banks to successfully transition towards sustainability initiatives and goals
2013	Establishment of Partnership for Action on Green Economy (PAGE)	Collaborates to reshape economic policies towards sustainability, transforming economies into sustainable engines of growth
2014	The launch of Green Bond Principles	To promote efficient green bond trading
2015	17 Sustainable Development Goals (SDGs) were agreed upon after its initiation in 2012	To address various social, economic and environmental challenges globally
2015	Paris Agreement was established	Necessitating actions and investments to limit global warming to below 2 or 1.5 degrees Celsius
2017	Hamburg Principles by the G20	To enhance private engagement in green financing
2017	The introduction of the Science-Based Targets Initiative (SBTI)	Empowers industries with emissions targets to limit global temperature rise below 1.5 °C, combating climate change

(continued)

Table 3.1 (continued)

Year	Development/policies	Aim/objectives
2017	Network for Greening the Financial System (NGFS) was established	To promote sustainable finance and address climate risks within the financial sector
2017	The launch of Climate Action 100+	This global initiative ensures corporate climate action, governance improvement and enhanced disclosure of climate risks/opportunities
2018	The launch of Integrated National Financing Frameworks (INFF) by the UN	It represents a coordinated effort to align financing strategies with sustainable development goals at the national level
2018	European Union Action Plan on Sustainable Finance	Acts as a guiding framework to foster sustainable investments and fund the European Green Deal
2018	Sustainable Blue Economy Finance Principles were launched	It outlines 14 principles guiding banks, insurers and investors in financing a sustainable blue economy
2019	Sustainability-Linked Loan Principles were introduced	These principles provide a framework for loans where the interest rate can be adjusted based on the borrower's achievement of predetermined sustainability performance targets
2019	Net-Zero Asset Owner Alliance was launched	It represents a coalition of institutional investors committed to transitioning their investment portfolios to net-zero greenhouse gas emissions by 2050
2019	Investing for Impact: Operating Principles for Impact Management was launched	These principles aim to provide a framework for investors to manage and measure the impact of their investments on social and environmental issues
2019	International Platform on Sustainable Finance was initiated	To promote the alignment of financial flows with sustainable development objectives globally, facilitating the sharing of best practices and standards among countries and stakeholders
2020	The Abu Dhabi Declaration was endorsed	It was a significant event where global leaders and stakeholders affirmed their commitment to various environmental and developmental goals

Source: Authors

Finance Disclosure Regulation (SFDR) and Taxonomy Regulation, which require asset managers and financial participants to disclose environmental, social and governance (ESG) information. Additionally, the Task Force on Nature-Related Financial Disclosures (TNFD) was launched to aid financial institutions and businesses in identifying and reporting nature and biodiversity risks, following the Climate-related Financial Disclosures (TCFD) framework. Lastly, France's 2021 green budget evaluated the environmental impacts of public expenditure, with a focus on climate change, pollution and biodiversity loss.

3.2.3 Green Finance Initiatives

Green finance initiatives are financial practices and investments aimed at supporting environmental sustainability, combating climate change and promoting ecological conservation. These initiatives involve allocating resources to projects and enterprises that have a positive environmental impact, integrating environmental concerns into financial decision-making and creating financial products and services that support a low-carbon and climate-resilient economy. While many green finance initiatives exist, this chapter concentrates on the most effective and widely adopted ones. These initiatives have been categorised into green bonds, climate finance, green loans and green grants.

3.2.3.1 Green Bonds

Green bonds are financial instruments designed to support environmentally beneficial projects while providing investors with stable or fixed income. Various entities, including development financial institutions (DFIs), corporations and banks, issue these bonds, often through entities created specifically for certain projects. To ensure

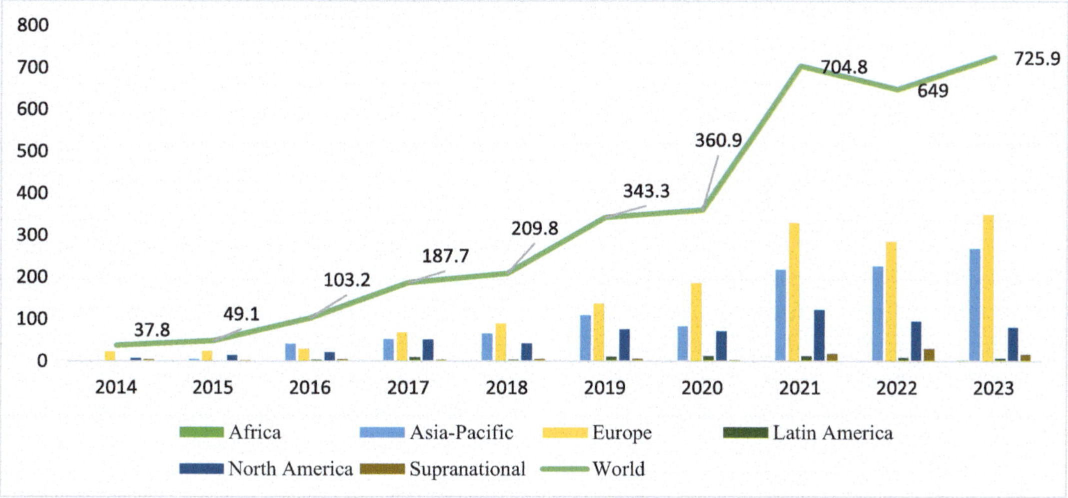

Fig. 3.1 Trend in green bonds by region (US$ billions), 2014–2023. (Source: Climate Bonds Initiative (2024))

consistency in this expanding market, standards such as the Climate Bond Standard and Green Bond Principles have been established. Adherence to these standards requires meeting criteria related to fund allocation, conducting thorough project assessment and selection, managing funds transparently and offering detailed reporting (World Economic Forum, 2020). In the past 14 years, green bonds have emerged as a vital resource in promoting a green environment. This initiative has become increasingly prevalent over the years. For instance, as illustrated in Fig. 3.1, the value of green bonds has surged from approximately US$37.8 billion in 2014 to about US$725.9 billion.

Despite China leading in total green bond issuance in 2023 (Myllyvirta, 2024), Europe stands out as the leading issuer at the regional level (see Fig. 3.1). It is not surprising that the European Investment Bank (EIB) was the first to issue a Climate Awareness Bond (CAB), making it the world's pioneer in green bonds in 2007.[2] Also, in response to demand from a group of Swedish pension funds in 2008, the World Bank became the first global institution to issue a green bond, aimed at supporting projects focused on mitigating climate change.[3] The World Bank remains a leading advocate for green bonds, with its subsidiary, the IFC issuing a US$1 billion green bond in March 2013. The IFC has also significantly influenced the global green bond market by aiding countries in the issuance of their own green bonds (see Box 3.1). Also, there have been some major corporations such as SNCF, Apple, ICBC and Credit Agricole who entered the green bond market. Notable developments include Tesla Energy's pioneering solar asset-backed security in 2013 and Massachusetts' launch of its first green municipal bond, marking significant milestones (OECD, 2015). By 2018, over US$13 billion had been raised through more than 150 green bonds globally (Kumar, 2022). Fiji issued the first sovereign green bond in 2016, and Fannie Mae emerged as the largest issuer in 2017, contributing to a total issuance of US$115.5 billion, with ambitions to reach US$1 trillion by 2020 (Wachira et al., 2023). Green bond issuance climbed to US$167.3 billion by 2018, supported

---

[2] https://www.eib.org/en/press/all/2022-308-15-years-of--eib-green-bonds-leading-sustainable-investment-from-niche-to-mainstream#:~:text=On%20July%205th%20 2007,of%20social%20and%20sustainability%20bonds. Accessed 18 April 2024

[3] https://www.worldbank.org/en/news/feature/2021/12/08/what-you-need-to-know-about-ifc-s-green-bonds#:~:text=In%20late%202007%2C%20a%20 group,for%20eligible%20climate%2Dfocused%20projects. Accessed 4 July 2024

by the Barclays/MSCI Green Bond Index introduced in 2014 to improve market liquidity (Kumar, 2022).

Box 3.1: International Finance Corporation Promote Green Bonds

International Finance Corporation (IFC) plays a crucial role in fostering the development of green bonds, particularly in emerging markets. In 2010, IFC introduced its first green bonds, catering to climate-conscious investors. In 2013, it issued the market's inaugural US$1 billion U.S. dollar-denominated green bonds, setting a precedent and strengthening the market. Acting as an anchor investor, IFC supports first-time issuers, facilitating their entry into the green bond market and preparing them for subsequent issuances. For instance, in August 2021, IFC invested US$100 million in Egypt's inaugural private sector green bond, aiding climate-smart projects and the nation's transition to a sustainable economy. Similarly, in Romania, IFC backed the first green bond issued by a financial institution, Raiffeisen Bank S.A. (RBRO), with a US$20 million investment. Additionally, IFC launched the Amundi Planet EGO Fund, the largest green bond fund in emerging markets, to invest in such bonds issued by financial institutions. Through initiatives like the Green Bond Technical Assistance Program (GB-TAP) and collaborations like the Real Economy Green Investment Opportunity Fund with HSBC Global Asset Management, IFC extends its support to expand the capacity of financial institutions and non-financial companies to issue green bonds, collectively raising over US$2.5 billion for investments in both sectors.

Source: World Bank (2021a, b).

3.2.3.2 Green Grants

A green grant is a financial support provided by governments, organisations or institutions to fund projects that promote environmental sustainability, conservation, or the transition to a low-carbon economy. These grants finance initiatives such as renewable energy projects, energy efficiency improvements, waste reduction, conservation efforts and other environmentally friendly activities. The objective is to encourage and facilitate actions that contribute to ecological benefits and sustainable development. This funding approach is adopted by various entities, including the World Bank, International Monetary Fund (IMF), United Nations (UN) and several universities. For example, New York University (NYU)'s Green Grants initiative supports projects enhancing environmental operations, promoting education and community involvement, advancing research and design, expanding sustainability education and showcasing effective practices and technologies (see Box 3.2).

Box 3.2: NYU Green Grant Projects

The Green Grants program at New York University (NYU) supports projects enhancing the university's environmental performance, promoting community engagement, advancing research, and expanding sustainability education. NYU is distinguished as a top global research institution recognized for its urban campuses in New York City, Abu Dhabi, and Shanghai, complemented by 12 study away sites. Renowned for its exceptional academic programs, NYU attracts students from 133 nations and champions social mobility, particularly for low-income and first-generation students. The university leads in research and patent awards, maintaining a vast global alumni network celebrated for accomplishments across various disciplines. The green grant project is open to all NYU students, faculty, administrators, and staff, the program offers grants up to US$20,000 for impactful projects with potential for institutionalisation. Applicants must submit an Expression of Interest (EOI) and undergo project consultation. Microgranta, up to US$3000, support immediate sustainability improvements and can evolve into larger Green Grants. Applications are reviewed by the Green Grants Committee (semesterly) or Office of Sustainability (rolling basis), ensuring projects align with NYU's sustainability goals and demonstrate innovative solutions for environmental stewardship.

Source: NYU (2024).

3.2.3.3 Climate Finance

Climate finance refers to the financial resources and instruments that are specifically designated to support efforts in mitigating and adapting to climate change. It encompasses a variety of funding sources and mechanisms aimed at reducing greenhouse gas emissions, enhancing resilience to climate impacts and promoting sustainable development (Wachira et al., 2023). Efforts to address climate change are bolstered by increased funding and investment. For instance, climate finance has surged from US$541 billion in 2018 to US$1415 billion in 2022, as depicted in Fig. 3.2. Climate finance has three main sources such as public, private sources and development partners. The public sources encompass government entities such as the State of Connecticut in the USA, New York MTA, the City of Gothenburg in Sweden, the Tokyo Metropolitan Government in Japan and national DFIs. Private sources include commercial banks, corporations, individual investors, philanthropic organisations, private equity firms and institutional investors. The development partners include funds from Multilateral Development Banks (MDBs), other Development Financial Institutions (DFI) and bilateral supports. I n addition to primary sources, various development institutions, including MDBs and partnerships with the sources mentioned earlier, establish specially designated funds. These are the Green Climate Fund (GCF), Climate Investment Fund (CIF), African Development Fund (ADF), Special Climate Change Fund (SCCF) and the Least Developed Countries Fund (LDCF), which are crucial contributors to climate finance.

While many institutions contribute to advancing climate finance, the United Nations Development Programme (UNDP) plays a significant role. It supports countries in designing programs to attract climate finance, as illustrated in Box 3.3.

Fig. 3.2 Global climate finance (US$ billion), 2012–2022. (Source: Climate Policy Initiative Report, 2023)

Box 3.3: United Nations Development Programme Role in Climate Finance

United Nations Development Programme (UNDP) is playing a crucial role in the Pacific Islands, where limited financial resources and frequent disasters challenge budgets. With its support, these nations are integrating disaster risk screening tools into budget assessments to enhance fiscal accountability for climate impacts. In Pakistan, ministries, including the Ministry of Water, are embedding climate considerations into budget planning, aligning over US$400 million with climate objectives. UNDP also fosters public-private partnerships for climate initiatives, such as aiding Indonesia's Ministry of Finance in issuing a US$1.25 billion Green Sukuk bond. Additionally, UNDP assists India, Vietnam, and Mongolia in accessing Green Climate Fund resources to mobilize private investments for climate adaptation.

Source: Beloe (2018) and UNDP (2022).

3.2.3.4 Green Loans

Green loans are a type of financing specifically designed to fund projects that have positive environmental benefits. These loans are structured to support initiatives aimed at improving environmental outcomes, such as reducing carbon emissions, enhancing energy efficiency and promoting resource conservation. Some examples include green credit cards, green car loans, green mortgages and municipal green loans, among others. The eligibility criteria for green loans ensure that projects meet the lender's environmental standards or certifications. Green lending began around 2005, with major U.S. banks like Wells Fargo and Bank of America allocating resources to sustainable ventures. Its usage has since surged, although the lack of regulatory mandates for disclosure makes it challenging to gauge the exact size of the green loan market. IFC has promoted a green loan agenda and has even supported countries to issue green loans to finance their green project (see Box 3.4).

Box 3.4: International Finance Corporation Facilitates Green Loans

IFC employs a bottom-up approach to estimate and track the size of green loan markets globally. This method identifies green loans based on project, corporate, and industry characteristics. The IFC combines various databases to estimate green activities, revealing that a significant portion of syndicated loans in 2014 incorporated green activities, totalling around $165 billion. Real estate comprises 41% of green loans by dollar value, while clean energy accounts for 15% In 2021, the International Finance Corporation (IFC) initiated several significant green loan projects across various countries, demonstrating its commitment to advancing sustainable development and climate action.

In Mexico, IFC structured a US$541 million, 15-year Green Loan facility for Infraestructura Energetica Nova (IEnova). This funding aims to support the construction of five solar plant projects with a combined capacity of 526 MW. These projects are pivotal in displacing carbon-intensive thermal generation in Mexico, expected to reduce approximately 793,000 tonnes of CO2 equivalent annually. By financing IEnova's inaugural solar power ventures, IFC seeks to facilitate the company's transition towards a more sustainable business model. Notably, this investment marked IFC's first certified Green Loan in Mexico following IEnova's adherence to the Green Loan Principles.

Moving to Brazil, IFC allocated up to US$120 million through a green loan to Sicredi, a cooperative financial institution with over 5 million members. This funding is designated to bolster Sicredi's climate finance program, particularly focusing on photovoltaic (PV) energy projects across Brazil. The initiative aims to diversify the country's energy mix, promote sustainability, and contribute to Brazil's climate objectives by facilitating investment in renewable energy infrastructure.

In South Africa, IFC made history by issuing Africa's inaugural certified green loan to Absa Bank Ltd., amounting to US$150 million. This financing is intended to increase support for biomass and other renewable energy projects, crucial for advancing South Africa's energy sector and economic recovery post-COVID-19. Absa Bank Ltd., one of Africa's largest financial services groups, will utilize the loan to expand its climate finance portfolio and help the country achieve its greenhouse gas reduction targets.

Finally, in Romania, IFC addressed the pandemic-stricken retail property sector by providing a €73.5 million green loan to NE Property BV, a subsidiary of NEPI Rockcastle, the largest retail property owner in Central and Eastern Europe. This financing will facilitate the development of sustainable infrastructure, fostering low-carbon economic growth in Romania and supporting the recovery of the retail property market.

Through these initiatives, IFC demonstrates its role as a catalyst for sustainable finance, leveraging green loans to drive impactful investments in renewable energy, climate-friendly projects, and resilient infrastructure across diverse global markets. These efforts not only promote environmental sustainability but also contribute to economic resilience and inclusive growth in the regions they serve.Source: IFC (2017) and World Bank (2021a).

The United States leads in total green loans, followed by the United Kingdom, Australia, France, Japan and China, reflecting the size of their financial markets (Gilchrist et al., 2021). In terms of the percentage of green credit in loan markets, Turkey has the largest share, exceeding 70% (Gilchrist et al., 2021).

3.2.4 Importance of Green Finance

Green finance becomes highly significant when a region or country adopts such financing options or policies to attract investment. The following are some importance of green finance.

3.2.4.1 Promote Sustainable Development

At the global, regional and national scales, efforts are underway to attain sustainable development goals. Therefore, there are some policies such as SDGs, EU Green Initiative, African Agenda 2063 and regional policies that aim for balanced sustainable development, integrating economic growth, social inclusion and environmental protection. They prioritise goals like poverty reduction, healthcare, education, climate action, clean energy and sustainable resource use, striving for an equitable, resilient world. However, financial constraints are causing slow and uncertain progress towards these goals. For example, achieving the SDGs requires an estimated annual investment of US$5–7 trillion across various sectors.[4] The Intergovernmental Panel on Climate Change (IPCC) recommends allocating between US$1.6 trillion and US$3.8 trillion towards the energy system from 2016 to 2050. Moreover, there is a US$2 trillion need for investments in green projects from 2021 to 2023, in the aftermath of the Coronavirus Disease 2019 (COVID-19) pandemic. The Paris Agreement underscores the necessity for continual, efficient funding and heightened global investment. However, the green finance initiative has made it possible to raise funds to finance such development agenda. Through this funding, it also helps in achieving sustainable development. For example, green finance initiatives enable investments in renewable energy projects such as solar, wind, hydro and geothermal power. These efforts promote the transition to clean energy sources, aligning with SDG 7, and contribute to reducing greenhouse gas emissions and addressing climate change goals outlined in SDG 13. Green finance initiatives are expected to lead to a 26% decrease in the combustion of fossil fuels (IEA, 2017). Green finance supports sustainable infrastructure initiatives (SDG 9), including public transportation, waste management and eco-friendly buildings.

[4]https://www.unepfi.org/positive-impact-2/#:~:text=An%20estimated%20%245%2D7%20trillion,address%20the%20SDGs%20financing%20gap. Accessed 18 Apiril, 2024

These investments enhance urban resilience, reduce environmental impacts and promote sustainable economic growth. They also aid in biodiversity conservation and ecosystem restoration, aligning with SDG 14 and SDG 15 goals. For instance, the Asian Development Bank (ADB) has used green finance to promote sustainable development in Asia (see Box 3.5).

Box 3.5: Asian Development Bank Used Green Finance to Promote Sustainable Development

ADB has committed US$600 million to a set of environmentally friendly projects aimed at transforming waste into clean energy, reducing CO2 emissions, enhancing eco-friendly transport, and preserving delicate wetland areas in rapidly growing second-tier cities within the People's Republic of China (PRC). A significant portion of the funding, amounting to US$200 million, will go towards developing biomass power plants in Anhui, Jiangsu, Shandong, and southern provinces. These plants will convert agricultural and municipal waste into clean energy, benefiting millions of people while reducing greenhouse gas emissions by 638,000 tons annually. Another US$150 million loan will be allocated to expanding district heating systems in Heilongjiang province, benefiting 270,000 households previously reliant on coal stoves for heating. This initiative will mitigate air pollution and respiratory diseases, particularly among vulnerable groups like women and children. In western Gansu province, US$100 million in ADB assistance will be directed towards improving infrastructure for pedestrians and cyclists, as well as enhancing road safety. Additionally, this project aims to restore 1400 hectares of damaged wetland habitats in Zhangye city, promoting sustainable ecotourism and creating job opportunities. Furthermore, US$150 million will be used to support the sustainable urbanization of emerging towns in Liaoning province by bolstering infrastructure to accommodate larger populations. This includes the construction of a wastewater treatment plant, upgrading heating networks, and expanding public green spaces, benefiting over 1.5 million people. ADB issued its first blue bond for US$300 million in September 2021 to promote blue sea. ADB is committed to boosting investment in healthy oceans to US$5 billion by 2024.

Source: ADB (2012).

3.2.4.2 Energy Security

Investing in renewable energy infrastructure, such as solar, wind, hydro and geothermal power, provides numerous strategic benefits for nations transitioning away from finite and environmentally harmful fossil fuels. These investments reduce reliance on non-renewable resources like coal, oil and gas, offering natural replenishment and lower environmental impact, thereby curbing greenhouse gas emissions and combatting climate change. Economically, such investments drive growth in renewable energy sectors, fostering job creation, local economic development and innovation. They attract private sector investment, enhancing economic resilience and prosperity. Diversifying energy sources through renewables bolsters energy security by reducing vulnerability to supply disruptions and price fluctuations associated with imported fossil fuels. This diversification ensures stable energy supplies for industries, households and essential services. In contrast, dependence on finite fossil fuels poses risks of geopolitical tensions and economic instability due to global competition. Renewable energy's abundance and widespread availability offer a sustainable and secure energy future. Overall, investing in renewable energy infrastructure strengthens nations' energy independence, supports sustainable development goals, promotes technological advancement and positions countries as leaders in the global transition to clean energy. For example, Box 3.6 highlights China's opportunity to enhance energy

Box 3.6: Renewable Energy Investment in China

China has actively promoted renewable energy, particularly wind and solar, through green finance initiatives. The People's Bank of China (PBOC) introduced a taxonomy defining eligible projects in the Green Bond Endorsed Project Catalogue in 2015, enhancing market integrity and boosting green bond issuances. From 2019 to 2023, China increased its renewable energy capacity by 185%. In 2023, substantial investments totalling 1 trillion yuan accelerated solar energy growth, driven by government programs like "whole county distributed solar" and "clean energy base," and supportive policies. Fifteen provinces prioritized solar development, adding approximately 200GW in capacity nationwide, setting a new record. China leads globally in solar PV technology and dominates wind energy production with nearly 329 GW installed capacity in 2021, bolstered by significant investments and government subsidies. Its leadership in onshore and offshore wind energy and electric vehicles (EVs), with 22% of 2022 passenger vehicle sales being all-electric, underscores its commitment to sustainable energy solutions, addressing energy security, poverty alleviation, and affordability challenges.

Source: IEA (2023), Macaire and Naef (2023) and Myllyvirta (2024).

security through substantial investments in green finance.

3.2.4.3 Reduce Greenhouse Gases

Green finance reduces greenhouse gases (GHGs) by financing projects that promote sustainable practices. It funds renewable energy sources like solar and wind, replacing fossil fuels and cutting greenhouse gases. Investments in energy-efficient industries, buildings and infrastructure lower energy use per output, reducing emissions. Financial tools like carbon pricing drive businesses to cut emissions or invest in offsets, further reducing carbon footprints (see Box 3.7).

Box 3.7: Carbon Pricing Mechanism

Carbon pricing is a fundamental tool aimed at integrating the external costs of greenhouse gas (GHG) emissions into economic frameworks. It assigns a monetary value to carbon dioxide (CO2) emissions, thereby holding emitters accountable for the environmental harm caused. This strategy empowers emitters to choose between reducing emissions through operational changes or continuing to emit and covering associated costs. The primary objective of carbon pricing is to achieve environmental goals in a flexible and cost-effective manner, encouraging investment in clean technologies and stimulating economic growth toward low-carbon alternatives. Governments find carbon pricing attractive not only for emissions reduction but also as a revenue source, crucial in financially constrained contexts.

Businesses employ internal carbon pricing to evaluate potential carbon-related expenses, assess climate risks, and explore revenue opportunities through sustainable initiatives. Investors utilize carbon pricing to gauge the impact of climate policies on their portfolios, prompting strategic shifts towards sustainable and climate-resilient investments. Globally, carbon pricing mechanisms vary and include carbon taxes, emissions trading systems (ETS), and carbon crediting mechanisms. ETS provide certainty in emissions reduction targets while maintaining price flexibility. In contrast, carbon taxes establish fixed carbon prices, ensuring predictability but with variability in environmental outcomes. Some regions implement carbon taxes with varying rates across different fuels, resembling indirect carbon pricing strategies. To

(continued)

Box 3.7: (continued)

effectively decarbonize economies and address climate challenges comprehensively, a diverse range of carbon pricing approaches must complement supportive policies and regulations. This holistic approach accelerates emission reductions and facilitates a sustainable economic transition. As international consensus on carbon pricing's significance grows, its role expands in shaping a future focused on decarbonization, influencing policy decisions and strategies across global sectors.

Source: World Bank (2024).

Finance supports constructing sustainable buildings and infrastructure with efficient systems, such as green roofs and insulation. It also backs electric vehicles, charging infrastructure and public transit to cut fossil fuel use in transportation. Funding conserves forests and wetlands to store carbon dioxide, easing climate impacts. For instance, in Box 3.5, ADB utilised green finance to reduce greenhouse gas emissions by 638,000 tonnes annually in Southern Asia.

3.3 Green Finance Initiative in Banking: Motivation and Potential Metrics

Green Finance Initiatives in Banking (GFIB) refer to efforts by financial institutions to integrate environmental considerations into their lending, investment and risk management practices. These initiatives aim to support environmentally sustainable projects and businesses while promoting responsible financial practices. Most banks have now integrated most of the green finance initiatives discussed into their investment and decision-making. Some of the initiatives include but are not limited to the following:

1. Offering green loans such as green car loans, green credit cards and green deposits
2. Provision of green bonds such as green bonds
3. Engage in carbon credits
4. Serve as venture capital for green projects and financing for client warehousing initiatives
5. Reducing investments in environmentally high-risk projects
6. Promotion of e-banking software and applications, enhancing customer engagement in electronic banking services
7. Installation of light-emitting diodes (LEDs) in bank offices to promote energy efficiency
8. Solar-powered ATMs and other renewable energy ATMs deployment
9. Adherence to global and national green banking principles and policies

3.3.1 Motivation

3.3.1.1 Regulatory Requirement and Policy Recommendation

There are numerous regulations in place that either enforce or incentivise banks to incorporate green finance initiatives into their decision-making processes. These regulations typically stem from government policies and international agreements aimed at addressing environmental challenges such as climate change and resource depletion. For instance, a regulatory requirement by the Bank of Ghana through its policy (i.e., Ghana Sustainable Banking Principle)

Box 3.8: Ghana Sustainable Banking Principles

In April 2015, the Bank of Ghana partnered with the International Finance Corporation (IFC) to establish a national platform aimed at promoting sustainable banking practices. This initiative culminated in the formation of the Sustainable Banking Committee, which was officially inaugurated on November 20, 2015. The committee's primary mandate was to develop guidelines and principles for sustainable banking. By

2017, significant progress was made as the committee drafted the first set of Sustainable Banking Principles and Sector Guidance Notes. These foundational documents laid the groundwork for integrating environmental and social risk management (ESRM) into the operations of Ghanaian banks.

A pivotal moment arrived in November 2019 when the Bank of Ghana officially launched the Ghana Sustainable Banking Principles (SBPs). These principles are designed to serve as a framework for effective ESRM policies within banks across the country. The SBPs encompass seven key principles:

1. Environmental and Social Risk Management (ESRM)
2. Integration of Environmental, Social, and Governance (ESG) factors in bank operations
3. Corporate governance and ethical standards
4. Gender equality
5. Financial inclusion
6. Resource efficiency, sustainable production, and consumption
7. Reporting requirements focused on sectors such as Agriculture & Forestry, Construction & Real Estate, Manufacturing, Oil & Gas, Mining, Power & Energy—sectors with significant environmental and social sensitivities and substantial portfolio exposure in banks

By 2020, 24 commercial banks in Ghana had committed to adopting and implementing the Sustainable Banking Principles. Notable institutions such as Barclays Bank of Ghana now Absa Bank Ghana Limited, Ecobank, and GCB Bank were among those that pledged to measure and publicly report their progress in adhering to these principles. This collaborative effort between the Bank of Ghana, IFC, and Ghanaian commercial banks underscores a concerted commitment to integrating sustainability into the financial sector. By promoting responsible banking practices through these principles, Ghana aims to enhance environmental stewardship, promote inclusive economic growth, and ensure sustainable development across key sectors of its economy.

Sources: Green Finance Platform (2024).

encouraged banks to integrate green finance initiatives into their business model (see Box 3.8).

3.3.1.2 Carbon Footprint Reduction

Green finance initiatives align with the Paris Agreement's objectives by steering investments towards environmentally sustainable initiatives. Therefore, banks that integrate these initiatives have facilitated the development of climate-resilient infrastructure, leading to reduced carbon emissions. Notably, such banks have spurred the uptake of low-carbon technologies like home geothermal power systems within communities. Expanding these initiatives could accelerate progress towards achieving net-zero emissions. To foster a low-carbon future, increased green finance initiatives in banks are essential to swiftly enhance the accessibility of community-level renewable energy systems. For example, Aspiration Green Bank introduced a product that seeks to minimise carbon footprint (see Box 3.9).

Other financial institutions, including the Bank of China (Hong Kong) Limited, have also introduced green time deposits. These deposits are designed to fund projects in green building, pollution prevention and environmentally sustainable agriculture, thereby supporting the transition to a low-carbon, climate-resilient economy (see Box 3.10).

3.3.1.3 Impact of Green Banks on Mainstream Banking Operations

Some countries have established green banks that specialise in financing environmentally sustain-

Box 3.9: Aspiration Green Bank Introduces Green Credit Card

Aspiration Bank has launched the Aspiration Zero Credit Card (AZCC), designed to promote environmental sustainability. This digital bank prioritizes reducing carbon footprints by offering cash back on purchases while planting trees. Aspiration allocates 10% of its profits to support struggling Americans and aims to plant 100 million trees globally. The AZCC distinguishes itself by funding only clean initiatives, thereby combatting "greenwashing." Despite a US$60 annual fee and foreign transaction fees, the card incentivizes eco-conscious spending by planting trees with each purchase and providing carbon-neutral rewards. It is also eco-friendly, crafted from plant-based materials. The ongoing tree planting acts as a carbon sink, gradually reducing carbon footprints.

Source: Holzhauer (2023).

Box 3.10: Bank of China (Hong Kong) Limited Issue Green Time Deposit

Bank of China (Hong Kong) Limited (BOCHK) is committed to advancing green and sustainability initiatives through its innovative green time deposit product for corporates. This deposit supports a wide range of green industries in Hong Kong and is notable for being the first green deposit certified by an independent third party in the region. Funds from this deposit are allocated to projects in green building, pollution prevention, and environmentally sustainable agriculture, fostering a transition to a low-carbon, climate-resilient economy. The certification is provided by Deloitte Touche Tohmatsu, with a post-issuance attestation report for 2023. The general terms of this product include the bank's discretion to amend, suspend, or terminate the products, services, and offers at any time without prior notice. The preferential interest rate applies only to the initial time deposit, with renewal rates subject to BOCHK's prevailing quotes. Any disputes are subject to the bank's final decision. Additionally, customers are responsible for any costs associated with using BOCHK Mobile Banking. In case of discrepancies between the English and Chinese versions of promotional materials, the Chinese version will take precedence.

Source: Bank of China (Hong Kong) Limited (2024).

able projects like renewable energy and conservation. These assess projects using strict environmental criteria, offering tailored financial solutions crucial for transitioning to a low-carbon economy. Green banks adhere to certifications such as B Corp Certification for high social and environmental standards. Many are part of the Global Alliance for Banking on Values (GABV), promoting sustainable development. They prioritise digital transactions to reduce environmental impact and emphasise technological efficiency. Some examples of these banks include Aspiration Bank (USA), Charity Bank (UK), Spring Bank (USA), vdk Bank (Belgium), Beneficial State Bank (USA), Triodos Bank (Netherlands), Sunrise Bank (USA), The Connecticut Green Bank (USA), DC Green Bank (USA), Montgomery County Green Bank (USA), New York Green Bank (USA) and among others.

The establishment of green banks has spurred existing commercial banks to incorporate GFIB into their offerings. This shift is driven by a desire to appeal to environmentally conscious customers and investors. To remain competitive, commercial banks have integrated these principles into their business models. For instance, one of the largest banks in India (Axis Bank) has widely integrated green finance initiatives (see Box 3.11). Also, Bank of China (Hong Kong) Limited has expanded its scope by including green finance initiative (see Box 3.10).

Box 3.11: Axis Bank Ltd. Integrate Green Finance Initiative

Axis Bank provides a wide range of financial services to diverse sectors including large and mid-corporates, Micro, Small, and Medium Enterprises (MSMEs), agriculture, and retail. This bank is the India's largest private bank which has a lot of customers. Demonstrating its commitment to sustainability, Axis Bank initiated a green finance initiative program in August 2011, converting dry waste from its Mumbai corporate office and 34 branches into notebooks, notepads, and envelopes, recycling over 100,000 kilograms of paper. Axis Bank's Mumbai corporate office is also Platinum LEED certified for its eco-friendly operations. The bank promotes carpooling to reduce carbon emissions, advocates for digital communications to reduce paper consumption, and sends annual reports via email. Moreover, Axis Bank has installed ten solar-powered ATMs in the Coimbatore circle as part of its sustainable practices and efforts to educate customers on environmental responsibility.

Source: Axis Bank (2024).

3.3.1.4 Green Corporate Image

GFIB enhances the reputation of banks by demonstrating and fulfilling its commitment to environmental stewardship. A well-executed green positioning strategy can positively influence how the brand is perceived. The corporate image of banks is shaped by interactions among the institution, its staff, customers and the community regarding environmental responsibilities. When a company's green products are reliable, align with environmental needs and exhibit strong environmental performance and reputation, they can foster a positive green image. According to Chen (2010), customers perceive a green brand image as indicative of environmental dedication and concerns, achieved through competence, sustainability and trustworthiness. This underscores that a positive green bank image significantly impacts customer loyalty by retaining existing customers, reclaiming lost ones and attracting new ones, ensuring the bank's long-term prosperity and sustainability. Research by Sharma and Choubey (2022) indicates that Indian banks integrating green finance initiatives tend to enhance their green corporate image, ultimately influencing customer loyalty positively over time. Also, Ghana has had a limited availability of green loans, but Standard Chartered Bank Ghana and Absa Bank Ghana limited stand out by having introduced a green loan to enhance its corporate image as an environmentally responsible institution (Box 3.12).

Box 3.12: Some Banks in Ghana Have Launched Green Loan to Gain Corporate Green Image

Standard Chartered Bank Ghana has recently launched a renewable energy loan, expanding its portfolio to support individuals and businesses interested in adopting renewable energy solutions and reducing their carbon footprint. This new loan is particularly focused on solar energy, a major renewable resource in Ghana. It features competitive interest rates and flexible repayment terms, designed to make solar energy more accessible to the public. In addition to the financial product, the bank provides expert guidance throughout the loan application process, ensuring clients receive comprehensive support. This initiative underscores Standard Chartered Bank Ghana's commitment to enhancing financial inclusion and promoting economic growth. By making renewable energy more accessible and affordable, the bank empowers Ghanaians to pursue their goals with confidence and embrace sustainable energy solutions. As the financial sector evolves, Standard Chartered Bank Ghana continues

(continued)

Box 3.12: (continued)

to prioritize innovative approaches that benefit both customers and communities, reinforcing its role in supporting sustainable development and economic advancement.

Ghana's effort to diversify its energy sources and embrace renewable energy has been significantly advanced by a US$24 million corporate green loan facilitated by Absa Bank Ghana Limited. This financing will be used to develop a 100 MW solar power facility for the Bui Power Authority, enhancing the country's energy infrastructure and supporting its sustainable development goals. The solar project is led by Meinergy Technology Limited, a well-regarded company specializing in renewable energy solutions. This initiative marks Absa Ghana's first venture into green lending, establishing the bank as a key proponent of eco-friendly energy and regional sustainability. Absa Bank Ghana limited's successful execution of this deal was characterized by a rigorous adherence to its sustainable lending criteria. The project underwent comprehensive evaluation to ensure it met high standards, thus earning its status as an exemplary green loan. This landmark transaction underscores Absa Bank Ghana limited's commitment to supporting renewable energy initiatives and promoting sustainable practices in the region.

Source: The Business and Financial Times (2024) and Chandak (2023).

3.3.2 Metrics for Evaluating Green Finance Effectiveness and Sustainability Contributions

As previously discussed, there are several motivations for banks to incorporate green finance initiatives into their business models. However, it is crucial for banks to establish key metrics for assessing the effectiveness and sustainability contributions of these initiatives. Therefore, we discuss some of these metrics: tonnes of CO_2 equivalent (tCO2e) reduced or avoided, percentage of funded projects adhering to recognised environmental standards or certifications, feedback and satisfaction levels from stakeholders, area of habitat preserved or restored (in hectares) and number of jobs created or communities impacted positively.

3.3.2.1 Tonnes of CO2 Equivalent (tCO2e) Reduced or Avoided

This metric quantifies the direct impact of green finance projects on mitigating greenhouse gas emissions by measuring the amount of carbon dioxide equivalent (CO2e) that is reduced or avoided as a result of the initiative. It provides a tangible measure of how effectively green finance projects contribute to the reduction of greenhouse gases, which is crucial for assessing their environmental benefits. By calculating tCO2e reductions, banks can evaluate the success of their green finance investments in lowering overall carbon emissions, thereby supporting climate goals and sustainability targets. This metric not only helps in tracking progress towards emission reduction objectives but also enhances transparency and accountability in reporting the environmental impact of green projects. Therefore, when banks want to integrate green finance initiatives, they must check how much tCO2e will be reduced.

3.3.2.2 Percentage of Funded Projects Adhering to Recognised Environmental Standards or Certifications

This metric gauges the proportion of financed projects that adhere to established environmental standards or hold recognised certifications. It is designed to ensure that the projects receiving financial support meet rigorous environmental criteria and align with best practices in sustainability. By monitoring this percentage, financial institutions can verify that their investments are contributing to environmentally responsible outcomes and are compliant with industry-recognised benchmarks. This metric helps in maintaining high standards for environmental performance, ensuring that funded projects not only achieve their intended environmental goals but also contribute to broader sustainability objectives. It also provides transparency and accountability in the green finance process, reinforcing confidence that financial support is directed towards projects that genuinely adhere to accepted environmental standards and practices.

3.3.2.3 Feedback and Satisfaction Levels from Stakeholders

This metric assesses the effectiveness of stakeholder engagement and the perceived value of green finance initiatives by gathering and analysing feedback from various stakeholders, including community members and environmental organisations. By measuring satisfaction levels and feedback, banks can gauge how well their green finance efforts align with stakeholder expectations and needs. This feedback provides valuable insights into the impact and reception of green projects, revealing whether they are meeting their intended goals and contributing positively to the community and environment. Understanding stakeholder perceptions helps banks to refine their strategies, enhance transparency and build stronger relationships with key parties involved. Ultimately, this metric supports the continuous improvement of green finance initiatives by ensuring they deliver meaningful benefits and foster greater trust and support from stakeholders.

3.3.2.4 Area of Habitat Preserved or Restored (in Hectares)

This metric assesses the positive impacts of green finance projects on local ecosystems and biodiversity by measuring the area of habitat preserved or restored, typically quantified in hectares, and monitoring changes in biodiversity indicators. It provides a clear measure of how effectively these projects contribute to the conservation and rehabilitation of natural environments. By tracking the extent of preserved or restored habitats, as well as shifts in biodiversity levels, banks and stakeholders can evaluate the environmental benefits of their green finance investments. This metric helps to ensure that green finance projects are not only addressing climate change but also supporting the resilience and health of ecosystems. It also highlights the contribution of these initiatives to broader environmental goals, such as protecting endangered species and enhancing ecological balance. Monitoring these outcomes is crucial for demonstrating the tangible benefits of green finance and for guiding future investments in environmental sustainability.

3.3.2.5 Number of Jobs Created or Communities Impacted Positively

This metric evaluates the broader socio-economic benefits of green finance initiatives by measuring the number of jobs generated and the extent of positive impacts on communities. It provides insights into how green finance projects contribute to economic development and social well-being. By tracking job creation, this metric highlights the role of green finance in fostering employment opportunities and supporting local economies. Additionally, assessing the positive impact on communities, such as improved infrastructure, enhanced quality of life, or increased access to essential services, demonstrates the wider benefits of these initiatives beyond environmental outcomes. This comprehensive evaluation helps to showcase the value of green finance in promoting sustainable development, addressing social inequalities and driving economic growth. It also underscores the alignment of green finance projects with broader societal goals, ensuring that they deliver holistic benefits and contribute to the overall resilience and prosperity of communities.

3.4 Policies and Principles That Support Green Finance Initiative in Banking

There have been various policies and principles at the global and national levels that support GFIB. This section discusses such policies.

3.4.1 Global Principles and Policies

The chapter specifically examines the direct principles, policies and practices that encourage banks to integrate green finance initiatives, distinguishing them from indirect aspects. It is noteworthy that all global policies and principles in this realm are voluntary. (see Table 3.2). This indicates that any bank can voluntarily adopt these policies.

An early global principle is the United Nations Environment Programme Finance Initiative (UNEP FI), which collaborates with the global financial sector, aiming to redirect private finance towards sustainable development. Partnering with banks, insurers, investors and supporting institutions, it promotes a financial sector prioritising societal and environmental well-being. Through inspiration and education, UNEP FI encourages institutions to improve lives while safeguarding the planet, leveraging the UN's influence to accelerate sustainable finance. Key initiatives include the development of industry-based principles such as the Principles for Responsible Banking (PRB), Principle for Sustainable Insurance (PSI), Principles for Responsible Insurance (PRI) and Sustainable Blue Economy Finance Principles (SBEFP). UNEP FI has helped a lot of countries to develop their national GFIB policies.

Also, EPs which were initiated in 2003 direct banks to evaluate environmental and social risks in project financing, focusing on sectors such as energy and infrastructure. It requires banks to account for community and biodiversity impacts, integrating sustainability into financing decisions to foster development while minimising adverse effects. Regular updates ensure alignment with evolving standards.

SBFN which was formerly known as SBN is another important policy that represents a distinctive gathering of financial sector regulators and banking associations hailing from emerging markets, united in their dedication to promoting sustainable finance. It emerges as a pivotal framework which provides fundamental principles, policies and operational strategies that support GFIB. Key stakeholders including central banks,

Table 3.2 Major global principles and policies for green finance initiative in banking

Year	Policy	Players	Conduct
1992	United Nations Environment Programme Finance Initiative (UNEP FI)	375 banks, 160 insurance, 6 other investors	Leveraging the UN's influence to accelerate sustainable finance, encouraging institutions to improve lives while safeguarding the planet
2003	Equator Principles (EPs)	101 Financial institutions	Facilitate responsible global investment, balancing economic development and environmental protection, while providing a standardised framework for financial institutions to manage project impacts
2009	Global Alliance for Banking on Values (GABV)	73 Banks	GABV is an international network of independent banks dedicated to leveraging finance for sustainable economic, social and environmental development
2012	Sustainable Banking and Finance Network (SBFN)	91 members: 55 regulators, 36 industry associations 80% banks, 18% capital markets, 13% insurance 13%, others 10%. 5% environmental agencies and 1% pension funds	Dedicated to promoting sustainable finance to address national development priorities, foster financial market deepening and ensure stability, SBFN collaborates closely with the International Finance Corporation (IFC), which serves as its secretariat and knowledge partner

Source: Zhang (2016), Barua (2020), Park and Kim (2020), Equator Principles (2024), SBFN (2024), PRI (2024) and UNEP FI (2024)

financial regulators and banking associations collaborate within this framework. Initiated in 2012, its core objective is to foster shared knowledge among members, empowering them to spearhead policy advancements and innovative initiatives essential for nurturing sustainable finance ecosystems. From its inception, the IFC has played a pivotal role, actively contributing as SBFN's secretariat, technical advisor and knowledge partner.

Furthermore, GABV also underpins the essence of GFIB. It is a coalition of autonomous banks leveraging finance to drive sustainable economic, social and environmental progress. It was founded in 2009 and has over 70 members across 40+ countries. GABV promotes fair and inclusive financial systems through initiatives like International Banking on Values Day and regional chapters. It advocates for regulated, value-based banking and provides tools like the GABV scorecard for banks to measure their social impact. GABV hosts summits to further its mission of sustainable finance and global impact.

3.4.2 National Policies and Principles

Table 3.3 outlines policies and principles regarding GFIB, revealing disparities in national approaches. While some countries focus on green bonds, others neglect products like sustainable insurance or green insurance. Private sector engagement is vital for green growth, with countries offering incentives such as awards or dedicated funds to encourage investment in green finance initiatives, enhancing sustainability efforts globally.

3.5 Challenges of Green Finance Initiative in Banking

Despite the relevance of green finance initiatives, banks may encounter some challenges when integrated into their activities. This section therefore provides discussions on some of these challenges.

3.5.1 Risk Assessment Challenges

Banks face challenges in accurately assessing the risks associated with green projects, such as technological risks in new renewable energy ventures or regulatory risks in emerging markets.[5] Green projects frequently involve innovative technologies, such as advanced solar photovoltaic systems, wind turbines and cutting-edge energy storage solutions. These technologies, while promising, may not have extensive performance histories, making it difficult to predict their long-term reliability and efficiency. Banks must grapple with the uncertainty of these technologies' lifecycles, potential technical failures and maintenance needs, which can impact the overall viability and success of the projects. Research conducted in India highlights that green banks face difficulties when assessing risks associated with environmental initiatives (Giridhar & Sudhakar, 2017; Rahman et al., 2023).

3.5.2 The Issue with Assessing Financial Viability

Evaluating the profitability and financial viability of green projects compared to traditional investments is challenging, affecting banks' willingness to finance them. Green projects, such as renewable energy or eco-friendly infrastructure, typically require substantial initial investment and have longer payback periods (Eti et al., 2024). They rely on regulatory support and market stability, which can be unpredictable. The lack of historical performance data further complicates risk assessment and return prediction, leading to cautious investment decisions. Traditional investments in sectors like fossil fuels offer more predictable and shorter payback periods with well-understood risk profiles, making banks more comfortable financing them. Balancing these factors, banks must weigh the higher perceived risks and potential long-term

[5] https://greenly.earth/en-us/blog/ecology-news/how-can-you-make-the-change-to-green-finance. Accessed 4 July 2024

Table 3.3 Some examples of national principles and policies of green finance initiative in banking

Year	Country	Principles/policy	Concepts
2007 and 2012	China	Green Credit Policy launched in 2007, followed by Green Credit Guidelines introduced by banks in 2012	Banks report loan balances in 12 green sectors, meeting global sustainability standards with monitoring systems
2008 and 2014	Brazil	Banking association launched Voluntary Green Protocol in 2008; FIs adopted Social and Environmental Responsibility Resolution in 2014	State-owned and commercial banks commit voluntarily to adhere to green guidelines
2009–2019	USA	National Green Banking Act 2019 (Proposed in 2009 but implemented in 2019	To regulate and promote environmentally sustainable practices within the banking sector
2011–2012	Bangladesh	Environmental Risk Management Guidelines	Banks are required to develop their environmental and social risk management framework and adopt an assessment tool
2012	Nigeria	Nigerian Sustainable Banking Principles	Mandates environmental, social and governance (ESG) policy adoption and provides three sector guidelines (oil and gas, power, agriculture) for sensitive sectors
2012	Colombia	Colombia Green Protocol	Voluntary implementation of E&S policy with clear performance standards, including examples and tools for guidance
2012	Lebanon	Green prudential regulation policy	Regulation to integrate climate, environmental and sustainability factors into banking practices
2014	Mongolia	Mongolian Sustainable Finance Principles	Assistance for banks in integrating E&S factors into lending decisions and product design to promote sustainability
2014	Turkey	Sustainability Guidelines for the Banking Sector	It outlines risk categorisation, mitigation strategies, monitoring plans and assessment methods
2014	South Africa	Principles for Managing Environmental and Social Risk	Mandate financial institutions to promote social, economic and environmental rights through sustainable practice
2014	Indonesia	Sustainable Finance Roadmap	Financial institutions are mandated to integrate sustainable finance into their business operations and activities
2015	Kenya	Sustainable Finance Guiding Principles	The five principles centre on economic viability, inclusive growth, social and environmental risk management, resource scarcity and ethical values
2015–2018	Vietnam	Directive on Promoting Green Credit Growth and E&S Risk Management	Credit institutions must formalise E&S risk policies, report to the central bank and promote lending to green projects.
2015	Peru	Regulation for Social and Environmental Risk Management	Peru introduces Regulations for Social and Environmental Risk Management in its Financial System, holding financial entities accountable for environmental damage
2015 and 2021	China	Green Bond Guidelines and Catalogue	In 2015, People's Bank of China (PBOC) issued guidelines for green bonds in China, outlining project criteria and management of proceeds. In April 2021, PBOC updated the green bond project catalogue, effective July 1, excluding coal and fossil fuels

2016	Morocco	Roadmap for Aligning the Moroccan Financial Sector with Sustainable Development, covering banking, capital markets and insurance	Align its financial sector with sustainable development, emphasising environmental, social and governance concerns, reviewed periodically
2016	Brazil	Guidelines for Issuing Green Bonds	Guidelines to boost the green bond market, leading to increased issuance and potential growth
2016	India	Disclosure Requirements for Issuing and Listing of Green Bonds	It promotes transparency, accountability and credibility in the green bond market, facilitating informed investment decisions and supporting sustainable development initiatives
2016	France	Energy Transition and Green Growth Law	This law ensures that listed companies, banks, credit providers and investors adhere to information disclosure requirements
2016	Morocco	Guidelines for the Issuance of Green Bonds	Moroccan Capital Market Authority and IFC collaborated to create guidelines promoting the Green Bonds market, clarifying roles and requirements
2016	Nigeria	Green Bond Guidelines	Standards and procedures for issuing green bonds, promoting environmental sustainability and attracting investments in eco-friendly projects
2017	South Africa	Green Segment/Amendment to the Johannesburg Stock Exchange (JSE) Debt Listing Requirements	Facilitate the listing of green bonds and promote transparency, credibility and investment in environmentally sustainable projects on the Johannesburg Stock Exchange (JSE)
2017	Bangladesh	Updated Environmental and Social Risk Management Guidelines	Enhance projects by integrating environmental and social factors, minimising negatives and maximising positives for sustainable development and well-being
2017	Indonesia	Sustainable Finance Regulation launched in 2014 to implement the roadmap for the entire financial sector	Promote sustainability by regulating finance, aligning with a roadmap, integrating ESG principles and fostering Islamic finance development
2018	Pakistan	Sustainable Banking Guidelines	Integrate environmental and social considerations into their operations and risk management practices
2018	Ghana	Sustainable Finance Roadmap	Integrate ESG principles in the financial sector
2018	Nepal	Environmental and Social Risk Management Guidelines	Integrate environmental and social factors into project phases to minimise harm and enhance positive outcomes
2018	Fiji	Sustainable Finance Framework and Green Bond Guidelines	Fiji's Green Bond Framework ensures transparency, integrity and quality, aligning with the 2017 Green Bond Principles
2018	Kenya	Green Bond Guidelines	Kenya's green bond guidelines streamline issuance, approved by the Capital Markets Authority and issued by Nairobi Securities Exchange
2018	Mexico	Green Bond Guidelines	Provide guidelines for issuers and certainty for investors regarding the environmental benefits of financed projects.
	Lebanon	Differential capital requirements for green projects	Incentivise banks to finance renewable energy and efficiency projects by lowering reserve requirements for these loans
2019	Ghana	Sustainable Banking Principles and Sector Guidance Notes	Highlight the banking sector's role in addressing climate change within Ghana's national and international commitments
2021	Pakistan	Green Bond Guidelines	Sustainable finance framework, offering guidance on ESG and climate risk

Source: Weber and Acheta (2016), Barua (2020), Park and Kim (2020), SBFN (2024) and Sharma and Choubey (2022)

benefits of green projects against the immediate, predictable returns of conventional investments.

3.5.3 Lack of Standardisation

The absence of standardised metrics and reporting frameworks for measuring and reporting the environmental impact of green projects complicates investment decisions by creating inconsistencies in how environmental benefits and risks are quantified and communicated (Radzewicz-Bak, 2024). Without a unified approach, banks and investors struggle to compare the environmental performance of different projects, assess their true sustainability impact and gauge potential returns accurately. This inconsistency leads to uncertainty and can deter investment, as stakeholders cannot confidently evaluate the efficacy or long-term viability of green projects. Moreover, it hampers transparency and accountability, making it difficult for banks to align their investment portfolios with their sustainability goals and for regulators to monitor and enforce environmental standards effectively. Standardised metrics would provide clarity, facilitate better risk management and encourage more informed and confident investment in green initiatives.

3.5.4 High Cost of Capital in Green Project

Green projects often entail higher upfront investment costs and longer payback periods compared to traditional ventures, which can deter banks that prioritise short-term returns. These projects, encompassing renewable energy, sustainable agriculture and eco-friendly infrastructure, demand significant initial capital and extended timelines to achieve financial returns (Radzewicz-Bak, 2024). This misalignment with the short-term financial goals commonly sought by banks can make green projects less appealing. Consequently, banks may hesitate to finance such initiatives due to the longer duration required to realise profitability, despite their potential long-term environmental and economic benefits. This focus on immediate returns over extended gains can limit the availability of funding for green projects, impacting the broader adoption of sustainable practices in the banking sector.

3.5.5 Changes in Regulatory and Policy Uncertainty

Shifts in environmental regulations and policies can create uncertainty for banks concerning the long-term viability of green investments. As governments and regulatory bodies frequently update standards to address evolving environmental challenges, these changes can impact the stability and predictability of green projects. Banks, therefore, face difficulty in assessing the future risks and returns of their green investments, as compliance requirements, financial incentives and market conditions may fluctuate. This regulatory uncertainty can lead to hesitancy in financing green initiatives, as banks strive to navigate the potential implications of policy changes on the financial sustainability and success of these projects over time.

3.5.6 Low Market Awareness and Demand

The growth of green finance initiatives in banks can be hindered by limited awareness and demand for green financial products among customers and investors, especially in rural areas. Many individuals and businesses may not fully understand the benefits or availability of green financial options, such as sustainable loans, green bonds, or eco-friendly investment funds. For example, in most areas of Ghana, most people will not understand the meaning of a green credit card. This lack of awareness translates into insufficient market demand, which can deter banks from developing or promoting such products. Without a strong customer base actively seeking green financial solutions, banks may find it challenging to justify the investment in these products, slowing the overall adoption and integration of green finance within the banking sector. This gap in knowledge and interest prevents the mainstreaming of green finance and limits its potential impact on environmental sustainability.

3.5.7 Inadequate Capacity Building

Banks may struggle with effectively assessing, managing and monitoring green investments due to a lack of necessary expertise and capacity (Giridhar & Sudhakar, 2017; Rahman et al., 2023). This deficiency can stem from limited experience with the unique risks and opportunities associated with sustainable projects, such as renewable energy ventures or eco-friendly infrastructure. The specialised knowledge required to evaluate the environmental impact, long-term viability and compliance with evolving green standards may not be as developed within traditional banking institutions. As a result, banks might find it challenging to integrate comprehensive green assessment frameworks, leading to hesitancy or suboptimal management of green finance initiatives. This expertise gap can impede their ability to fully support and scale sustainable investments, ultimately affecting the growth of green finance in the sector.

3.6 Future Trends and Prospects of Green Finance Initiative in Banking

3.6.1 Expansion of Sustainable Financial Products

Banks are likely to continue expanding their range of sustainable financial products to cater to growing customer demand and regulatory pressures. This includes more innovative green loans, green bonds and sustainability-linked loans that incentivise environmentally responsible behaviour. The development of specialised investment funds focusing on renewable energy, sustainable agriculture and low-carbon technologies will also become more prominent, providing more avenues for customers to invest in green initiatives. For example, since their inception in some Asian countries in 2020, green deposits have expanded to others, including Thailand, which has recently implemented this financial initiative (see Box 3.13).

Box 3.13: Green Deposit

Green Deposits have gained traction globally, enabling corporate and retail customers to invest in projects with positive environmental impacts via time deposits. These deposits are used to fund areas like alternative energy, sustainable agriculture, and clean technology, allowing savers to earn interest while supporting green initiatives.

Various banks have adopted green deposits with tailored frameworks. The Bank of China (Hong Kong) Limited (BOCHK) targets mainly corporate customers, funding pollution prevention and green building projects between 2020 and 2023. ICBC (Asia) has offered green deposits since 2022. In Japan, Sumitomo Mitsui Banking Corporation (SMBC) has been offering green deposits since April 2021, requiring significant minimum deposits of USD 50 million for 30-day terms and US$25 million for 180-day terms, all in US dollars. MUFG, another major Japanese bank, introduced green deposits in April 2022, available to corporate and retail customers for 1 to 12-month terms and USD US$1 million to US$25 million deposits, also in US dollars. Deutsche Bank launched green deposits for corporate clients in Germany in 2021, supporting renewable energy, energy efficiency, ICT, clean transportation, and green buildings, with €516 million allocated primarily to green buildings between January 2021 and June 2022.

India's Reserve Bank of India (RBI) established a framework in April 2023 for green deposits, setting regulations to prevent greenwashing and ensure proper fund allocation. Eligible projects mirror Japan's

(continued)

Box 3.13: (continued)

categories but exclude fossil fuels, nuclear power, and other environmentally harmful sectors. Indian banks like IndusInd Bank and HDFC Bank have enticed savers with higher interest rates and shorter terms for green deposits. In Thailand, the Bank of Ayudhya Public Company Limited has recently introduced green deposits, pioneering this financial product in the country's market.

Source: Satafang (2024).

3.6.2 Integration of Advanced Technology and Rise in Fintech

Advanced technologies like artificial intelligence (AI), blockchain and big data analytics will play a crucial role in enhancing the effectiveness of green finance (Radzewicz-Bak, 2024). AI can improve the assessment and management of green projects by providing more accurate risk analysis and predicting environmental impacts. Blockchain technology can increase transparency and traceability in green finance transactions, ensuring that funds are used as intended (Mirza et al., 2023). Big data analytics can help banks better understand the environmental impact of their investments and identify new opportunities for sustainable finance.

The fintech sector will increasingly offer solutions that facilitate green finance, such as platforms for carbon offsetting, green investment robo-advisors and tools for tracking personal carbon footprints. These innovations will provide consumers and businesses with easier access to green financial products and services, driving broader adoption and integration of green finance in daily financial activities (Mirza et al., 2023). For instance, many countries are now exploring the issuance of central bank digital currencies, which could be facilitated through commercial banks to finance green projects (Bouza et al., 2024).

3.6.3 Strengthening Regulatory Frameworks

The regulatory landscape for green finance is expected to become more robust and standardized. Governments and international bodies will likely introduce more comprehensive guidelines and mandatory reporting requirements for sustainable finance, reducing uncertainty and increasing accountability. Enhanced regulations will encourage banks to integrate ESG criteria into their core operations and investment decisions, promoting greater consistency and reliability in green finance practices. This indicates that regulatory frameworks are continually evolving to further advance green finance initiatives. For instance, the Greenhouse Gas Reduction Act (GGRA) was first enacted in 2009. It has been updated multiple times, with significant amendments in 2016 and the most recent update in 2022, now called the Climate Solutions Now Act of 2022.[6] The legislation mandates the reduction of greenhouse gas emissions and outlines measures to achieve specific targets.

3.6.4 Increased Collaboration and Partnerships

Banks will increasingly collaborate with governments, non-governmental organisations (NGOs) and private sector entities to promote green finance. These partnerships can facilitate knowledge sharing, resource pooling and the development of large-scale green projects. Joint initiatives will help banks navigate the complexities of green finance and contribute to achieving broader sustainability goals. For example, Ecobank Ghana Limited, in partnership with African Development Bank (AfDB), has secured US$20 million from the GCF to improve affordable financing for women-led Micro, Small, and

[6] https://www.chesapeakeclimate.org/maryland/greenhouse-gas-reduction-act/#:~:text=In%202009%2C%20Maryland%20became%20one,greenhouse%20gas%20emissions%20by%202020. Accessed 4 July 2024

Box 3.14: Ecobank Ghana and AfDB Secure US$20 Million from the Green Climate Fund for Sustainable Projects

Ecobank has collaborated with the African Development Bank (AfDB) to secure approval for the utilization of US$20 million from the Green Climate Fund (GCF). This funding is aimed at enhancing access to affordable financing specifically for women-led micro, small, and medium-sized enterprises (MSMEs), as well as farmer-based associations operating in Ghana's most vulnerable agro-ecological zones. The initiative seeks to address the financial barriers faced by these groups, who are crucial for both agricultural productivity and economic development in Ghana. By providing targeted financial support, the project aims to empower women entrepreneurs and smallholder farmers, helping them adopt climate-resilient practices and technologies. This, in turn, contributes to broader goals of sustainable development and climate adaptation within the country. Through this partnership, Ecobank and AfDB are working to improve the resilience of agricultural systems in areas highly susceptible to climate change impacts, ensuring that these communities have the necessary resources to thrive despite environmental challenges.

Source: EUROMONEY (2024).

Medium Enterprises (MSMEs) and farmer associations in vulnerable agro-ecological zones (see Box 3.14).

3.6.5 Growing Focus on Climate Risk Assessment

As climate risks become more pronounced, banks will prioritise the integration of climate risk assessments into their financial decision-making processes. This trend involves evaluating the potential impacts of climate change on the financial health of borrowers and investments, incorporating climate scenario analysis and developing strategies to mitigate climate-related risks. This proactive approach will help banks safeguard their portfolios against climate-induced disruptions and align with global climate goals.

3.6.6 Adoption of Global Standards and Certifications

The adoption of global standards and certifications related to green finance will continue to gain traction. Certifications such as B Corp, ISO 14001 and adherence to frameworks like the TCFD will become more prevalent, enhancing credibility and transparency in banks' sustainability efforts. These standards will help banks demonstrate their commitment to green finance and attract environmentally conscious investors.

3.6.7 Focus on Impact Measurement and Reporting

There will be a growing emphasis on measuring and reporting the environmental and social impact of green finance initiatives. Banks will develop more sophisticated metrics and frameworks to quantify the outcomes of their sustainable investments and communicate these impacts to stakeholders. Enhanced impact measurement will improve transparency, accountability and the overall effectiveness of green finance strategies.

3.6.8 Evolution of Consumer Preferences

Consumer preferences are shifting towards more sustainable and ethical financial products. Banks will need to adapt by aligning their offerings with these evolving expectations, incorporating sustainability into their brand values and enhancing customer engagement on environmental issues. This trend will drive banks to innovate and expand their green finance portfolios to meet the growing demand for responsible financial solutions.

3.6.9 Global Alignment and Cooperation

The global financial community will increasingly align on common goals and cooperative efforts to address climate change and promote sustainable development. International initiatives and agreements will foster greater cooperation among banks, regulators and other stakeholders, leading to more coordinated and effective green finance practices worldwide. This collective approach will help scale green finance efforts and accelerate the transition to a low-carbon economy.

3.7 Conclusion

Globally, policymakers and scholars are exploring strategies to achieve sustainable development without compromising environmental integrity, highlighting the necessity for financing solutions. This has spurred the emergence of green finance, aimed at directing funds towards environmentally beneficial projects that support sustainable development goals. This chapter examines the concept of green finance and its various initiatives, particularly those adopted by the banking sector. It discusses policies and principles that foster the integration of green finance initiatives in banking, along with potential challenges encountered in this process. Additionally, the chapter outlines future trends and prospects that encourage banks to persist in integrating these initiatives despite the obstacles they may face.

The chapter identifies a variety of green finance initiatives that mobilise financing from public, private and multilateral development organisations towards sustainable development goals. These initiatives encompass green bonds, climate finance, green loans and green grants. It also notes the widespread integration of these initiatives across most sectors of the economy, including significant adoption by banks, which play a pivotal role in financing projects, whether eco-friendly or not. Banks are increasingly incorporating green finance into their activities and decision-making processes driven by regulatory requirements, policy recommendations, efforts to reduce ecological footprints and the desire to cultivate a green corporate image. The chapter also finds that global policies and principles encouraging banks to integrate green finance initiatives include UNEP FI, Equator Principles, GABV and SBFN, complemented by national policies guiding the banking sector in adopting green finance. However, integrating green finance presents challenges such as difficulties in risk assessment, assessing financial viability, lack of standardised metrics for green projects, high initial capital costs, regulatory uncertainties, low market awareness and demand in emerging economies and insufficient capacity building. Despite these challenges, integrating green finance in banking holds future trends and prospects, including expanding sustainable financial projects, leveraging new technological innovations, enhancing regulatory frameworks, fostering increased collaborations and partnerships, focusing on climate risk assessment, adopting global standards and certifications, embracing green fintech solutions, prioritising impact measurement and reporting, responding to evolving consumer preferences and promoting global alignment and cooperation.

The chapter provides policy implications based on the findings. First, given that many banks are driven to adopt green finance initiatives due to regulatory pressures, it is suggested that policymakers, including central banks and environmental regulators, implement policies that incentivise banks for genuine green initiatives and penalise those involved in greenwashing practices. Second, because of constraints in capacity and resources, numerous banks encounter difficulties in pursuing green investments. Thus, policymakers should establish institutions focused on training banking sector personnel to adeptly comprehend, engage in and oversee green investments. Third, nations lacking national development banks should endeavour to establish such entities. These banks could act as intermediaries between private sectors—including commercial banks—and green projects, facilitating smoother financing and fostering sustainable development. Fourth, policymakers should strengthen collaboration between domestic banks and MDBs to fund green projects. Such partnerships can facilitate technology and expertise transfer, thereby enhancing the capacity of domestic banks to engage in green financing initiatives. Fifth, since banks encounter issues when accessing risk attached to green projects and the identification of greenwashing, policymakers should develop and adopt standardised definitions and metrics for what constitutes a 'green' project. This helps in setting clear benchmarks and reduces ambiguity in project claims. Banks can also mandate detailed disclosure of project objectives, methodologies and outcomes. Transparency in reporting helps stakeholders assess the genuine environmental benefits of projects. Policymakers should engage independent third-party experts to assess and verify the risk profiles of green projects. This can provide an unbiased evaluation of potential risks and enhance credibility.

References

Abor, J. Y., Gyeke-Dako, A., Fiador, V. O., Agbloyor, E. K., Amidu, M., & Mensah, Lord. (2019). *Sustainable banking* (pp. 311–331). https://doi.org/10.1007/978-3-319-77458-9_15

ADB. (2012, September 27). *ADB provides $600 million for eco-friendly projects in People's Republic of China*. ADB. https://www.adb.org/news/adb-provides-600-million-eco-friendly-projects-peoples-republic-china

Axis Bank. (2024). *We're just the bank you need*. https://www.axisbank.com/about-us

Bank of China (Hong Kong) Limited. (2024). *Green Time Deposit*. https://www.bochk.com/en/corporate/ESG/greendeposit.html

Barua, S. (2020). Chapter 6. The principles of green banking. In *Principles of green banking* (pp. 65–78). De Gruyter. https://doi.org/10.1515/9783110664317-006

Beloe, T. (2018, December 5). *Financing the response to climate change—We all need to play our part*

Bouza, S., Miccoli, M., & Mircheva, B. (2024, June 18). *Central bank digital currencies can boost middle east's financial inclusion, payment efficiency: 19 countries in the Middle East and Central Asia are exploring issuing a CBDC*. https://www.imf.org/en/Blogs/Articles/2024/06/18/central-bank-digital-currencies-can-boost-middle-easts-financial-inclusion-payment-efficiency#:~:text=Central%20Bank%20Digital%20Currencies%20Can%20Boost%20Middle%20East's%20Financial%20Inclusion%2C%20Payment%20Efficiency,-19%20countries%20in&text=Almost%20two%2Dthirds%20of%20countries,efficiency%20of%20cross%2Dborder%20payments

Chandak, P. (2023, August 16). *Absa bank facilitates $24 million green loan to Propel Ghana's renewable energy transition*. https://solarquarter.com/2023/08/16/absa-bank-facilitates-24-million-green-loan-to-propel-ghanas-renewable-energy-transition/#:~:text=Absa%20Bank%20Facilitates%20%2424%20Million%20Green%20Loan%20To%20Propel%20Ghana's%20Renewable%20Energy%20Transition,-By&text=Ghana's%20ambitious%20drive%20to%20diversify,loan%20facilitated%20by%20Absa%20Ghana

Chen, Y.-S. (2010). The drivers of green brand equity: Green brand image, green satisfaction, and green trust. *Journal of Business Ethics, 93*(2), 307–319. https://doi.org/10.1007/s10551-009-0223-9

Climate Bonds Initiative. (2024). *Interactive data platform*. Climate Bonds Initiative. https://www.climatebonds.net/green-infrastructure-investment-opportunities-giio-programme

Eti, S., Yüksel, S., Dinçer, H., Kalkavan, H., Hacioglu, U., Mikhaylov, A., Shah Danish, M. S., & Pinter, G. (2024). Assessment of technical and financial challenges for renewable energy project alternatives. *Cleaner Engineering and Technology, 18*, 100719. https://doi.org/10.1016/j.clet.2023.100719

EUROMONEY. (2024, June 19). *Awards for excellence country winners: Ghana*. https://www.euromoney.com/article/2dbnwmzqqdy6ej7jfqz9d/awards/awards-for-excellence/awards-for-excellence-country-winners-ghana#:~:text=During%20the%20review%20period%2C%20Ecobank,vulnerable%20agro%2Decological%20zones%20in

Gilchrist, D., Yu, J., & Zhong, R. (2021). The limits of green finance: A survey of literature in the context of green bonds and green loans. *Sustainability, 13*(2), 478. https://doi.org/10.3390/su13020478

Giridhar, K. V., & Sudhakar, K. G. (2017). Green Banking: Issues and Challenges in India. *International Journal of Latest Technology in Engineering, Management & Applied Science, VI*, 1–13.

Green Finance Platform. (2024). *Ghana sustainable banking principles*. https://www.greenfinanceplatform.org/policies-and-regulations/ghana-sustainable-banking-principles

Holzhauer, B. (2023, April 11). Aspiration launches first credit card*: Earn cash back and plant trees with every purchase*. Select. https://www.cnbc.com/select/aspiration-launches-first-credit-card/.

IEA. (2017). World Energy Outlook 2017. https://www.iea.org/reports/world-energy-outlook-2017

IEA. (2023). World Energy Outlook 2023. https://www.iea.org/reports/world-energy-outlook-2023

IFC. (2017). *Green finance: A bottom-up approach to track existing flows*. https://www.cbd.int/financial/gcf/ifc-greentracking.pdf

Kumar, B. R. (2022). *Project finance: structuring, valuation and risk management for major projects*. Springer International Publishing. https://doi.org/10.1007/978-3-030-96725-3

Macaire, C., & Naef, A. (2023). Greening monetary policy: Evidence from the People's Bank of China. *Climate Policy, 23*(1), 138–149. https://doi.org/10.1080/14693062.2021.2013153

Mirza, N., Umar, M., Afzal, A., & Firdousi, S. F. (2023). The role of fintech in promoting green finance, and profitability: Evidence from the banking sector in the euro zone. *Economic Analysis and Policy, 78*, 33–40. https://doi.org/10.1016/j.eap.2023.02.001

Myllyvirta, L. (2024, January 25). *Analysis: Clean energy was top driver of China's economic growth in 2023*. Carbon Brief. https://www.carbonbrief.org/analysis-clean-energy-was-top-driver-of-chinas-economic-growth-in-2023/#:~:text=Back%20to%20top-,Solar%20power,%25%20year%2Don%2Dyear

NYU. (2024). *Green Grants*. https://www.nyu.edu/life/sustainability/programs/green-grants.html

OECD. (2015). *Infrastructure financing instruments and incentives*. https://www.oecd.org/finance/private-pensions/Infrastructure-Financing-Instruments-and-Incentives.pdf

OECD. (2020). *Green finance and investment: Measuring green finance flows in Kazakhstan*. OECD Publishing. https://doi.org/10.1787/3ef6618c-en

Oil Change International. (2023). *Banking on climate chaos 2023: Fossil fuel finance report*. https://www.banktrack.org/download/banking_on_climate_chaos_2023/bocc_2023_0627.pdf

Ozili, P. K. (2022). Green finance research around the world: A review of literature. *International Journal of Green Economics, 16*(1), 56. https://doi.org/10.1504/IJGE.2022.125554

Park, H., & Kim, J. D. (2020). Transition towards green banking: Role of financial regulators and financial institutions. *Asian Journal of Sustainability and Social Responsibility, 5*(1), 5. https://doi.org/10.1186/s41180-020-00034-3

PRI. (2024). About the PRI. In PRI. https://www.unpri.org/about-us/about-the-pri

Radzewicz-Bak, B. (2024). Preparing financial sectors for a green future. *Departmental Papers, 2024*(002), 1. https://doi.org/10.5089/9798400255250.087

Rahman, M. H., Rahman, J., Tanchangya, T., & Esquivias, M. A. (2023). Green banking initiatives and sustainability: A comparative analysis between Bangladesh and India. *Research in Globalization, 7*, 100184. https://doi.org/10.1016/j.resglo.2023.100184

Satafang, N. (2024, March 8). *Green deposits for greater sustainability*. https://www.krungsri.com/en/research/research-intelligence/green-deposits-2024.

SBFN. (2024, April 16). Latest News—April 16, 2024: SBFN launched its 2024 Global Progress Brief and new Data Portal. SBFN. https://www.sbfnetwork.org/

Sharma, M., & Choubey, A. (2022). Green banking initiatives: A qualitative study on Indian banking sector. *Environment, Development and Sustainability, 24*(1), 293–319. https://doi.org/10.1007/s10668-021-01426-9

The Business & Financial Times. (2024, May 10). *Standard Chartered Bank offers the lowest rate on individual loans—BOG report*. https://thebftonline.com/2024/05/10/standard-chartered-bank-offers-the-lowest-rate-on-individual-loans-bog-report/#:~:text=In%20addition%20to%20their%20existing,thereby%20reduce%20their%20carbon%20footprint

UNDP. (2022). *The UNDP approach to risk informed development*.

UNEP FI. (2024). About Us. UNEP FI. https://www.unepfi.org/about/about-us/

Wachira, M. M., Iddrisu, K., Abban, G., & Abor, J. Y. (2023). Sustainable financing in developing economies. In J. Y. Abor (Ed.), *Sustainable and responsible investment in developing markets* (pp. 159–171). Edward Elgar Publishing. https://doi.org/10.4337/9781803927060.00019

Weber, O., & Acheta, E. (2016). *The Equator Principles: Do they make banks more sustainable*. (16/05; Inquiry Working Paper).

World Bank. (2021a, November 4). *What you need to know about green loans*. Climate Explainer Series. https://www.worldbank.org/en/news/feature/2021/10/04/what-you-need-to-know-about-green-loans

World Bank. (2021b, December 8). *What you need to know about IFC's green bonds*. https://www.worldbank.org/en/news/feature/2021/12/08/what-you-need-to-know-about-ifc-s-green-bonds#:~:text=In%20late%202007%2C%20a%20group,for%20eligible%20climate%2Dfocused%20projects

World Bank. (2024). *Carbon pricing*. https://carbonpricingdashboard.worldbank.org/what-carbon-pricing#:~:text=Carbon%20pricing%20is%20an%20instrument,to%20their%20sources%20through%20a

World Economic Forum. (2020, November 9). *Climate change: What is green finance and why is it important?* https://www.weforum.org/agenda/2020/11/what-is-green-finance/

Zhang, R. (2016). *Sustainable banking network* (110539; IFC E&S). http://documents.worldbank.org/curated/en/623531480317898290/Sustainable-banking-network.

FinTech Innovations for Sustainable Banking

4

Khadijah Iddrisu, Ibrahim Nandom Yakubu, and Simplice A. Asongu

4.1 Introduction

Financial technology (i.e. FinTech) is rapidly becoming essential in consumer purchasing and investment behaviours, integrating diverse technologies into daily life. It has transformed how banks deliver financial services, prompting financial institutions to constantly innovate in integrating banking solutions and financial features for their customers. The rise of financial technology (FinTech) and its plethora of companies has brought forth a wave of innovations and user-friendly applications designed to enhance banking activities (Kohtamäki et al., 2019). Furthermore, the coronavirus disease 2019 (COVID-19) pandemic has significantly accelerated the adoption and usage of FinTech. In particular, the use of pure-play lending applications (apps) has surged in emerging markets and developing economies during this period (Fu & Mishra, 2022). The total value of digital payment transactions rose from US$4.1 trillion in 2019 to US$5.2 trillion in 2020. Therefore, can we say FinTech promotes sustainable banking?

Sustainable development has captured worldwide focus, leading nations to weave sustainability into their economic practices (Eisenmenger et al., 2020). The banking sector, just like the other sectors, has embraced the idea of sustainability, which has affected the core activities of banking (Wachira et al., 2023). Examples include policies and frameworks like the triple bottom line banking, the United Nations Environment Programme Finance Initiative (UNEP FI), the Global Alliance for Banking on Values (GABV), the Equator Principles (EPs), and the Sustainable Banking and Finance Network (SBFN), which steer banking practices towards sustainability (Equator Principles, 2024; GABV, 2024; SBFN, 2024; UNEP FI, 2024). While these policies were first introduced by a handful of central banks and individual banks, the increasing emphasis on sustainability has led more banks to adopt them. The GABV began with just ten banks in 2009 and has since expanded to over 70 members across more than 45 countries (GABV, 2024). The SBFN, established by the International Finance Corporation (IFC) of the World Bank in 2012, has grown to include 85 members (SBFN, 2024). Therefore, looking at the effort made by banks and FinTech firms to promote FinTech innovations, it is necessary to discuss how these innova-

K. Iddrisu (✉)
School of Business, Simon Diedong Dombo University of Business and Integrated Development Studies, Wa, Ghana
e-mail: khadijahiddris@iug.edu.gh

I. N. Yakubu
School of Business, University for Development Studies, Tamale, Ghana
e-mail: ibrahim.yakubu@uds.edu.gh

S. A. Asongu
School of Economics, University of Johannesburg, Johannesburg, South Africa

I. N. Yakubu (ed.), *Strategic Approaches to Banking Business and Sustainable Development Goals*, Sustainable Development Goals Series, https://doi.org/10.1007/978-3-031-80744-2_4

tions can promote sustainable banking. Hence, the overarching objective of this chapter is to provide a discussion on FinTech and how it can promote sustainable banking and bring into bear the key issues that are driving and hindering it.

The rest of this chapter is structured as follows: Section 4.2 shows an overview of FinTech, particularly the concept, the evolution, and the available innovations. Section 4.3 provides theoretical and empirical evidence as to how FinTech enhances or minimizes sustainable banking. Section 4.4 discusses regulatory and ethical considerations. Section 4.5 outlines the limitations of FinTech and possible opportunities associated with the technology. Finally, this chapter ends with conclusions, which entail recommendations.

4.2 FinTech: Concepts and Evolution

4.2.1 Concept of FinTech

FinTech lacks a universally accepted definition but is characterized by some common terminologies. The Financial Stability Board (2023) defines FinTech as innovation in financial services driven by technology, potentially leading to new business models, applications, processes, or products that significantly impact financial markets, institutions, and the delivery of financial services. According to Batunanggar (2019), FinTech has evolved alongside developments in information technology (IT), such as the Internet, smartphones, and big data analytics, facilitating the more efficient and cost-effective delivery of financial services. Navaretti et al. (2018) expressed FinTech as innovative processes and products in financial services made possible by advancements in digital technology. A recent perspective from the World Bank and the International Monetary Fund (IMF) broadly defines FinTech as technological advancements with the potential to revolutionize financial services, driving the creation of innovative business models, applications, processes, and products (World Bank, 2022).

The concepts discussed by the aforementioned authors highlight two key terms: digital technology and financial services. From these, we can infer that FinTech involves leveraging digital technology or information and communication technology (ICT) to improve financial services and economic activities. Today, FinTech has gained global attention as countries intend to go cashless. Investment in FinTech has shown a consistent upward trend over time, despite its inherent volatility, as illustrated in Fig. 4.1. For instance, between 2010 and 2019, investment in FinTech surged from US$9 billion to US$216.8 billion, dipped in 2020 amidst increasing relevance, and rebounded in 2021. However, from 2021 to 2023, investment declined from US$225.8 billion to US$113.7 billion.

4.2.2 Evolution and Trends of FinTech

This subsection discusses the evolution and trends (statistics of FinTech usage) of FinTech. The evolution of FinTech is summarized in Table 4.1. The FinTech industry has come a long way since its inception in the nineteenth century. Its roots can be traced back to 1866 when the use of telegraphs and railroads enabled rapid transmission of financial information across distances (Fessenden, 2018). This innovation laid the groundwork for the US Federal Reserve's creation of the Federal Reserve Wire Network (Fedwire) in 1918, facilitating the shift from physical cash and gold transactions in 1915 to electronic fund transfers (Garbade & Silber, 1979). Fedwire connected all 12 Federal Reserve Banks, the Federal Reserve Board, and the US Treasury through telegraph lines, using Morse code to communicate and process transactions electronically. This platform has been updated over the years to capture the current needs of users. For instance, currently, the Fedwire transfer system has evolved from using telegraph lines to leveraging high-speed data networks, resulting in substantial improvements in speed, capacity, and efficiency.[1] In 1988, successful transfers via

[1] See https://www.federalreservehistory.org/essays/fedwire accessed 22 June 2024

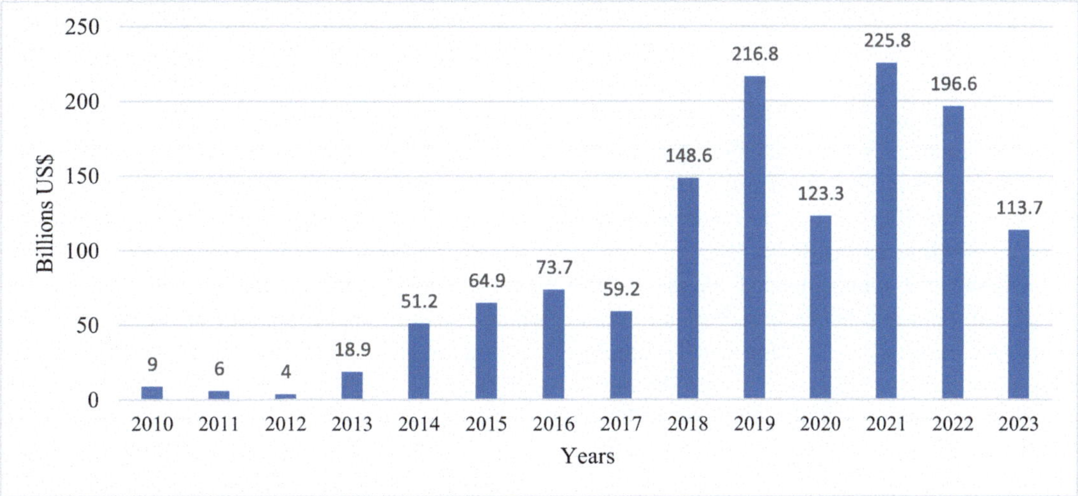

Source: Statista Research Department (2024)

Fig. 4.1 FinTech investment in US$ billion, 2010 to 2023. (Source: Statista Research Department (2024a)

Table 4.1 Key developments in the FinTech industry

Year	Development or innovations introduced
1866	Establishment of transatlantic telegraph cable for rapid financial communication
1950	Launch of Diners Club, the first universal credit card
1967	Introduction of the first automated teller machine (ATM) by Barclays Bank
1973	Creation of the Society for Worldwide Interbank Financial Telecommunication (SWIFT) for standardized international transactions
1981	Implementation of the first fully online banking services by the Bank of Scotland
1994	First online banking by Stanford Federal Credit Union
1998	Founding of PayPal, a key player in online payments
2001 to 2011	Mobile money (MoMo) financial platforms
2009	Introduction of Bitcoin, the first decentralized cryptocurrency
2011	Launch of Google Wallet, enhancing mobile payment capabilities
2014	Apple Pay brings digital wallets to the mainstream
2020	Rapid growth in digital banks and open banking initiatives

Source: Diners Club International (2024), Fessenden (2018), and Salampasis and Mention (2018)

Fedwire totalled around US$1.6 million, whereas by 2023, the volume had grown significantly to approximately US$1.087 billion.[2]

The Diners Club Card, launched in 1950, became the first universal credit card, inspired by Frank McNamara's experience of forgetting his wallet at a New York restaurant (Diners Club International, 2024). To avoid such situations, McNamara developed the card, using it at Major's Cabin Grill with his partner Ralph Schneider a year later (Diners Club International, 2024). Known as the "First Supper," this event revolutionized the credit industry by introducing the world's first multipurpose charge card. By 1951, membership grew to 42,000 as the card gained acceptance in major US cities. In 1953, Diners Club expanded internationally to the UK, Canada, Cuba, and Mexico, and by 1955, its network spanned Europe, Asia, and the Middle East. It entered the travel market in 1958 and reached a million members by 1959, listing on the New York Stock Exchange (NYSE). The card

[2]See https://www.frbservices.org/resources/financial-services/wires/volume-value-stats/annual-stats.html accessed 23 June 2024

was accepted in 130 countries by 1967 and started in Russia in 1969. Diners Club introduced the first rewards programme in 1984 and a multinational corporate card in 1990. Acquired by Discover Financial Services in 2008, Diners Club celebrated its 70th anniversary in 2020, offering access to over 1000 airport lounges worldwide (Diners Club International, 2024).

Customers of banks needed access to cash outside of banking hours and beyond branch locations. Hence, on 27 June 1967, Barclays Bank revolutionized banking by introducing the world's first automated teller machine (ATM) in Enfield, London, and Switzerland received its first ATMs the same year. This innovation, the De La Rue Automatic Cash System, was developed by John Shepherd-Barron from the De La Rue company. Although early attempts like Luther Simjian's Bankmatic machine in 1939 failed, the 1960s saw successful ATM advancements. Notably, Scottish inventor James Goodfellow patented a similar machine in 1966. Shepherd-Barron's ATM, inspired by his frustration with limited banking hours, allowed 24/7 cash access. British actor Reg Varney conducted the first transaction, marking a significant milestone in banking history. The global ATM market size reached US$20 billion in 2020 and is projected to grow to US$28.82 billion by 2026[3]. In 2020, there were approximately 3.2 million ATMs worldwide, with 41.24 machines per 100,000 people[3]. The Asia–Pacific region leads in ATM usage. Despite growth, cash transactions are declining due to security concerns and competition from online banking.[3]

Financial transactions were prone to errors and insecurity due to varied communication methods like telex. The lack of standardization led to inefficiencies and increased fraud risks. Manual processing made costs high, and traditional systems were vulnerable to interception and tampering, necessitating a secure, reliable network. Therefore, in 1973, the Society for Worldwide Interbank Financial Telecommunication (SWIFT) was introduced. The SWIFT payment system allows banks and financial institutions to securely send electronic messages regarding international transactions. It offers a uniform messaging platform that streamlines communication among these entities. Now, SWIFT supports messaging services for more than 10,000 financial institutions across 212 countries worldwide, significantly enhancing global business operations. On 30 November 2021, SWIFT set a new record with 50.3 million messages exchanged, marking an increase of 8.5% compared to the previous peak on 26 February 2021.[4]

During the early 1980s, various New York City (NYC) banks, such as Citibank, Chase Manhattan, Chemical Bank, and Manufacturers Hanover, embarked on trials of computerized banking to innovate the financial sector. They provided customers with home banking access to assess its potential success. To enhance efficiency and easy access to financial services, the Bank of Scotland fully implemented online banking services in 1981. Although this bank was known as the first bank to adopt centralized computer processing for its accounts in the UK banking sector in 1959, this activity was only limited to the workers of the bank. Therefore, the scope was expanded by including customers through electronic home banking services. The idea of online banking service received much attention when Stanford Federal Credit Union introduced its banking services website. This was a pivotal moment in the evolution of FinTech. This initiative quickly spread beyond its origins in the UK and NYC. By 1994, Stanford Federal Credit Union had established Internet banking as a mainstream solution for its clients. The next year, the Presidential Bank became the first in the country to provide customers with online access to their accounts, further advancing the adoption of digital banking. In today's market, online banking services are essential for banks to remain competitive—those without risk of losing customers. In 2020, the Far East and China had more than 805 million active users of online banking, a

[3]https://www.enterpriseappstoday.com/stats/atm-statistics.html

[4]https://tipalti.com/payments-hub/swift-payments/#:~:text=SWIFT%20is%20a%20member%2Downed,the%20SWIFT%20network%20comes%20in

number expected to approach one billion by 2024. While Asia led in overall user numbers, European countries had the highest penetration rates for online banking.[5]

PayPal, initially founded as Confinity in 1998 by Peter Thiel and Max Levchin, aimed to revolutionize digital payments with ease and affordability (Soni, 2022). Based in Mountain View, California, the company seized an emerging market for online transactions amidst the Internet's growing influence. By simplifying payment processes and incentivizing users to share their email and financial information, PayPal quickly gained traction. Within three years, it processed over US$3 billion from 10.2 million consumers and 2.6 million businesses (Laine, 2021). Recognizing the opportunity, the founders rebranded as PayPal in 2000, focusing exclusively on Internet payments, a niche previously overlooked. This strategic shift propelled PayPal's rapid ascent, addressing a gap left by traditional payment methods like checks and money orders. PayPal was acquired by eBay for US$1.5 billion in July 2002, boosting its growth as eBay's preferred payment method. In 2013, PayPal acquired Braintree for US$800 million.[6] Activist investor Carl Icahn pushed for PayPal's split from eBay in 2014, leading to PayPal becoming a separate public company in 2015.[7] In 2020, PayPal's stock price more than doubled, elevating its market capitalization to over US$274 billion (Laine, 2021). During the third quarter of 2020, PayPal reported net revenues of US$5.46 billion, up significantly from US$4.38 billion in the same period of 2019—a 16% year-over-year increase. Additionally, PayPal's customer base grew from 295 million in 2018 to over 360 million by the end of 2020 (Laine, 2021). As of early 2022, PayPal's market cap reached around US$210 billion, making it a top FinTech company globally[6]. The FinTech landscape underwent a notable shift when network service providers collaborated with information technology (IT) firms to offer financial services, which is called mobile money (MoMo). Originating in the Asian financial sector, this form of FinTech has recently garnered interest in various emerging markets.[8] Since the launch of the first mobile money provider in East Asia and the Pacific in 2001, M-Pesa's introduction in Kenya in March 2007 marked a pivotal moment that popularized the model globally[8]. Now, a lot of rural areas have access to financial services because of MoMo (Iddrisu et al., 2022).

In 1983, David Chaum, an American cryptographer, proposed the idea of anonymous electronic money, known as e-Cash, which he later implemented through DigiCash in 1989 (Dreier et al., 2015). Despite initial efforts in a single bank, the concept struggled to gain traction, leading to the company's dissolution in 1998 after being acquired by a major credit card issuer. Bitcoin emerged in 2009 as the first decentralized cryptocurrency, gaining recognition in 2013 as the best investment by Forbes and the worst in 2014 by Bloomberg.[9] In October 2021, the Security Exchange Commission (SEC) approved the ProShares Bitcoin Strategy Exchange-Traded Funds (ETFs), while Futures Exchange (FTX) declared bankruptcy in November 2022. Finally, in January 2024, the SEC approved 11 spot Bitcoin ETFs, enhancing investor access to crypto markets[9]. The quest to promote more cashless transactions led to the creation of Google Wallet in 2011 and Apple Pay in 2014 (Liu et al., 2015). Google Wallet, introduced in 2011 by Google, is a mobile payment platform designed to innovate the payment experience. Rather than using conventional credit or debit cards, the app enables users to securely store their card details on their smartphones, aiming to transform the way payments are made. Apple officially entered the payments industry in 2014 after years of

[5] https://www.statista.com/statistics/1228757/online-banking-users-worldwide/

[6] https://fintechmagazine.com/digital-payments/story-paypal-worlds-most-valuable-fintech-firm

[7] https://www.americanbanker.com/payments/news/carl-icahn-who-pushed-paypals-independence-from-ebay-dumps-his-stake#:~:text=Icahn%20originally%20pushed%20for%20eBay,dumping%20his%20position%20in%20AIG

[8] https://www.uncdf.org/article/7904/where-in-the-world-is-mobile-money-prominent

[9] https://money.usnews.com/investing/articles/the-history-of-bitcoin

preparation, acquisitions, and patent filings.[10] By partnering with Visa, Mastercard, and American Express, Apple aimed to replace traditional transaction systems with secure digital tokens. The company leveraged its strong brand and extensive hardware and software ecosystem to introduce Apple Pay. This new payment method, unveiled by CEO Tim Cook in September 2014, allowed users to make purchases using their devices through Apple Wallet[10]. The initiative quickly gained global traction, with major retailers like Starbucks, KFC, and JCPenney adopting Apple Pay within a few years of its launch[10].

The emergence of COVID-19 in 2020 highlighted significant advancements in financial services through FinTech. Digital banks such as Revolut, Chime, and Monzo gained popularity for their digital-only banking models. At the same time, artificial intelligence (AI) and machine learning saw a surge in applications like fraud detection, credit assessment, and customer service using chatbots and predictive analytics. Regulatory changes like the revised Payment Services Directive (PSD2) in Europe promoted open banking, allowing third-party applications to integrate with traditional banking systems. Furthermore, there was a growing trend of integrating financial services into non-financial platforms such as e-commerce apps, enhancing the seamless delivery of financial experiences. These innovations are increasingly crucial in optimizing FinTech utilization, particularly in emerging markets. Efforts are underway to enhance current innovations, with some FinTech firms focusing on building new solutions based on these innovations. For example, the global number of FinTech entities has risen from 12,131 in 2018 to 29,955 in 2024 (see Fig. 4.2).

4.2.3 FinTech Innovations and Forms

FinTech innovations vary and have been increasing over time. For instance, innovations have made it possible for the unbanked in rural areas, especially in most emerging economies like Africa, to be banked (Iddrisu et al., 2022). These innovations have been classified into different forms by institutions and scholars. For example, Aba and Linardy (2021) categorized FinTech innovations into (i) payment, settlement, and clearing; (ii) electronic aggregator; (iii) risk and investment management; and (iv) peer lending. Bank Indonesia (2017) also classified it into a payment system; market support; investment and risk; lending, financing, funding, and capital raising; and other means of FinTech services. Rahmi (2019) also grouped it into four main forms: savings and investment; transfer and payment; borrowing; and insurance. Building upon these ideas, this chapter presents different forms of FinTech innovations to capture the new trends of innovations: e-savings and e-investment; e-transfer and e-payment; e-lending and e-borrowing; e-insurance; e-financial security and e-regulation; and other e-banking activities.

4.2.3.1 E-savings and E-investment

E-savings, or electronic savings accounts, are bank accounts managed online through digital platforms, such as mobile apps or websites. They provide a convenient way for users to save money without needing to visit physical bank branches. E-investment, or electronic investment, refers to the management and execution of investment activities through digital platforms, including online brokers, mobile apps, and automated investment services. This approach leverages technology to provide easy access to financial markets, enabling users to invest in various assets such as stocks, bonds, ETFs, and mutual funds without needing traditional financial intermediaries. In a nutshell, FinTech has enabled users to save or invest comfortably from home using tools like robo-advisors, micro-investing platforms, mobile apps, and websites. Most people in emerging economies have been using MoMo for savings and investment purposes (see Box 4.1).

4.2.3.2 E-transfer and E-payment

Electronic transfer (e-transfer) streamlines the digital movement of money or data between accounts, eliminating physical cash or cheques

[10] https://www.bigcommerce.com/blog/apple-pay-tips-mobile-payments/

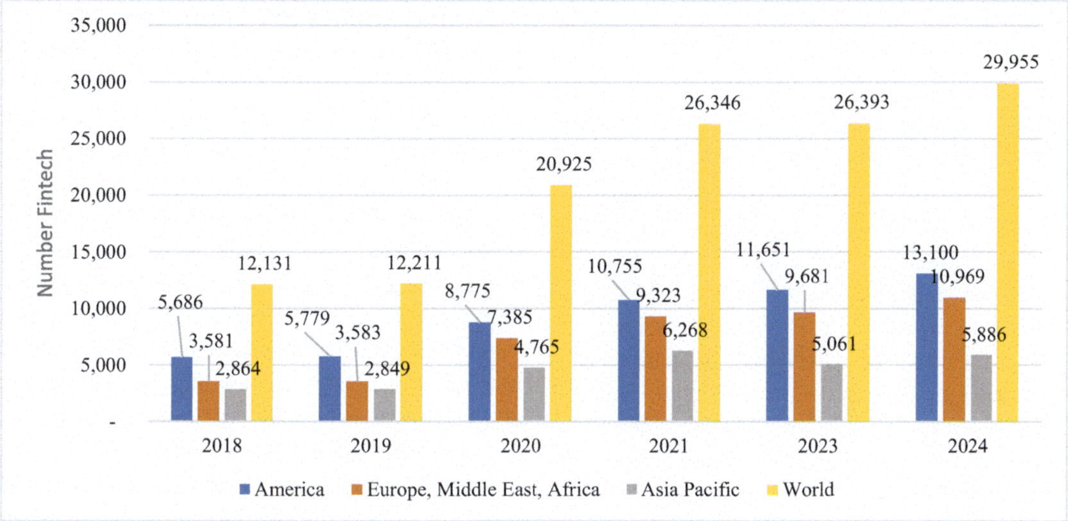

Source: Statista Research Department (2024)

Fig. 4.2 Number of FinTech entities worldwide from 2018 to 2024, by region. (Source: Statista Research Department (2024b)

(Tut, 2023). It boosts transaction speed and convenience for both domestic and international transfers via bank transfers, mobile apps, or email/phone identifiers. Examples include Interac e-Transfer in Canada, Zelle in the United States, Single Euro Payments Area (SEPA) in Europe, and MTN Mobile Money in Africa (GSM Association, 2024; Iddrisu et al., 2022, 2023; Salampasis & Mention, 2018; Tut, 2023). E-transfers offer ease of use, secure encryption protocols, multi-factor authentication, and rapid transaction speeds, allowing for scheduled future transfers and efficient remittance handling.

Electronic payment (e-payment) utilizes digital systems to facilitate the payment of goods, services, or bills (Tut, 2023). This method conducts transactions online or via electronic devices, serving as an alternative to cash, checks, or physical credit cards. E-payment can be performed through websites, e-commerce platforms, mobile apps, mobile wallets, and contactless methods like quick-response (QR) codes in physical stores. Payments can involve entering card details for online or mobile transactions, storing card details or funds in an app (e.g. Apple Pay and Google Wallet), or using digital currencies like Bitcoin. This method integrates seamlessly with online shopping sites and point-of-sale systems in stores, providing quick checkout options on e-commerce platforms and enabling automatic payments for subscriptions or bills. It ensures the secure transmission of payment data, replaces sensitive card details with unique transaction tokens, and includes security measures like biometric verification or personal identification numbers (PINs).

Box 4.1: Mobile Money

Mobile money (MoMo) leverages digital technology to facilitate financial transactions via mobile phones. Users can send and receive funds, pay for goods and services, settle bills, and withdraw cash using a digital wallet linked to their phone. By extending financial services to previously underserved regions, especially rural areas, mobile money has enhanced convenience and accessibility. This shift has not only empowered individuals but also spurred economic growth by streamlining transactions for both businesses and consumers.

(continued)

Box 4.1: (continued)

Globally, mobile money has seen remarkable growth, with 1.6 billion registered accounts in 2022—a 13% increase year-over-year. Africa alone accounts for 781 million of these, with the region's accounts rising by 17% annually. Transaction values surged 22% from 2021 to 2022, reaching US$1.26 trillion. Between 2013 and 2022, services contributed over US$150 billion to the region's gross domestic product (GDP), adding an extra 3.7%. As of 2024, sub-Saharan Africa leads the world in mobile money adoption. M-Pesa, Orange Money, MTN Mobile Money, and Airtel Money are leading mobile money services in Africa, boasting millions of users continent-wide. While over 144 mobile money providers operate in sub-Saharan Africa, these four dominate the market, enabling users to transfer money, pay bills, and purchase airtime and digital products. This innovation has revolutionized financial access in Africa, where over 60% of people who lack traditional banking services now have access to financial services. By the end of 2023, there were about 435 million active mobile money accounts, showing a 9% annual growth rate, which was lower than the 13% seen in 2022 and 15% in 2021.

MoMo has evolved to include insurance, lending, investment, and savings services through partnerships with banks and other institutions. In Ghana, MTN MoMo, Telecel Cash, and AT Cash collaborate with organizations like Ecobank, BIMA, AFB Bank, miFuture, Hollard Insurance, and miWay Insurance to offer integrated solutions for loans, insurance, and investments. This innovation has also enabled users to request transaction statements and seamlessly link their MoMo accounts to formal bank accounts. This integration facilitates easy transfers, payments, and withdrawals between the bank and their mobile money (MoMo account.

Source: Awanis et al. (2022), GSM Association (2024), Iddrisu et al. (2022), IFC (2011), and Jack and Suri (2011).

4.2.3.3 E-lending and E-borrowing

E-lending and e-borrowing have revolutionized the financial sector by leveraging technological advancements. These digital processes simplify traditional lending and borrowing, making them more efficient, accessible, and user-friendly. Innovations like peer-to-peer (P2P) lending, online lenders, digital bank loans, and invoice financing enable quicker application and approval processes, provide 24/7 access to loan management, and offer transparent terms online. These platforms also facilitate lending for individuals and businesses who may not have easy access to conventional banking services, though they must be carefully managed to address security and regulatory concerns. MoMo also serves as an e-lending and e-borrowing form of FinTech as it provides a ready market for lenders and ready bank loans as shown in Box 4.1.

4.2.3.4 E-insurance and Insurtech

E-insurance represents a significant evolution in the insurance industry, mirroring the broader trend of digital transformation across various sectors. This innovation enhances accessibility, efficiency, and transparency in managing insurance needs (Toukabri & Ettis, 2021). Key developments include online applications, automated underwriting, policy management through digital portals, and mobile apps for accessing policy details and making payments (Toukabri & Ettis, 2021). Emerging technologies, such as Insurtech, leverage cloud-based platforms and "as-a-service" business models, integrating artificial intelligence (AI), big data, and open application programming interfaces (APIs) to enhance speed and efficiency.[11] Examples include digital claims submissions, AI-driven claims assessments, and customer support via AI, online chat, and email[11]. Additionally, innovations like monitoring devices in vehicles, wearable activity trackers, and mobile applications such as TRUDI and Policy Buddy are being utilized (Del-Valle-Soto et al., 2024; Toukabri & Ettis, 2021).

[11] https://www.ey.com/en_us/insights/insurance/insurtech-trends-and-investment-landscape

4.2.3.5 E-financial Security and E-regulation

E-financial security utilizes encryption, multi-factor authentication, and real-time monitoring to protect digital financial transactions and assets from cyber threats (Rahmi, 2019; Tut, 2023). Innovations include AI-driven risk management tools for monitoring and mitigating risks, AI-based credit scoring using alternative data to promote inclusive lending, and machine learning algorithms for real-time fraud detection by analysing transaction patterns. E-regulation oversees regulatory frameworks ensuring consumer protection, data privacy, and compliance with financial laws, facilitated by technological advancements that streamline Know Your Customer (KYC) and anti-money laundering (AML) compliance (Rahmi, 2019).

4.2.3.6 Other Online Banking Activities

There are other innovations that enhance financial and banking services. Users can manage account preferences and update personal information like contact details and beneficiaries. Online banking provides e-statements for viewing transaction history and printing if necessary. Some banks enable remote cheque deposits via mobile apps, allowing cheques to be scanned for deposit. For instance, in Ghana, Guaranty Trust Bank (GTBank) (Ghana) Limited introduced remote cheque deposits (see Box 4.2). Budgeting tools and financial planning resources are offered to aid in effective money management. Customer support is available through chatbots, email, or phone for assistance with transactions and inquiries. Credit score monitoring tools provide insights and suggestions for improving credit health. Additionally, options for recurring and scheduled bill payments ensure timely payments.

Box 4.2: Guaranty Trust Bank (Ghana) Limited Introduced GTRemote Cheque Capturing

Guaranty Trust Bank (GTBank) (Ghana) Limited introduced GTRemote Cheque Capturing, enabling remote cheque deposits without branch visits. This secure, three-step online system involves scanning, verifying, and depositing cheques from anywhere. Benefits include time savings, cost reduction, and enhanced convenience and security. Suitable for corporate clients, it simplifies cheque processing and integrates with the Ghana Interbank Payment and Settlement Systems. The bank also launched a pioneering third-party transfer service for online money transfers. Recognized for its electronic banking innovations, GTBank has won numerous awards, including Bank of the Year and Best Mobile Bank in Ghana in 2014.

Source: GTBank Ghana Ltd. (2014).

4.3 Sustainable Banking: The Role of FinTech Innovations

Sustainable banking involves banking practices that support the transition to a low-carbon, sustainable economy, addressing modern global challenges such as climate change and new environmental risks (see Box 4.3). FinTech innovations have played a major role in promoting or minimizing sustainable banking. This section provides two schools of thought: (i) those who argue that FinTech has promoted sustainable banking and (ii) those with the view that FinTech has minimized sustainable banking.

Box 4.3: What Is Sustainable Banking?

Sustainable banking is the provision of financial products and services designed to fulfil people's needs, protect the environment, and generate profits. It can also be defined as banking institutions that prioritize environmental and social responsibility at the heart of their business and mission. The progression of sustainable banking commenced as banks assumed responsibility

(continued)

Box 4.3: (continued)

towards society, giving rise to concepts like ethical banking, green banking, and corporate social responsibility (CSR). Green banking represents a modern financial trend where banks realign their investment strategies to prioritize environmentally sustainable projects like renewable energy and conservation efforts. Ethical banking is a conscientious banking model that prioritizes fairness and sustainability, aiming to have a positive impact on all stakeholders, including clients, employees, the communities they serve, and the environment. CSR of banks is their commitment to operate ethically and contribute positively to society beyond their primary financial objectives.

Sustainable banking has spurred the adoption of practices like establishing smart branches and reducing investments in environmentally risky projects. Banks also offer green banking products such as green credit, green bonds, carbon credits, green credit cards, and green deposits, while serving as venture capitalists for green projects. They innovate by installing solar-powered automated teller machines (ATMs) and other renewable energy solutions and promote energy efficiency by adopting light-emitting diodes (LEDs) in bank offices. Banks employing sustainable banking practices offer benefits to both themselves and other stakeholders. For the bank, these include enhancing their green corporate image, meeting regulatory requirements, reducing operational costs, and gaining trust in sustainability efforts. Meanwhile, stakeholders benefit from reduced carbon footprints, increased environmental awareness among customers, and enhanced support for environmentally friendly project investments.

Sources: Abor et al. (2019), Nosratabadi et al. (2020), Wachira et al. (2023), and Wang et al. (2022).

4.3.1 Theoretical Framework

There are a lot of theoretical arguments that underpin the link between FinTech and sustainable banking. Some of these arguments include the technology acceptance model (TAM), sustainable development theory, diffusion of innovations theory, stakeholder theory, institutional theory, triple bottom line, resource-based view (RBV), and risk management theory. However, TAM was further discussed since it is a widely and popularly known theoretical argument that has been updated over time to capture changes in technology and users' attitudes towards technology (Marikyan & Papagiannidis, 2023).

Davis (1989) developed the TAM, building on the theory of reasoned action (TRA) due to the limitations of other arguments. During the 1980s, the rise of personal computers brought increased attention to research on technology adoption (Marikyan & Papagiannidis, 2023). This era marked the beginning of early studies in this field, which concentrated on user involvement in information system design and the development of system characteristics. However, these studies did not establish a comprehensive framework addressing factors that enhance user engagement in technology (Benbasat et al., 1986; Robey & Farrow, 1982). Therefore, TAM was developed to solve such an issue. Unlike TRA, TAM specifically identified factors influencing technology use, emphasizing that behavioural intention is driven by beliefs related to technology rather than a general attitude (Davis, 1989). TAM aimed to provide a comprehensive yet straightforward framework for understanding technology user behaviour. TAM summarized the technology user behaviour into perceived ease of use, perceived usefulness, and behaviour intention.

However, due to changes in technology and users' attitudes, TAM has been upgraded into TAM2 and TAM3 (Marikyan & Papagiannidis, 2023). TAM2 consists of five additional exogenous variables (such as subjective norms, image, job relevance, output quality, and result demonstrability) and two moderators (such as experience and voluntariness). Due to critiques of TAM's limited practical guidance (Lee et al., 2003), Venkatesh and Bala (2008) integrated the antecedents of perceived usefulness and ease of use into a single model, establishing a nomological network to comprehensively explain technology adoption, which gave rise to TAM3. TAM3 is an upgrade to TAM2, which also includes an additional six exogenous variables such as computer self-efficacy, perception of external control, computer anxiety, computer playfulness, perceived enjoyment, and objective usability (Marikyan & Papagiannidis, 2023). Despite the various extensions, the aim of TAM (which is how perceived ease of use and perceived usefulness influence technology adoption) has not changed but rather has been improved upon.

This theory can be used to examine the factors influencing the adoption of FinTech solutions in banking, such as perceived ease of use, perceived usefulness, and the impact of these technologies on customer satisfaction and banking efficiency.

4.3.2 Positive Effect of FinTech on Sustainable Banking

FinTech has enhanced sustainable banking in many ways as discussed subsequently.

4.3.2.1 Innovative Banking Products

FinTech companies develop green financial products like digital wallets for carbon credits, investment platforms for renewable energy projects, and sustainable lending options. With the support of FinTech, several banks have successfully introduced innovative products such as green deposits and green credits. These products cater to environmentally conscious consumers and businesses. For example, Aspiration Bank launched a green credit card that gives cash to customers to be used on green projects (see Box 4.4). Hidayat-ur-Rehman and Hossain (2024) also found that in Pakistan, FinTech promotes sustainable banking through innovative products.

Box 4.4: The Introduction of Aspiration Zero Credit Card by Aspiration Bank

Aspiration Bank has launched the Aspiration Zero Credit Card (AZCC), a green credit card that rewards purchases with cash back while actively contributing to environmental sustainability by planting trees to offset carbon footprints. As a digital bank committed to environmental impact, Aspiration Bank allocates 10% of its profits to assist financially struggling Americans and aims to plant 100 million trees worldwide. Unlike other cards, AZCC ensures its funds exclusively support clean initiatives, thereby combating "greenwashing." While the card incurs a US$60 annual fee and foreign transaction charges, it promotes eco-friendly spending by planting trees for each purchase and offering carbon-neutral rewards. The card is constructed from sustainable materials derived from plants, further enhancing its environmental credentials. Increased tree planting not only supports carbon sequestration efforts but also contributes to reducing overall carbon footprints.

Source: Holzhauer (2023).

4.3.2.2 Promoting Financial Inclusion: A Pillar of Sustainable Banking

The expansion of FinTech solutions has promoted sustainable banking by enhancing access to financial services in underserved areas. This accessibility empowers individuals and businesses in these regions to engage in sustainable banking practices such as green investments, eco-friendly financing options, and participation in environmental initiatives. By bridging gaps in financial inclusion, FinTech fosters economic empowerment and facilitates the adoption of sustainable practices across diverse communities, thereby contributing to broader environmental and social sustainability goals.

4.3.2.3 FinTech Collaborates with Traditional Banks to Promote Sustainable Banking

FinTech encourages partnerships between traditional banks and sustainable FinTech startups, accelerating the adoption of sustainable practices and facilitating industry-wide knowledge sharing. For example, in Ghana, through its partnership with MTN MoMo, Ecobank extends loans to users of MTN MoMo, including underserved areas where individuals may lack the typical requirements needed to secure loans from traditional banks (see Box 4.1). This is supported by an empirical study conducted by Iddrisu et al. (2022), which shows that in Africa, the presence of banks, particularly foreign ones, alongside FinTech, promotes financial inclusion.

4.3.2.4 Easy Identification of Environmental Risk

FinTech leverages advanced analytics to provide banks with comprehensive insights into environmental risks and opportunities. By integrating vast datasets and applying sophisticated analysis techniques, FinTech enables banks to make more informed decisions regarding sustainable investments and effective environmental impact management. This not only helps in mitigating risks and maximizing opportunities but also aligns financial practices with broader sustainability goals. For instance, a FinTech innovation like ComplyAdvantage offers an application that enables banks to detect fraud and manage risk more effectively (see Box 4.5). Also, FinTech companies are revolutionizing the environmental finance market by streamlining the process for individuals and organizations to invest in environmental projects, enhancing accessibility and efficiency in sustainable investing (Mertzanis, 2023).

Box 4.5: ComplyAdvantage AI-driven Fraud and AML Risk Detection

ComplyAdvantage was founded in 2014, which enhances the efficiency of banks' compliance processes by reducing false positives by up to 70% and cutting onboarding times by up to 50% through its artificial intelligence (AI)-driven fraud and anti-money laundering (AML) risk detection solution. The solution uses advanced automation, graph network detection, identity clustering, and dynamic thresholds to deliver three key capabilities:

1. *Customer Screening:* It offers real-time risk assessments with flexible parameters, integrating data feeds, case management, and customer relationship management (CRM) systems for continuous monitoring.
2. *Adverse Media Screening:* The system triggers alerts based on a comprehensive AML and combating the financing of terrorism (CFT) taxonomy aligned with regulatory guidelines.
3. *Transaction Monitoring:* A cloud-based platform enables real-time risk monitoring and maximizes straight-through processing using configurable risk-based rules.

It is tailored for digital, regional, mid-market, and large banks needing scalable AI solutions; this system harmonizes compliance and business objectives with key features. It leverages machine learning, identity clustering, and graph analytics to uncover emerging AML risks, while unsupervised ML algorithms and peer group data identify hidden threats. The platform offers hourly updates from global watchlists with customizable rules and prioritizes alerts with rule customization for different customer tiers. It supports a variety of payment rails and non-transaction events, streamlines case creation and workflow integration, uses identity clustering to pinpoint accounts controlled by a single entity, and ensures transparency with easy access to transaction records for regulatory cooperation. As of 2023, ComplyAdvantage was utilized by more than 1000 financial institutions and banks worldwide, including prominent clients like Allianz, OakNorth Bank, and Santander.

Source: Valk (2023).

4.3.2.5 Promote Regulatory Framework

FinTech has contributed to the evolution of regulatory frameworks by introducing technologies that enhance transparency, efficiency, and compliance in the financial sector. Through innovations like blockchain and AI-driven analytics, FinTech enables real-time monitoring of transactions and risk assessment, thereby helping regulatory bodies enforce stricter guidelines and prevent financial misconduct. For example, bank regulators in most developed economies have a FinTech innovation known as a regulatory sandbox, which assists in easy monitoring and regulations (see Box 4.6).

Box 4.6: Regulatory Sandbox

A regulatory sandbox is a structured method endorsed by regulators that permits the real-time, limited-duration experimentation of new financial products, technologies, and business models. This controlled environment operates under specific rules, supervision requirements, and safeguards, fostering innovation by providing a platform where both established entities and emerging players can explore ideas that may push the boundaries of existing regulatory frameworks. Successful trials within these sandboxes can lead to various outcomes, such as official approval of the innovation with tailored conditions, adjustments to regulatory standards, or even directives to halt certain activities. Originating in the UK in 2015, the concept has since gained global traction, with over 20 jurisdictions by early 2018 actively adopting or exploring its implementation.

Source: Zeidy (n.d.)

4.3.3 FinTech Hampers Sustainable Banking

Although FinTech has revolutionized various sectors of the financial industry, it introduces several challenges that may impede the advancement of sustainable banking. The following are challenges FinTech poses to sustainable banking.

4.3.3.1 Environmental Impact of Technology

The infrastructure that underpins FinTech, encompassing data centres and blockchain technology, often carries substantial environmental footprints due to their high energy consumption, posing a challenge to achieving sustainability goals. This trend is particularly pronounced in emerging economies with limited existing infrastructure, necessitating the construction of new infrastructure, which often compromises sustainability efforts. Empirically, Pawłowska et al. (2022) showed that despite FinTech's help in reducing the ecological footprint, it also affects the environment since most of the FinTech infrastructure consumes a large amount of energy.

4.3.3.2 Displacement of Traditional Banking Models

The disruption of traditional banking by FinTech can result in job losses and economic changes that do not always support the principles of sustainable development or ethical banking, which strive for fair and balanced growth.

4.3.3.3 Inequality

While FinTech holds the potential to improve financial inclusion, it also risks widening digital disparities, potentially marginalizing those without access to technology and impeding Sustainable Development Goals. Some FinTech practices, such as utilizing large datasets and machine learning, have introduced discriminatory outcomes that hinder financial inclusion and small- and medium-sized enterprises' (SMEs') credit access. Biases in machine learning models, often stemming from historical data biased towards white and male subjects, result in inaccuracies like facial recognition failures for non-whites and lower credit scores for females. Addressing these biases necessitates AI systems explicitly designed to detect and mitigate bias, supported by regulatory frameworks ensuring fairness. However, achieving unbiased data collection and usage presents significant challenges for technology firms, increasingly mandated by regulators. Moreover, FinTech's data accumulation may lead to data monopolies, limiting consumer choice and imposing high switching costs on SMEs locked into specific platforms.

4.3.3.4 Exposure to Risk and Vulnerability

FinTech has introduced significant risks and vulnerabilities in emerging economies. While they expand access to financial services like payments and investments, the extensive collection of consumer data raises concerns about privacy, consumer protection, and effective data governance. FinTech, especially larger firm business models, centred on aggregating large datasets, heightens the exposure of personal information to security threats. Moreover, customers often lack ownership and control over their data, exacerbating these risks and potentially compromising financial sector stability. This exposes customers to fraudsters. For example, in Ghana, despite fewer reported incidents, Mobile Money (MoMo) fraud losses in 2022 doubled compared to 2021. Losses rose from GHc14.2 million to GHc27 million, a 47% increase. Despite some recovery efforts, net losses amounted to about GHc26 million. Payment service providers reported 12,166 MoMo fraud incidents in 2022, slightly fewer than in 2021, primarily involving wrong transfers and fraudulent cash reversals.[12]

4.3.3.5 Promote Crime and Money Laundering

Despite providing significant advantages like enhanced financial inclusion and efficiency, FinTech can unintentionally facilitate crime and money laundering due to various inherent vulnerabilities. For example, cryptocurrencies have been identified as tools for money laundering (Allen et al., 2020). Emerging cryptocurrencies enable transactions through novel methods that offer both heightened privacy (Kareem et al., 2020) and substantial anonymity (Faccia et al., 2020). Concurrently, the increasing theft of cryptocurrencies complicates and intensifies the technical demands of anti-money laundering efforts (Allen et al., 2020).

[12] https://gna.org.gh/2023/06/momo-fraud-declines-but-losses-increase-to-ghs27m-in-2022/

4.4 Regulatory and Ethical Considerations

The detrimental impact of FinTech primarily arises from FinTech firms disregarding regulations and ethical standards. Therefore, the integration of FinTech into sustainable banking requires careful consideration of both regulatory and ethical aspects to ensure that the potential benefits are realized while mitigating risks.

4.4.1 Regulatory Considerations

FinTech firms must comply with various regulations to ensure they positively impact sustainable banking. These include anti-money laundering and counter-terrorism financing (CTF), customer protection, operational risk management and governance, and authorization and vetting requirements.

4.4.1.1 Anti-money Laundering and Counter-Terrorism Financing

FinTech must comply with anti-money laundering and counter-terrorism financing regulations such as the Financial Action Task Force (FATF) Recommendations, the Bank Secrecy Act (BSA) in the United States, and the European Union's (EU) Anti-Money Laundering Directive (AMLD). This can be achieved when there are robust KYC and transaction monitoring systems in the digital environment to prevent illegal activities. The FATF Recommendations provide a global framework for countries to combat money laundering, terrorist financing, and proliferation financing (see Box 4.7).

Box 4.7: Financial Action Task Force (FATF) Recommendations

The Financial Action Task Force (FATF), established in 1989 and headquartered in Paris, is the global watchdog against money laundering and terrorist financing. The

FATF Recommendations, first adopted in 2012 and last updated in November 2023, provide a robust framework for countries to combat money laundering, terrorist financing, and the proliferation of weapons of mass destruction. The FATF sets international standards to prevent these illicit activities and encourages national legislative and regulatory reforms through political advocacy.

With 40 member countries and over 200 global jurisdictions committed to its standards, the FATF ensures coordinated efforts to address serious crimes such as drug trafficking, illicit arms trade, and cyber fraud. Countries implement these standards through measures tailored to their specific legal, administrative, and financial systems. The FATF's framework includes Recommendations, Interpretive Notes, and a Glossary to guide national authorities. Notably, FATF suspended Russia's membership on 24 February 2023. The organization's efforts aim to enhance the global response to organized crime, corruption, and terrorism by fostering international cooperation and adherence to its standards.

Source: FATF (2023).

BSA mandates businesses to keep records and file reports aiding investigations, helping law enforcement combat money laundering, terrorism, and tax evasion (see Box 4.8).

Box 4.8: Bank Secrecy Act in the United States

The Bank Secrecy Act (BSA), passed by Congress in 1970, was the first US law to combat money laundering. It mandates businesses to maintain records and file reports useful in criminal, tax, and regulatory investigations. These records assist law enforcement in identifying and deterring money laundering linked to crimes, terrorism, and tax evasion. The Internal Revenue Service, part of the US National Money Laundering Strategy, balances enforcement with education. Key BSA requirements include filing Form 8300 for cash payments over US$10,000, reporting foreign financial accounts, and guidelines for money services businesses. A helpline offers compliance assistance.

Source: Internal Revenue Service (2024).

AMLD aims to combat money laundering and terrorist financing to enhance global security, maintain the integrity of the financial system, promote financial stability, and foster sustainable economic growth (see Box 4.9).

Box 4.9: European Union Anti-money Laundering Directive

The Anti-Money Laundering Directive (AMLD) is a framework of regulations issued by the European Union (EU) to combat money laundering and terrorist financing within member states. Originating in 1990 and aligned with FATF guidance, AMLD mandates that EU countries integrate these rules into their national laws. Key requirements include customer due diligence, transaction monitoring, and reporting suspicious activities to prevent the misuse of the financial system for illicit purposes.

The directive is consistently updated to address emerging risks, including those related to virtual assets and crowdfunding. It complements other regulations like the Markets in Crypto-Assets Regulation (MiCA). The Regulation on the Traceability of Transfers of Funds (TFR), effective from December 2024, enhances the traceability and authentication of crypto-asset transfers, aligning with international standards.

(continued)

Box 4.9: (continued)

The European Commission oversees the implementation and compliance of AMLD by working with member states and relevant authorities to ensure effective enforcement and adaptation of these regulations to evolving threats. This comprehensive and evolving legislative framework aims to create a uniform regulatory environment across the EU to safeguard against money laundering and terrorist financing.

Source: European Commission (2024a).

4.4.1.2 Consumer Protection

FinTech firms must prioritize consumer protection by incorporating consumer protection laws into their decision-making processes and ensuring compliance with these regulations (World Bank, 2022). This approach helps foster sustainable banking by reducing fraud and enhancing financial inclusion. These laws include the EU Payment Services Directive (PSD2) as shown in Box 4.10, regulations from the Consumer Financial Protection Bureau (CFPB) in the United States (see Box 4.11), and various country-specific regulations, which are supervised by central banks and other financial regulatory authorities. Regulators can also embark on awareness campaigns to encourage consumers to engage only with licensed or registered entities.

Box 4.10: Payment Services Directive

The Payment Services Directive (PSD2) grants the European Commission the authority to adopt delegated and implementing acts to guide compliance by authorities and market participants with PSD2's obligations. Key acts include:

1. *Commission Delegated Regulation (EU) 2019/411* (Art. 15(4)): This regulation sets technical standards for the development, operation, and maintenance of the electronic central register for payment services and provides access to its information.
2. *Commission Implementing Regulation (EU) 2019/410* (Art. 15(5)): It establishes technical standards detailing the structure and content of information that competent authorities must notify the European Banking Authority regarding payment services.
3. *Commission Delegated Regulation (EU) 2017/2055* (Art. 28(5)): This regulation provides standards for cooperation and information exchange between competent authorities related to the establishment and freedom of service provision by payment institutions.
4. *Commission Delegated Regulation (EU) 2021/1722* (Art. 29(6)): It specifies the framework for cooperation and information exchange between competent authorities of the home and host member states in supervising cross-border payment institutions and electronic money institutions.

These regulations ensure a cohesive framework for implementing PSD2 across the EU, facilitating regulatory compliance, efficient supervision, and enhanced cooperation among member states.

Source: European Commission (2024b).

Box 4.11: US Consumer Financial Protection Bureau

The Consumer Financial Protection Bureau (CFPB) was established by the Dodd–Frank Act in 2010 following the 2007–2008 financial crisis to address predatory and deceptive practices in the financial industry. Senator Elizabeth Warren played a key role in its development, and the bureau officially began operations in July 2011 with Richard Cordray as its first director.

The CFPB's core mission is to protect consumers in financial matters through:

- Creating and enforcing rules to prevent deceptive, unfair, and abusive practices
- Handling consumer complaints about financial companies
- Monitoring financial markets for risks to consumers
- Offering financial education tools and resources

The CFPB is led by a director, appointed by the president and confirmed by Congress, currently Rohit Chopra as of May 2024. The agency comprises six divisions focusing on different aspects of consumer protection and financial regulation.

Funded by quarterly transfers from the Federal Reserve, the CFPB has secured over US$20 billion in consumer relief and collected US$4 billion in penalties. Despite a Supreme Court ruling in 2024 affirming its funding structure, the CFPB faces criticism for perceived overreach and lack of accountability. Critics argue its directors' independence and special court deference complicate oversight. Nonetheless, the CFPB remains a vital resource for consumer financial protection and education.

Source: Marquit (2024).

4.4.1.3 Operational Risk Management and Governance

Regulators are increasingly imposing risk management and governance requirements on FinTech entities that are typically applied to traditional financial service providers (World Bank, 2022). Therefore, FinTech firms must comply with these regulations. This is because these obligations are designed to be adaptable, establishing standards for FinTech entities that align with their business characteristics and operational context. For instance, FinTech firms operating in the UK must adhere to overarching requirements like the "Principles for Business,[13]" which mandate responsible and effective organizational and risk management practices. The UK's Financial Conduct Authority (FCA) has further elaborated on these principles through detailed guidelines on risk management. Similarly, Mexico's Financial Technology Institutions Law emphasizes the importance of demonstrating effective operational risk controls and fraud prevention measures as essential criteria for FinTech operators seeking authorization (Staschen & Kerse, 2021).

4.4.1.4 Authorization and Vetting Requirement

Pre-licensing or registration vetting of FinTech entities helps exclude those prone to fraud (World Bank, 2022). This process, including background checks on past misconduct, ensures that entities and their management can effectively handle risks. Therefore, FinTech firms should undergo thorough vetting before obtaining licensing. As regulators increasingly promote awareness campaigns encouraging consumers to choose reputable FinTech firms, adhering to vetting processes before registration becomes essential. Countries like Australia and Portugal mandate licensing for all consumer credit providers, including digital microcredit. Licensing or registration for P2P lending is increasingly common globally, with requirements in the EU, China, and the UK (World Bank, 2022). Crowdfunding regulations vary, with specific categories for operators in the EU and the United States, while others use existing firm categories.

4.4.2 Ethical Considerations

To ensure the effective function and operation of FinTech solutions, firms must adhere to essential ethical principles. These include efforts to reduce inequality, promote customer education, maintain transparency and accountability, and minimize environmental impact, particularly carbon footprint.

[13] https://www.handbook.fca.org.uk/handbook/PRIN/2/1.html

4.4.2.1 Effort to Reduce Inequality

One of the fundamental goals of FinTech is to democratize access to financial services and activities (Aba & Linardy, 2021; Iddrisu et al., 2022). This entails ensuring that FinTech solutions are not only technologically advanced but also inclusive and accessible to all segments of the population, particularly the unbanked and underbanked (Batunanggar, 2019; Navaretti et al., 2018). These populations often face barriers such as limited access to traditional banking services, financial literacy challenges, or geographical isolation. FinTech firms play a crucial role in bridging these gaps by developing solutions that cater to diverse needs and circumstances (Batunanggar, 2019; Navaretti et al., 2018). This may involve offering mobile banking apps with intuitive interfaces, enabling simplified account opening processes, providing microfinance options, or facilitating P2P lending platforms. By making financial services more accessible through digital channels, FinTech solutions empower individuals and businesses to manage their finances more effectively, save money, and access credit opportunities. Mitigating the digital divide is another critical aspect. FinTech firms should strive to ensure that their solutions are accessible across different devices and Internet connectivity levels, including in rural or underserved areas. This involves designing user-friendly interfaces, supporting multiple languages, and providing robust customer support to assist users in navigating and utilizing these technologies. Ultimately, by promoting inclusivity and accessibility, FinTech firms not only expand their market reach but also contribute to broader economic empowerment and financial inclusion goals, driving positive social impact and sustainable growth in communities worldwide.

4.4.2.2 Customer Education

Many FinTech solutions are technologically intricate and can be challenging to grasp without IT expertise. Consequently, it is essential for FinTech firms to prioritize educating users about various aspects of their solutions (Batunanggar, 2019). This education should cover how to effectively use the FinTech services, understanding their functions and capabilities, recognizing limitations and potential risks involved, and understanding the periods during which the services will be accessible (Doku et al., 2023). In emerging markets like African countries, where formal education on technical matters such as reading terms and conditions or user manuals may be less common (Evans & Mendez Acosta, 2021), this educational effort becomes particularly critical. FinTech firms need to bridge this knowledge gap by providing accessible and clear information through user-friendly interfaces, interactive guides, and possibly localized support. By doing so, they empower users to make informed decisions about adopting and using FinTech solutions effectively. This proactive educational approach not only enhances user experience and satisfaction but also promotes trust and confidence in FinTech services, fostering greater adoption and sustainable growth in these markets (Abdeldayem & Al Dulaimi, 2020).

4.4.2.3 Transparency and Accountability

Maintaining transparency in FinTech operations is crucial, particularly in today's landscape where data usage, algorithmic decision-making, and AI-driven processes play significant roles (Aggarwal et al., 2024). Transparency ensures that users, regulators, and stakeholders understand how FinTech firms collect, handle, and utilize data (Aggarwal et al., 2024; Aldboush & Ferdous, 2023). It involves disclosing the purposes for which data is collected, how algorithms make decisions, and the ethical considerations behind AI applications. This transparency builds trust and allows users to make informed decisions about using FinTech services. It also enables regulators to monitor compliance with data protection laws and ethical standards, fostering a fair and accountable FinTech ecosystem. By promoting openness in these areas, FinTech firms can mitigate risks associated with privacy breaches, algorithmic bias, and unintended consequences of AI technologies, ultimately enhancing their credibility and long-term sustainability in the market.

4.4.2.4 Minimizing Carbon Footprints

FinTech firms evaluate their environmental impact by assessing energy use, emissions, resource consumption, and waste from operations like data centres (Puschmann et al., 2020). They promote green finance through sustainable practices like renewable energy adoption and energy-efficient technologies. Integrating environmental considerations into financial products involves offering green investments and eco-friendly financing. Collaboration with experts, regulators, and stakeholders is key for adhering to global sustainability standards, enhancing credibility, and attracting socially responsible investors. Public advocacy and education further support these efforts, emphasizing the importance of environmentally sustainable practices in finance to clients and communities.

4.5 Future Prospects of FinTech

FinTech has been noted for its risks and limitations as outlined earlier. However, considering international, regional, and national policy developments, there is a promising future for FinTech. Given that countries aim to integrate nearly all sectors, particularly finance, technology is essential for achieving this integration. This section explores various prospects in FinTech.

4.5.1 Central Bank Digital Money

Many governments are actively pursuing the issuance of central bank digital currencies (CBDCs). For instance, 19 countries across the Middle East and Central Asia are currently researching CBDC implementation (Bouza et al., 2024). Some countries, like Bahrain, Georgia, Saudi Arabia, and the United Arab Emirates, have progressed to the advanced "proof-of-concept" stage (Bouza et al., 2024). Kazakhstan has made significant strides with two pilot programmes for the digital tenge, indicating a continuing necessity and relevance for FinTech advancements.

4.5.2 Review and Implementations of FinTech Policies

Policymakers at international, regional, and national levels are actively revising existing regulations to accommodate new FinTech innovations and implementing new measures to enhance policy effectiveness. For example, the European Commission has updated the FATF guidelines to include the forthcoming Regulation on the Traceability of Transfers of Funds (TFR), set to take effect in December 2024 (see Box 4.9). TFR is a policy to regulate the activities of FinTech in Europe.

4.5.3 The Quest to Achieve Financial Inclusion

The United Nations has emphasized financial inclusion as a crucial strategy for alleviating poverty, prompting countries worldwide to implement policies aimed at enhancing access to financial services (Iddrisu et al., 2023). Financial inclusion is recognized not just as a tool for poverty alleviation but as a fundamental aspect of achieving broader economic and social development goals. Empirical studies strongly support the notion that FinTech can significantly advance financial inclusion by overcoming traditional barriers between urban and rural areas and bridging the gap between the affluent and the economically disadvantaged (Demir et al., 2022; Iddrisu et al., 2022, 2023; Pawłowska et al., 2022; Rahmi, 2019; Salampasis & Mention, 2018). Also, the Alliance for Financial Inclusion (AFI) has been implementing Inclusive Green Finance (IGF) policies to assist countries in extending access to finance to all populace (see Box 4.12). AFI has catalysed the adoption of digital finance solutions in numerous countries, enabling greater access to financial services for underserved populations. For example, India's move from indirect to direct payments, deposited into bank accounts with associated debit cards, reduced corruption and saved US$7 billion in 2.5 years (Pazarbasioglu et al., 2020). As an active member of AFI, the Bank of Ghana launched its first Mobile Money interoperability system in May 2018, significantly advancing financial inclusion. By 2019, mobile money transactions reached GH₵308 million, establishing Ghana as a leading mobile money market in Africa.[14] In 2021, mobile money accounts in Ghana increased from 32.7 million to 40.9 million.[15]

[14] https://qz.com/africa/1662059/ghana-is-africas-fastest-growing-mobile-money-market. Accessed 23 April 2024

[15] https://www.bog.gov.gh/wp-content/uploads/2021/09/Summary-of-Economic-Financial-Data-September-2021-1.pdf. Accessed 23 April 2024

Box 4.12: Alliance for Financial Inclusion

The Alliance for Financial Inclusion (AFI), a legally recognized entity in Malaysia under the International Organizations (Privileges and Immunities) Act of 1992, is primarily focused on advocating for and developing evidence-based policies to improve the lives of impoverished communities. AFI functions as a coalition of member central banks and financial regulatory bodies, united in their mission to advance financial inclusion on local, regional, and global levels. Through partnerships with regulators, international organizations, and the private sector, AFI aims to implement impactful policy reforms by leveraging a collaborative approach centred on peer learning, knowledge exchange, and proactive initiatives. Founded to create a global forum for sharing knowledge, AFI has consistently developed this vision over the years, supported by significant investments from donors and partners, resulting in tangible progress. With membership including central banks and financial regulators from 81 emerging and developing countries, AFI has committed more than a decade to addressing financial exclusion, achieving notable advancements in promoting financial inclusion. It provides a framework of six foundational principles for micro, small, and medium enterprise (MSME) finance policy, drawn from the best practices and international standards of the AFI network. The Sharm El Sheikh Accord, adopted at the 2017 AFI Global Policy Forum, commits members to work together in implementing financial inclusion strategies that also benefit the environment, especially vulnerable communities. Additionally, AFI's Inclusive Green Finance (IGF) workstream, part of the International Climate Initiative supported by Germany's Federal Ministry for Economic Affairs and Climate Action, has seen active climate commitments from members such as Bangladesh Bank, Reserve Bank of Fiji, and Central Bank of Egypt.

Source: AFI (2024).

4.5.4 Globalization and Social Media

The rapid advancement of globalization, the widespread influence of social media, and increasing urbanization have significantly boosted the prominence of e-commerce and digital payments (Ciupac-Ulici et al., 2022). Today, consumers are frequently exposed to products and services via social media platforms, transforming the way they shop and make payments (Aldboush & Ferdous, 2023). With the convenience of digital payment systems, buyers can seamlessly purchase items directly from advertisements on their social media feeds (Salampasis & Mention, 2018). This shift in consumer behaviour has not only led to a surge in demand for digital payment solutions but has also highlighted the urgent need for innovative FinTech developments. These FinTech solutions are essential to support the expanding e-commerce ecosystem, ensuring secure, efficient, and user-friendly transaction processes. The integration of social media and digital payments is reshaping the retail landscape, making it imperative for businesses and financial institutions to adapt and innovate. As a result, the demand for advanced FinTech solutions continues to grow, driving the need for new technologies and platforms that can accommodate this evolving marketplace.

4.6 Conclusion

FinTech has garnered global attention, prompting many countries to implement policies that promote its growth. The COVID-19 pandemic accelerated the shift from traditional financial systems to FinTech solutions. Today, numerous FinTech innovations aim to enhance financial and banking activities. This chapter has explored FinTech and its solutions, examining the evolution and current trends in the industry. It also evaluated how FinTech can contribute to sustainable banking and under what conditions it might hinder it. To ensure the smooth operation of FinTech, regulatory and legal considerations were discussed. Finally, this chapter examined the future prospects of FinTech.

While there is no universally accepted definition of FinTech, the concept broadly revolves around leveraging information technology to enhance financial services and activities. The evolution of FinTech dates back to the nineteenth century, starting with the establishment of the transatlantic telegraph cable, which revolutionized rapid financial communication. Over time, FinTech has advanced significantly, leading to the development of digital banking and open banking initiatives, which are central to modern financial innovation. FinTech innovations can be grouped into six main categories: e-savings and e-investment; e-transfer and e-payment; e-lending and e-borrowing; e-insurance; e-financial security and e-regulation; and other e-banking activities. This chapter has also identified that FinTech has the potential to advance sustainable banking by offering innovative financial products, enhancing financial inclusion, facilitating the identification of environmental risks, improving banking regulations, and collaborating with traditional banks to deliver efficient services. However, despite these benefits, FinTech also poses challenges to sustainable banking. These include potential negative environmental impacts, the displacement of traditional services, increased exposure to customer risk and vulnerability, the exacerbation of inequality, and the facilitation of crime and money laundering.

Given these adverse effects, it is crucial for FinTech firms to adhere to regulatory and ethical standards. Compliance with licensing and registration requirements, customer protection regulations, and Anti-Money Laundering and Counter-Terrorism Financing (AML/CTF) measures is essential. Ethically, firms should strive to minimize inequality, educate customers, maintain transparency and accountability, and consider their environmental impact. The future of FinTech appears promising, driven by factors such as the implementation of central bank digital currencies, the development and review of FinTech policies, the ongoing quest to enhance financial inclusion, and the influences of globalization and social media.

The findings of this chapter carry several important policy implications for governments, regulatory bodies, and industry stakeholders: First, policymakers must develop and adapt regulatory frameworks that balance innovation with the need for financial stability and consumer protection. This includes updating existing financial regulations to accommodate the unique aspects of FinTech, particularly in areas like digital banking, e-payments, and e-lending. Clear guidelines for compliance with Anti-Money Laundering and Counter-Terrorism Financing (AML/CTF) measures are essential to prevent the misuse of FinTech platforms. Second, to harness the potential of FinTech in promoting sustainable banking, policies should encourage the development of financial products and services that prioritize environmental sustainability and social equity. Governments could consider incentivizing FinTech solutions that contribute to financial inclusion and environmental risk management, while also imposing regulations to mitigate any negative environmental impacts associated with digital financial services. Third, as FinTech continues to expand, there is a growing need for policies that protect consumers from risks such as fraud, data breaches, and financial exploitation. Policymakers should mandate transparency and accountability in FinTech operations and ensure that firms provide adequate customer education, particularly in underserved or vulnerable populations. Fourth, this chapter highlights the ethical considerations FinTech firms must navigate, such as minimizing inequality and addressing the potential for increased vulnerability among certain customer segments. Policymakers should promote the adoption of ethical standards within the FinTech industry, possibly through codes of conduct or industry certifications that emphasize responsible business practices and social responsibility. Fifth, to maximize the benefits of FinTech while minimizing disruptions to traditional financial services, policies should encourage collaboration between FinTech firms and traditional banks. This could include facilitating partnerships that enhance service delivery and financial inclusion, as well as supporting open banking initiatives that foster innovation while maintaining financial stability. Last, with the rapid evolution of FinTech, it is crucial for policymakers to remain agile, continuously reviewing and updating financial policies to reflect new developments such as central bank digital currencies and emerging global trends. This forward-looking approach will help ensure that the regulatory environment remains conducive to innovation while safeguarding the broader financial ecosystem.

References

Aba, F., & Linardy, D. (2021). *Financial Technology in Financial Inclusions and its Implications on Poverty from Indonesia* (2021110061). https://doi.org/10.20944/preprints202111.0061.v1.

Abdeldayem, M. M., & Al Dulaimi, S. H. (2020). Trends of global FinTech education practices and the gcc perspective. *International Journal of Advanced Science and Technology, 29*, 7150–7163.

Abor, J. Y., Gyeke-Dako, A., Fiador, V. O., Agbloyor, E. K., Amidu, M., & Mensah, Lord. (2019). *Sustainable banking* (pp. 311–331). https://doi.org/10.1007/978-3-319-77458-9_15

AFI. (2024). *Inclusive green finance*. AFI. https://www.afi-global.org/thematic-areas/inclusive-green-finance/

Aggarwal, K. K., Sharma, A., Kaur, R., & Lakhera, G. (2024). *Algorithmic FinTech pioneering the financial landscape of tomorrow* (pp. 27–45). https://doi.org/10.4018/979-8-3693-1746-4.ch002.

Aldboush, H. H. H., & Ferdous, M. (2023). Building trust in FinTech: An analysis of ethical and privacy considerations in the intersection of big data, AI, and customer trust. *International Journal of Financial Studies, 11*(3), 90. https://doi.org/10.3390/ijfs11030090

Allen, F., Gu, X., & Jagtiani, J. (2020). A survey of FinTech research and policy discussion. *Review of Corporate Finance, 1*(3–4), 259–339.

Awanis, A., Lowe, C., Andersson-Manjang, S. K., & Lindsey, D. (2022). *State of the Industry Report on Mobile Money 2022*.

Bank Indonesia. (2017). *Bank Indonesia Regulation No. 19/12/PBI/2017 regarding the Organisation of Financial Technology*. https://www.dataguidance.com/legal-research/bank-indonesia-regulation-no-1912pbi2017

Batunanggar, S. (2019). *FinTech development and regulatory frameworks in Indonesia*. (1014; ADBI Working Paper Series).

Benbasat, I., Dexter, A. S., & Todd, P. (1986). An experimental program investigating color-enhanced and graphical information presentation: an integration of the findings. *Communications of the ACM, 29*(11), 1094–1105. https://doi.org/10.1145/7538.7545

Bouza, S., Miccoli, M., & Mircheva, B. (2024, June 18). *Central Bank Digital Currencies Can Boost Middle East's Financial Inclusion, Payment Efficiency: 19 countries in the Middle East and Central Asia are exploring issuing a CBDC*. https://www.imf.org/en/Blogs/Articles/2024/06/18/central-bank-digital-currencies-can-boost-middle-easts-financial-inclusion-payment-efficiency#:~:text=Central%20Bank%20Digital%20Currencies%20Can%20Boost%20Middle%20East's%20Financial%20Inclusion%2C%20Payment%20Efficiency,-19%20countries%20in&text=Almost%20two%2Dthirds%20of%20countries,efficiency%20of%20cross%2Dborder%20payments

Ciupac-Ulici, M., Beju, D.-G., Bresfelean, V. P., & Zanellato, G. (2022). Which factors contribute to the global expansion of M-commerce? *Electronics, 12*(1), 197. https://doi.org/10.3390/electronics12010197

Davis, F. D. (1989). Perceived usefulness, perceived ease of use, and user acceptance of information technology. *MIS Quarterly, 13*(3), 319. https://doi.org/10.2307/249008

Del-Valle-Soto, C., López-Pimentel, J. C., Vázquez-Castillo, J., Nolazco-Flores, J. A., Velázquez, R., Varela-Aldás, J., & Visconti, P. (2024). A comprehensive review of behavior change techniques in wearables and IoT: Implications for health and well-being. *Sensors, 24*(8), 2429. https://doi.org/10.3390/s24082429

Demir, A., Pesqué-Cela, V., Altunbas, Y., & Murinde, V. (2022). FinTech, financial inclusion and income inequality: A quantile regression approach. *The European Journal of Finance, 28*(1), 86–107. https://doi.org/10.1080/1351847X.2020.1772335

Dinners Club International. (2024). *The Diners Club Journey*. https://www.dinersclub.com/about-us/history/

Doku, J. N., Iddrisu, K., Bortey, D. N. A. S., & Ladime, J. (2023). Impact of digital financial technology on financial inclusion in sub-Saharan Africa: The moderation role of institutional quality. *African Finance Journal, 25*(2).

Dreier, J., Kassem, A., & Lafourcade, P. (2015). Formal analysis of e-cash protocols. *12th International Joint Conference on E-Business and Telecommunications (ICETE)*, 65–75.

Eisenmenger, N., Pichler, M., Krenmayr, N., Noll, D., Plank, B., Schalmann, E., Wandl, M.-T., & Gingrich, S. (2020). The Sustainable Development Goals prioritize economic growth over sustainable resource use: A critical reflection on the SDGs from a socio-ecological perspective. *Sustainability Science, 15*(4), 1101–1110. https://doi.org/10.1007/s11625-020-00813-x

European Commission. (2024a). *Anti-money laundering and countering the financing of terrorism at EU level*. https://finance.ec.europa.eu/financial-crime/anti-money-laundering-and-countering-financing-terrorism-eu-level_en

European Commission. (2024b). *Implementing and delegated acts—PSD 2*. https://finance.ec.europa.eu/regulation-and-supervision/financial-services-legislation/implementing-and-delegated-acts/payment-services-directive_en

Evans, D. K., & Mendez Acosta, A. (2021). Education in Africa: What are we learning? *Journal of African*

Economies, 30(1), 13–54. https://doi.org/10.1093/jae/ejaa009

Equator Principles. (2024). About the equator principles. In *Equator principles*. https://equator-principles.com/about-the-equator-principles/

Faccia, A., Moşteanu, N. R., Cavaliere, L. P. L., & Mataruna-Dos-Santos, L. J. (2020). *Electronic money laundering, the dark side of FinTech* (pp. 29–34). Proceedings of the 2020 12th International Conference on Information Management and Engineering. https://doi.org/10.1145/3430279.3430284

FATF. (2023, November). *The FATF recommendations*. https://www.fatf-gafi.org/en/publications/Fatfrecommendations/Fatf-recommendations.html

Fessenden, H. (2018). The great telegraph breakthrough of 1866. *Econ Focus, 2*(Q), 28–30.

Financial Stability Board. (2023, December 4). *FinTech*. https://www.fsb.org/work-of-the-fsb/financial-innovation-and-structural-change/fintech/

Fu, J., & Mishra, M. (2022). FinTech in the time of COVID−19: Technological adoption during crises. *Journal of Financial Intermediation, 50*, 100945. https://doi.org/10.1016/j.jfi.2021.100945

GABV. (2024, July 21). *The regional chapters*. GABV. https://www.gabv.org/declarations/the-regional-chapters/

Garbade, K. D., & Silber, W. L. (1979). The payment system and domestic exchange rates: Technological versus institutional change. *Journal of Monetary Economics, 5*(1), 1–22. https://doi.org/10.1016/0304-3932(79)90021-7

GSM Association. (2024). *The State of the Industry Report on Mobile Money 2024*.

GTBank Ghana Ltd. (2014, May 12). *GTBank Ghana Introduces Online Cheque Deposit System*.

Hidayat-ur-Rehman, I., & Hossain, M. N. (2024). The impacts of FinTech adoption, green finance and competitiveness on banks' sustainable performance: digital transformation as moderator. *Asia-Pacific Journal of Business Administration*. https://doi.org/10.1108/APJBA-10-2023-0497

Holzhauer, B. (2023, April 11). *Aspiration launches first credit card: Earn cash back and plant trees with every purchase*. Select. https://www.cnbc.com/select/aspiration-launches-first-credit-card/

Iddrisu, K., Abor, J. Y., & Banyen, K. T. (2022). FinTech, foreign bank presence and inclusive finance in Africa: Using a quantile regression approach. *Cogent Economics & Finance, 10*(1). https://doi.org/10.1080/23322039.2022.2157120

Iddrisu, K., Doku, J. N., Abor, J. Y., & Dziwornu, R. (2023). Financial inclusion and inclusive growth in Africa: What is the moderation role of financial stability? *Cogent Economics & Finance, 11*(2). https://doi.org/10.1080/23322039.2023.2267857

Internal Revenue Service. (2024, April 26). *Bank Secrecy Act*. https://www.irs.gov/businesses/small-businesses-self-employed/bank-secrecy-act#:~:text=Congress%20passed%20the%20Bank%20Secrecy,%2C%20tax%2C%20and%20regulatory%20matters

International Finance Corporation. (2011). *Mobile money study 2011*.

Jack, W., & Suri, T. (2011). *Mobile Money: The Economics of M-PESA*. https://doi.org/10.3386/w16721

Kohtamäki, M., Parida, V., Oghazi, P., Gebauer, H., & Baines, T. (2019). Digital servitization business models in ecosystems: A theory of the firm. *Journal of Business Research, 104*, 380–392. https://doi.org/10.1016/j.jbusres.2019.06.027

Kareem, H. M., Duhaidahawi, A., Zhang, J., & Abdulreza, M. S. (2020). Analysing the effects of FinTech variables on cybersecurity: Evidence form Iraqi Banks. *International Journal of Research in Business and Social Science, 9*(6), 123–133.

Laine, A. (2021, November 15). *Insights from inside the infamous PayPal referral program*. GROWSURF. https://growsurf.com/blog/paypal-referral-program

Lee, Y., Kozar, K. A., & Larsen, K. R. T. (2003). The technology acceptance model: past, present, and future. *Communications of the Association for Information Systems, 12*. https://doi.org/10.17705/1CAIS.01250

Liu, J., Kauffman, R. J., & Ma, D. (2015). Competition, cooperation, and regulation: Understanding the evolution of the mobile payments technology ecosystem. *Electronic Commerce Research and Applications, 14*(5), 372–391. https://doi.org/10.1016/j.elerap.2015.03.003

Marikyan, D., & Papagiannidis, S. (2023). Technology acceptance model: a review. In S. Papagiannidis (Ed.), *Theory Hub Book*. Theory Hub. https://open.ncl.ac.uk/

Marquit, M. (2024, June 17). *What is the Consumer Financial Protection Bureau (CFPB)?* https://www.britannica.com/money/Consumer-Financial-Protection-Bureau

Mertzanis, C. (2023). FinTech finance and social-environmental performance around the world. *Finance Research Letters, 56*, 104107. https://doi.org/10.1016/j.frl.2023.104107

Navaretti, G. B., Calzolari, G., Mansilla-Fernandez, J. M., & Pozzolo, A. F. (2018). *FinTech and banking. Friends or foes?* Friends or Foes.

Nosratabadi, S., Pinter, G., Mosavi, A., & Semperger, S. (2020). Sustainable banking; evaluation of the European business models. *Sustainability, 12*(6), 2314. https://doi.org/10.3390/su12062314

Pawłowska, M., Staniszewska, A., & Grzelak, M. (2022). Impact of FinTech on sustainable development. *Financial Sciences, 27*(2), 49–66. https://doi.org/10.15611/fins.2022.2.05

Pazarbasioglu, C., Mora, A. G., Uttamchandani, M., Natarajan, H., Feyen, E., & Saal, M. (2020). *Digital financial services*. https://pubdocs.worldbank.org/en/230281588169110691/Digital-Financial-Services.pdf

Puschmann, T., Hoffmann, C. H., & Khmarskyi, V. (2020). How green FinTech can alleviate the impact of climate change—The case of Switzerland. *Sustainability, 12*(24), 10691. https://doi.org/10.3390/su122410691

Rahmi, M. (2019). *FinTech for financial inclusion: Indonesia case* (pp. 805–807). In 1st International Conference on Economics, Business, Entrepreneurship, and Finance (ICEBEF 2018).

Robey, D., & Farrow, D. (1982). User involvement in information system development: A conflict model and empirical test. *Management Science, 28*(1), 73–85. https://doi.org/10.1287/mnsc.28.1.73

Salampasis, D., & Mention, A.-L. (2018). FinTech: Harnessing innovation for financial inclusion. In *Handbook of blockchain, digital finance, and inclusion* (Vol. 2, pp. 451–461). Elsevier. https://doi.org/10.1016/B978-0-12-812282-2.00018-8

SBFN. (2024, April 16). Latest News—April 16, 2024: SBFN launched its 2024 Global Progress Brief and new Data Portal. SBFN. https://www.sbfnetwork.org/

Soni, J. (2022). *The founders: The story of paypal and the entrepreneurs who shaped silicon valley*. Simon and Schuster.

Staschen, S., & Kerse, M. (2021, January 20). *Is Mexico's "FinTech Law" leading a new trend in FinTech regulation?* CGAP. https://www.cgap.org/blog/is-mexicos-fintech-law-leading-new-trend-in-fintech-regulation

Statista Research Department. (2024a, July 7). Investment activity in FinTech globally 2010-2024. Statista. https://www.statista.com/statistics/719385/investments-into-fintech-companies-globally/

Statista Research Department. (2024b, May 22). Number of fintechs worldwide from 2018 to 2024, by region. Statista. https://www.statista.com/statistics/893954/number-fintech-startups-by-region/

Toukabri, M. T., & Ettis, S. A. (2021). The acceptance and behavior towards e-insurance. *International Journal of E-Business Research, 17*(2), 1–16. https://doi.org/10.4018/IJEBR.2021040102

Tut, D. (2023). FinTech and the COVID-19 pandemic: Evidence from electronic payment systems. *Emerging Markets Review, 54*, 100999. https://doi.org/10.1016/j.ememar.2023.100999

UNEP FI. (2024). About us. UNEP FI. https://www.unepfi.org/about/about-us/

Valk, L. (2023, November). *Comply advantage alternatives & competitors*. https://salv.com/blog/complyadvantage-alternatives-competitors/

Venkatesh, V., & Bala, H. (2008). Technology acceptance Model 3 and a research agenda on interventions. *Decision Sciences, 39*(2), 273–315. https://doi.org/10.1111/j.1540-5915.2008.00192.x

Wachira, M. M., Iddrisu, K., Abban, G., & Abor, J. Y. (2023). Sustainable financing in developing economies. In J. Y. Abor (Ed.), *Sustainable and responsible investment in developing markets* (pp. 159–171). Edward Elgar Publishing. https://doi.org/10.4337/9781803927060.00019

Wang, K.-H., Zhao, Y.-X., Jiang, C.-F., & Li, Z.-Z. (2022). Does green finance inspire sustainable development? Evidence from a global perspective. *Economic Analysis and Policy, 75*, 412–426. https://doi.org/10.1016/j.eap.2022.06.002

World Bank. (2022). Financial consumer protection and FinTech: An overview of new manifestations of consumer risks and emerging regulatory approaches.

Zeidy, I. A. (n.d.). *The role of financial technology (FINTECH) in changing financial industry and increasing efficiency in the economy*. Retrieved June 25, 2024, from https://www.comesa.int/wp-content/uploads/2022/05/The-Role-of-Financial-Technology.pdf

5 Islamic Banks' Investment in Green Sukuk and Sustainable Development in the UAE

Abdul Jalil Mahama and Ibrahim Nandom Yakubu

5.1 Introduction

The global push toward sustainable development has gained significant momentum, with nations worldwide striving to balance economic growth, environmental protection, and social equity (United Nations Development Programme, 2021). The United Arab Emirates (UAE) has positioned itself as a pioneering force in this arena, embedding sustainability into its national vision and strategic plans. As part of its ambitious Vision 2021 and the UAE Energy Strategy 2050, the nation seeks to diversify its economy away from oil dependence, promote renewable energy, and foster a green and sustainable future (Barchiche, 2023).

A critical aspect of this transformation is the innovative use of financial instruments that align with sustainability goals. Among these, Green Sukuk has emerged as a significant tool, particularly within the framework of Islamic finance, to mobilize capital for sustainable projects (Liu & Lai, 2021; Alkadi, 2024). Green Sukuk, a Sharia-compliant bond-like instrument, is specifically designed to raise funds for projects that have a positive environmental impact, such as renewable energy, energy efficiency, and sustainable infrastructure (Araminta & Timur, 2022).

The inception of Islam is deeply rooted in ethics that govern every facet of business finance. Hence, the introduction of Sukuk as a Sharia-based investment tool is aimed at promoting the development of environmentally and socially responsible projects (Cortallini & Panetta, 2021; Liu & Lai, 2021). Green Sukuk is geared toward sustainability with strict compliance with Sharia law, making all its investment portfolios halal, contrary to the conventional finance-oriented model of green bonds (Alam et al., 2016a, b).

Islamic banks, which operate in accordance with Sharia principles, have increasingly recognized the potential of Green Sukuk as a means to support environmentally friendly initiatives (Utermarck et al., 2024). By investing in Green Sukuk, Islamic banks not only fulfill their ethical and religious obligations but also contribute to the broader agenda of sustainable development. This dual alignment with financial returns and social responsibility makes Green Sukuk an attractive option for Islamic banks seeking to expand their portfolio of sustainable investments (Sisdianto et al., 2024).

The impact of Islamic banks' investment in Green Sukuk on sustainable development in the UAE is multifaceted. These investments channel much-needed capital into green projects, facilitating the transition toward a low-carbon

A. J. Mahama (✉)
Faculty of Economics and Finance, National University of Uzbekistan, Tashkent, Uzbekistan

I. N. Yakubu
School of Business, University for Development Studies, Tamale, Ghana
e-mail: ibrahim.yakubu@uds.edu.gh

I. N. Yakubu (ed.), *Strategic Approaches to Banking Business and Sustainable Development Goals*, Sustainable Development Goals Series, https://doi.org/10.1007/978-3-031-80744-2_5

economy. Projects funded by Green Sukuk can range from solar and wind energy plants to sustainable real estate and water conservation initiatives. These investments help reduce the nation's carbon footprint, promote energy security, and create green jobs, thus contributing to environmental sustainability and economic diversification.

The UAE's strategic embrace of Green Sukuk and the active participation of Islamic banks in this market highlight the country's commitment to aligning its financial sector with Sustainable Development Goals (SDGs). In recent years, the UAE Green Sukuk, bonds, and loans combined have reached USD17 billion (WAM, 2022). The UAE holds 19.2% of the global environmental, social, and governance (ESG) Sukuk market, considered the largest in the Gulf Cooperation Council (GCC) market with a 52% share (Andy Sambidge, 2023).

Following the hosting of the 28th United Nations Climate Change Conference (COP28) and the debut of the Dirham-oriented Green Sukuk issuance by the First Abu Dhabi Bank (FAB), there has been a surge in Green Sukuk by 41.6% in the UAE banking industry. Reputable firms such as Majid Al Futtaim, Aldar Properties, Masdar, and Mubadala have raised significant amounts through green bonds and Sukuk, aimed at achieving the emirate's sustainability goals (Pramod Kumar, 2023; Alkesh Sharma, 2023).

Despite these advancements, challenges remain in the progress of Green Sukuk issuances. These include the narrow linkage of projects to only renewable energy and green real estate, difficulties in classifying green assets and projects by financial institutions, and steep learning curves for new issuers coupled with high costs of review and reporting (The World Bank, 2020).

This study aims to explore the role of Islamic banks' investment in Green Sukuk in driving sustainable development in the UAE. While the topic of Green Sukuk has gained prominence in the literature (Abubakar & Handayani, 2020; Abdullah & Keshminder, 2022; Araminta et al., 2022; Musari & Hidayat, 2023), there is a notable gap in research examining its impact on sustainable development, especially in the Middle Eastern context. Therefore, this study will contribute to bridging this gap not only in the broader Middle Eastern context but also on a global scale, particularly in countries practicing Islamic finance.

In the next section, an overview of Green Sukuk and Islamic finance is provided. Section 5.3 reviews the literature on green finance. The research methodology is presented in Sect. 5.4. In Sect. 5.5, the results are discussed, while the study is concluded in Sect. 5.6.

5.2 Overview of Green Sukuk and Islamic Finance

Green Sukuk represents a unique intersection between Islamic finance principles and sustainable investment objectives, offering an innovative approach to financing environmentally friendly projects while adhering to Sharia-compliant principles. Islamic finance principles emphasize ethical and responsible financial practices, including the prohibition of Riba (interest) and the promotion of socially beneficial investments (Iqbal & Mirakhor, 2017). Green Sukuk aligns with these principles by financing projects that have positive environmental impacts, such as renewable energy infrastructure, green buildings, and climate adaptation initiatives (Ahmed, 2021).

Through the integration of Islamic finance principles into the issuance and structuring of Green Sukuk, issuers ensure compliance with Sharia requirements while mobilizing capital toward Sustainable Development Goals (Harahap et al., 2023). Thus, Green Sukuk serves as a manifestation of the ethical and sustainable principles inherent in Islamic finance, providing investors with opportunities to participate in socially responsible investments that promote environmental conservation and societal well-being. Studies have shown various ways Sukuk contracts could be managed, which include Mudarabah, Musharakah, Ijarah, Istisna, and Wakalah, as well as the integration of these as unified contracts (Abdullah & Nayan, 2020).

Furthermore, the linkage between Green Sukuk and Islamic finance extends beyond adherence to Sharia-compliant principles, encompassing broader concepts of social justice, wealth redistribution, and equitable economic development. Islamic finance principles such as Zakat (mandatory charitable giving) and Sadaqah (voluntary charity) underscore the importance of wealth redistribution and social welfare, which are integral aspects of sustainable development (Iqbal & Mirakhor, 2011). Green Sukuk can leverage these principles to allocate a portion of Sukuk proceeds toward social welfare initiatives, community development projects, and environmental conservation efforts, thereby fostering inclusive and equitable economic growth (Ali et al., 2023).

Moreover, Islamic finance principles promote risk-sharing and cooperative financing arrangements, which are conducive to sustainable development (Iqbal & Mirakhor, 2011). Through Green Sukuk, issuers can mobilize private capital for environmentally sustainable projects, leveraging partnerships and profit-sharing arrangements to promote inclusive and environmentally sustainable economic development (Akin & Mirakhor, 2016).

5.2.1 Islamic Finance Principles and Sustainability

Islamic finance principles and sustainability are deeply intertwined, with Islamic finance principles emphasizing ethical and responsible financial practices that promote social welfare and environmental stewardship. One of the fundamental principles of Islamic finance is the prohibition of Riba (interest), which aligns with sustainability goals by discouraging speculative and exploitative financial practices that can harm society and the environment (Iqbal & Mirakhor, 2011; Islam, 2015). The prohibition of Gharar (excessive uncertainty) and Maysir (gambling) further supports ethical financial practices (Uddin, 2015). In the context of Green Sukuk in the UAE, Islamic finance principles guide the structuring and issuance of Sukuk instruments that adhere to Sharia-compliant investment criteria while financing environmentally sustainable projects. Scholars argue that Green Sukuk represents a convergence of Islamic finance principles and sustainability objectives, facilitating investments in projects that promote environmental conservation, renewable energy, and climate resilience while ensuring compliance with Islamic finance principles (Akin & Mirakhor, 2016). Adhering to Sharia-compliant investment principles demonstrates how Green Sukuk in the UAE upholds ethical and sustainable financial practices that contribute to the country's Sustainable Development Goals.

Moreover, Islamic finance principles such as Zakat (mandatory charitable giving) and Sadaqah (voluntary charity) underscore the importance of wealth redistribution and social justice, which are integral aspects of sustainability (Iqbal & Mirakhor, 2011). In the context of Green Sukuk in the UAE, these principles can be leveraged to allocate a portion of Sukuk proceeds toward social welfare initiatives, community development projects, and environmental conservation efforts. Scholars argue that by incorporating Zakat and Sadaqah principles into Green Sukuk structures, the UAE can enhance the social impact and sustainability of Sukuk issuances, thereby fostering inclusive and equitable economic development (Sadiq & Mushtaq, 2015; Habib, 2017). Findings from similar studies posit that Islamic financial institutions, capital markets, and the social sector play a vital role in fostering robust economic expansion, improving access to financial services for the wider population, and facilitating the flow of funds. This, in turn, mitigates risks and vulnerabilities among marginalized communities, contributing significantly to overall financial stability and advancement (Ahmed et al., 2015).

Additionally, Islamic finance principles emphasize the concepts of Mudarabah (profit-sharing) and Musharakah (partnership), which promote risk-sharing and cooperative financing arrangements conducive to sustainable development (Iqbal & Mirakhor, 2011; Abdul-Rahman & Nor, 2016). The extension of green finance to cover integrated Islamic finance products aimed at solving both social and commercial needs is explored as an alternative in the case of Indonesia, considered the largest Muslim nation. Initial proposals developed on Cash Waqf Linked Sukuk (CWLS) through Green Sukuk were considered a significant finance strategy (Khairunnisa, 2022). Musari (2022) further explored the integration of Esham principles, Green Sukuk, and CWLS. Hence, the evaluation of the model and its potential introduction in the UAE could serve as a tool for overcoming the challenges of climate change, complementing existing frameworks. The extent to which investors in the Green Sukuk capital markets voluntarily accept lower yields in their investments demonstrates the overwhelming trust they have in environmental sustainability (Pirgaip & Ayaydin, 2024). In relation to these findings, the increase in the issuance and extension of Green Sukuk products to viable potential sectors indicates that the UAE could harness these principles to mobilize private capital for sustainable infrastructure projects, renewable energy initiatives, and environmental conservation programs, thereby contributing to the country's transition toward a green and sustainable economy.

5.2.2 Global Trends and Developments of Green Sukuk

The market for Green Sukuk, also known as sustainable or environmentally friendly Islamic bonds, has experienced notable growth and development globally in recent years. This growth could be attributed to increasing awareness of environmental sustainability and the growing demand for ethical investment instruments globally. Countries and financial institutions around the world have shown interest in issuing Green Sukuk to finance environmentally sustainable projects such as renewable energy, green infrastructure, and climate adaptation initiatives (Liu & Lai, 2021). The recent collaboration between the International Center for Private Sector in Development (ICPSD) and the Islamic Development Bank (IsDB) aims to expand the operations of Islamic finance as an enabler to achieving the United Nations Sustainable Development Goals (SDGs) under the United Nations Development Programme (UNDP). The apparent launch of the Green Sukuk Initiative under the Global Islamic Finance and Impact Investing Platform (GIFIIP) is significant. Through this impactful project, the Islamic Corporation for the Development of Private Sector under the IsDB has supported countries like Uzbekistan, Kuwait, Pakistan, Bahrain, and Turkey to help grow their private sectors. The main priority areas include (a) enhancing access to funding for green energy initiatives in developing nations globally; (b) offering financial solutions to address environmental issues; and (c) increasing the involvement of Islamic finance in meeting the Sustainable Development Goals (SDGs) (UNDP, 2024). Moreover, there has been a surge in investor interest in socially responsible investments, from both conventional and Islamic financial systems, leading to greater demand for Green Sukuk (Bennett & Iqbal, 2013). Studies in Malaysia indicated that education sector investment is a key driver and motive for issuers and investors of Green Sukuk (Rahman et al., 2020). The diverse nature of the Green Sukuk investment (GSI) portfolio reduces the level of its risk exposure (Araminta & Timur, 2022). As a result, the market for Green Sukuk has witnessed a significant increase in issuance volume and diversity of projects financed, indicating a shift toward more sustainable and responsible investment practices globally.

5.2.3 Growth and Expansion of Green Sukuk

The Green Sukuk market has seen rapid expansion and diversification, with issuances occurring in various regions across the globe. This growth can be attributed to several factors, including favorable regulatory environments, increasing investor demand for sustainable investments, and advancements in Green Sukuk structuring and issuance frameworks (Kareem, 2023). Additionally, governments and multilateral institutions have played a pivotal role in promoting the growth of the Green Sukuk market by providing incentives, support, and regulatory frameworks to facilitate issuances in the UAE (MOCCAE, 2022). Furthermore, the incorporation of technology and innovative financial instruments has enabled the expansion of the Green Sukuk market by enhancing transparency, efficiency, and accessibility for investors (Wahab & Mohamed Naim, 2019).

The growth and expansion of the Green Sukuk market signify its increasing significance as a viable tool for financing sustainable development projects globally. The market has seen another corporate body (Tabreed) gearing up to expand and diversify its investment portfolio across the Indian economy by raising Sukuk worth US$1 billion (Gulf News, 2024). Notable state institutions or stakeholders in the UAE championing the Green Sukuk regime include First Abu Dhabi Bank (FAB), Masdar, Aldar Properties, Majid Al Futtaim, and Abu Dhabi Islamic Bank, while Emirates NBD Bank, HSBC Bank, Sharjah Islamic Bank, and Standard Chartered Banks serve as bookrunners for Green Sukuk issuers (ISFI, 2023). These developments serve as evidence of the growth potential of Sukuk in the UAE.

Despite the growth potential of Green Sukuk, the emergence of Covid-19 led to a 9.1% contraction globally during the peak of the pandemic, reducing issuance by 44.4% in 2020. Reports from the World Bank further revealed the lack of green specifications in terms of what constitutes "Green" as a significant challenge to its issuance across various sectors of the economy. This hurdle calls for the identification of projects outside the renewable energy brackets that could be considered green (Mardi et al., 2020).

The emergence of Green Sukuk has transformed the bond market, attracting numerous investors from both conventional and Islamic finance enclaves, consequently increasing the total market cap of this capital market. The significance of Green Sukuk has also helped investors initially interested in Sharia-compliant products diversify their investments from the conventional bond system due to religious lineage or beliefs that seem to conflict with Islamic values. This dynamic shift has shown the growth potential of this market, coupled with a positive perception of Islamic finance as a major rival to the interest-based conventional system.

In terms of growth figures, the inception of Green Sukuk and its subsequent issuance around the globe, especially in Muslim-majority countries, have been significant, with double-digit growth figures compared to World Bank-led conventional bond issuances. According to official figures, from 2008, US$13 billion was raised from green bonds, followed by US$37 billion in 2014. In the case of a recent advisory given by the Islamic Development Bank (IsDB), the London Stock Exchange (LSEG) to Emirates Islamic Bank led to a massive oversubscription of the bank's first Green Sukuk issuance of $750 million in April, while Dubai Islamic Bank issued USD 1 billion, attracting significant resources from various investors across different economies (AlGhoul & Esmail, 2024). These new developments signify the efforts of the UAE government to leverage the potential of Islamic finance toward achieving net-zero climate change goals by 2050. The adoption of Islamic finance in expanding green initiatives could be replicated by other developing economies employing the frameworks used by the UAE to further establish resilient energy infrastructure.

Reports from Crédit Agricole indicate that green bonds are projected to account for approximately 64% of ESG-based issuances in 2024, showcasing their dominant role in the sustainable finance market (Global Finance, 2024). Furthermore, data from S&P Global highlights that the UAE is leading the green bond market with 40% of issuances, followed by Malaysia at 25% and Saudi Arabia at 20% (International Finance Corporation, 2024). This trend underscores the growing importance of Green Sukuk and bonds as essential instruments for financing sustainable projects globally.

5.2.4 Regulatory Framework

The growth and expansion of the Islamic banking and finance industry have extended to more than 75 countries, with apparent total asset accumulation reaching US$200 billion from 200 Islamic financial institutions globally (Rahman & Riyadh, 2016). The regulatory framework surrounding Green Sukuk issuances plays a crucial role in shaping the market's development and growth. Studies by El-Hawary et al. (2007) and Aziz (2008) posited that the creation of the Islamic Financial Services Board propelled the establishment of regulations and standards guiding the operations of Islamic finance.

Regulatory bodies and standard-setting organizations have introduced guidelines, principles, and frameworks to ensure the integrity, transparency, and credibility of Green Sukuk issuances (IFSB, 2020). These regulatory frameworks provide guidance on the eligibility criteria for Green Sukuk projects, the use of proceeds, reporting requirements, and disclosure standards (Securities Commission Malaysia, 2019). Moreover, regulatory bodies such as central banks and security commissions play a supervisory role in overseeing Green Sukuk issuances to ensure compliance with Sharia principles, environmental standards, and disclosure requirements.

The existence of strong leadership in the energy sector in establishing robust regulatory frameworks aims to enhance investor confidence, promote market transparency, and foster the growth of the Green Sukuk market as a sustainable financing instrument for addressing global environmental challenges.

Experts in this field have long questioned the risk associated with Sharia Standard 62 on Sukuk, developed by the Accounting and Auditing Organization for Islamic Financial Institutions. This standard includes requirements that support the distribution of risk embedded in assets among Sukuk owners during business headwinds. Such a situation could discourage potential investors from increasing their contributions, knowing the negative influence that the performance of their assets might pose to their returns. Hence, countries that practice Islamic finance and its related investment components might need to potentially review the requirements of Standard 62 to ensure it does not adversely affect investor interest (Shibli, 2024).

Furthermore, based on the strategic vision of the UAE's Climate Change and Sustainability Framework 2021–2031, the government seeks to incentivize sustainability initiatives aimed at increasing investment. Therefore, the development of global standards that could affect the potential of this strategy might further require amendment or revision at the state level.

5.3 Literature Review

5.3.1 Theoretical Framework

This study seeks to employ the ESG or triple bottom line (TBL) theory, which emphasizes the significance of considering three dimensions of sustainability—economic, social, and environmental—in decision-making processes (Elkington, 1998). This theory has been reemphasized in corporate sustainability management (Ozanne et al., 2016), examining the dilemma of tensions most public firms face in managing these key pillars of sustainability.

Applied to the context of Green Sukuk in the UAE, the TBL theory underscores the significance of evaluating the social and environmental impacts of green projects financed through Sukuk issuances alongside their financial performance. By adopting a holistic approach to sustainability, the UAE can ensure that Green Sukuk contributes not only to environmental conservation but also to social development and economic prosperity.

The concept of sustainable responsible investment comes into play in this context, where investment decisions are considered in line with not only financial returns (profits) but also the social and environmental benefits that could be realized. This creates alternative sources of funding for green financing that extend beyond traditional bonds to cover Islamic-oriented finance instruments with the potential of bridging income inequalities while reducing the impacts of environmental pollution, increasing investment in the circular economy, waste and sustainable water management practices, terrestrial conservation, energy-efficient practices, and clean transportation (Liu & Lai, 2021).

The recent partnerships between UAE governments and their subsidiary firms in the areas of clean energy maritime transportation in emerging economies represent a significant step toward reducing carbon emissions currently recorded in this sector. The joint production of green methanol fuels in Egypt could be considered a justifiable example, with the potential of this alternative fuel confirmed by the International Renewable Energy Agency (Alvin, 2024). Through Green Sukuk, the UAE can fund projects that address societal needs, such as affordable housing, healthcare, and education, while simultaneously promoting environmental sustainability. By integrating TBL principles into the design and implementation of Green Sukuk, the UAE could achieve a balanced and inclusive approach to sustainable development that benefits both present and future generations.

Ecological modernization theory (EMT) posits that environmental protection and economic development can be mutually reinforcing, leading to sustainability through technological innovation, market mechanisms, and regulatory frameworks (Mol & Sonnenfeld, 2014). This has been further explored on the premises of an environment–development constellation (Weber & Weber, 2020a, b), under which we seek to analyze the impacts of green finance in attaining the Sustainable Development Goals (SDGs). Previous literature has delineated the potential areas that have received Green Sukuk investment in line with SDGs. Weber and Weber (2020a, b) further uncovered the direct linkage of EMT and the SDGs, their studies factoring science and technology, ecology, social, economic, and political change into the circle of influence in the realization of key Sustainable Development Goals.

In the context of Green Sukuk in the UAE, EMT can be applied to understand how the integration of environmental considerations into financial instruments like Green Sukuk can foster sustainable development. The theory suggests that the adoption of Green Sukuk facilitates the transition toward a more environmentally sustainable economy by incentivizing investments in green projects, such as renewable energy infrastructure and sustainable transportation systems. Recent studies have further shown a positive correlation between Sukuk's potential in reducing risk levels and volatilities of green bond investments (Billah et al., 2023). Through the issuance of Green Sukuk, the UAE could encourage private sector participation in environmentally friendly initiatives while promoting economic growth and diversification. By aligning financial incentives with environmental objectives, Green Sukuk could contribute to the country's transition toward a greener and more sustainable future.

5.3.2 Empirical Literature Review

Green Sukuk, often considered a Sharia-compliant financial instrument developed to finance environmentally friendly projects around the globe, has gained interest as an essential tool that could help promote sustainable development. Empirical studies on Green Sukuk highlight its potential in achieving multiple Sustainable Development Goals (SDGs) across different sectors of the economy. Biancone et al. (2022) emphasized that Green Sukuk could be used to mobilize large amounts of capital for renewable energy projects, directly contributing to SDG 7 (affordable and clean energy). The benefits of these projects on rural electrification have been extensively explored through the application of a green Waqf model as a component of Green Sukuk (Ibrahim, 2023). A study by Ahmed (2017) indicated that countries in Southeast Asia such as Malaysia and Indonesia, considered pioneers of Green Sukuk issuance, have seen notable investments in solar and wind energy, resulting in increased access to alternative renewable energy and reduced carbon footprints. Ahmed (2017) argued that Green Sukuk not only enhances energy infrastructure but also stimulates green jobs, thereby contributing to SDG 8 (decent work and economic growth), as well as financial inclusion (Ebrahim et al., 2021) and poverty alleviation (Habibullah & Hassan, 2023). Furthermore, Green Sukuk's role in addressing environmental challenges extends to SDG 13 (climate action). Aassouli et al. (2018) asserted that funds raised through Green Sukuk are often earmarked for potential projects aimed at mitigating climate change impacts, especially in the area of green infrastructure in sub-Saharan Africa. According to their study, this financial mechanism helps bridge the funding gap for climate resilience projects in developing countries, where traditional financing tools may be scarce. According to Obaidullah (2017), the transparency and accountability inherent in Green Sukuk, considered a Sharia investment tool, bolster investor confidence, attracting a broader base of ethical investors and amplifying the impact on various SDGs.

In the case of the UAE, respective government institutions in the emirates of Dubai and Abu Dhabi have fully supported the development of an alternative energy mix through concerted efforts by Dubai Electricity and Water Authority (DEWA) and Masdar Abu Dhabi in establishing solar parks, including the Mohammed bin Rashid Al Maktoum Solar Park, considered one of the largest in the world. Recent issuances of Green Sukuk by various stakeholders in the country have been directed toward renewable energy and other carbon reduction projects, exemplifying how Green Sukuk mobilizes private capital to support large-scale renewable energy infrastructure, thereby reducing carbon emissions and enhancing energy security (Mahama & Mumin, 2023). These developments not only demonstrate progress toward clean energy but also align with the UAE's Green Agenda 2030, which aims to increase the contribution of clean energy to the total energy mix.

5.4 Methodology

5.4.1 Data and Variables

This study employs a country-level dataset spanning the period from 2018Q1 to 2022Q4. This period is selected based on data availability. The Sustainable Development Goals Index (SDGI) serves as the dependent variable. The independent variable is Green Sukuk Investment (GSI), measured by the total investment in Sukuk by the Islamic banking sector in UAE dirham (AED). This study also controls for the impact of financial development (FDEV) and foreign direct investment (FDI). Financial development is measured using domestic credit to the private sector by banks as a percentage of gross domestic product (GDP), while FDI is measured by the net inflows of foreign direct investment as a percentage of GDP (Yakubu, 2020). Data sources include the 2023 UAE Islamic Finance Report, the SDG Transformation Center, and the World Development Indicators of the World Bank.

5.4.2 Model Specification and Analytical Technique

The empirical model employed in this study is specified as follows:

$$\begin{aligned} \text{lnSDGI}_t &= \propto_0 + \beta_1 \text{lnGSI}_t + \beta_2 \text{lnFDEV}_t \\ &\quad + \beta_3 \text{lnFDI}_t + \epsilon_t \end{aligned} \tag{5.1}$$

The sign $\propto_0$ represents the intercept term, indicating the baseline level of sustainable development when all independent variables are zero. β represents the coefficient values of the explanatory or independent variables, while ϵ is considered the error term. t represents the time period. The acronyms SDGI, GSI, FDEV, and FDI are previously defined. The variables are set to natural logarithm form to linearize the relationships among them.

In analyzing the data, this study employs the robust least squares (RLS) technique to enhance the accuracy and reliability of its findings. Traditional ordinary least squares (OLS) regression is highly sensitive to outliers and leverage points, which can distort results by giving undue influence to large residuals (Paleologu et al., 2016). RLS, however, is designed to minimize this sensitivity, providing more reliable parameter estimates when dealing with anomalies or non-normal error distributions. Given the country-level data spanning multiple years, the likelihood of encountering extreme values or reporting inconsistencies is high (Sun et al., 2006). By using RLS, this study ensures that its estimates accurately reflect the true underlying patterns in the data, making the results more robust and reliable.

5.5 Results and Discussion

5.5.1 Descriptive Statistics

The descriptive statistics in Table 5.1 provides a comprehensive overview of the variables in their non-logarithm form. The mean SDGI score is 69.530, indicating a moderate overall level, with a median close to the mean, suggesting a symmetrical distribution. GSI exhibits a higher mean (88.200) compared to SDGI, with noticeable skewness (0.524) indicating a slight rightward tail in its distribution. FDEV and FDI have means of 75.221 and 4.372, respectively, with FDEV showing low skewness and FDI displaying a negative skew, suggesting a leftward tail. The variables exhibit varying levels of dispersion: GSI has a notably higher standard deviation (24.222) compared to the others, indicating greater variability. Kurtosis values for all variables suggest generally moderate peakedness relative to a normal distribution. The Jarque–Bera tests for normality show that the variables are normally distributed.

5.5.2 Multicollinearity Analysis

The multicollinearity analysis in Table 5.2 reveals the relationships between the natural logarithms of the variables. The pairwise correlations show that lnGSI is negatively correlated with lnFDEV (−0.652) and positively correlated with lnFDI (0.349), while lnFDEV and lnFDI are positively correlated (0.341). The variance inflation factor (VIF) values are all below 5, with lnGSI and lnFDEV having similar VIFs of 4.875 and 4.848, respectively, and lnFDI having a VIF of 3.171. These VIF values suggest the absence of multicollinearity among the variables, as they are below the commonly used threshold of 10 (Gujarati, 2003, cited in Yakubu, 2019). The tolerance values further confirm this, as they are all above 0.1, indicating that the regression analysis is not impacted by multicollinearity.

5.5.3 Regression Results

The regression results from Table 5.3 show the impact of the various independent variables on the Sustainable Development Goals Index (SDGI) as analyzed using the robust least squares (RLS) technique. Each coefficient represents the

Table 5.1 Descriptive statistics

	SDGI	GSI	FDEV	FDI
Mean	69.530	88.200	75.221	4.372
Median	70.430	71.000	75.146	4.484
Maximum	71.310	124.000	87.919	5.690
Minimum	66.830	68.000	64.097	2.432
Standard deviation	1.647	24.222	7.790	1.114
Skewness	−0.641	0.524	0.304	−0.749
Kurtosis	1.961	1.429	2.487	2.496
Jarque–Bera	2.269	2.972	0.527	2.084
Probability	0.322	0.226	0.768	0.353

Source: Authors' Computation

Table 5.2 Multicollinearity analysis

	*ln*GSI	*ln*FDEV	*ln*FDI	VIF	Tolerance (1/VIF)
*Ln*GSI	1.000			4.875	0.205
*ln*FDEV	−0.652	1.000		4.848	0.206
*ln*FDI	0.349	0.341	1.000	3.171	0.315

Source: Authors' Computation

Table 5.3 Robust least squares (RLS) estimates

Variables	*ln*SDGI
*ln*GSI	0.044***
	(0.012)
*ln*FDEV	0.074**
	(0.031)
*ln*FDI	0.054***
	(0.009)
C	1.585***
	(0.076)
R-squared	0.808
Adjusted R-squared	0.772
Standard error of regression	0.002

Source: Authors' Computation
Values in ( ) are standard errors; *** and ** are statistical significance at 1% and 5%, respectively

elasticity of the SDGI with respect to the independent variables, with all variables transformed into their natural logarithms for a linear relationship.

The results show that the coefficient for Green Sukuk investment is 0.044 and is statistically significant at the 1% level. This positive coefficient suggests that a 1% increase in Green Sukuk investment by Islamic banks is associated with approximately a 0.044% increase in the SDGI. This result highlights the positive contribution of Green Sukuk investments to sustainable development in the UAE, reinforcing the idea that environmentally focused financial instruments can drive progress toward sustainability goals.

Financial development has a positive significant relationship with SDGs. This implies that a 1% increase in financial development, measured as domestic credit to the private sector by banks (percentage of GDP), corresponds to a 0.074% increase in the SDGI. This finding underscores the importance of a well-developed financial sector in facilitating investments that promote sustainable development. A robust financial sector can provide the necessary capital for sustainable projects and ensure the efficient allocation of resources toward initiatives that support the SDGs.

Foreign direct investment shows a coefficient of 0.054 and is also statistically significant at the 1% level. This indicates that a 1% increase in FDI net inflows (percentage of GDP) leads to a 0.054% increase in the SDGI. The positive and significant impact of FDI suggests that foreign investments are instrumental in advancing Sustainable Development Goals in the UAE. FDI brings not only capital but also technology transfer, managerial expertise, and innovation, all of which are crucial for achieving sustainable economic growth and development.

5.6 Conclusion and Policy Recommendations

This study empirically analyzes the impact of Green Sukuk investment by Islamic banks on sustainable development in the UAE over the period 2018–2022, while controlling for the impact of financial development and foreign direct investment. Using robust least squares (RLS) regression, the results reveal that Green Sukuk investment positively and significantly contributes to the Sustainable Development Goals Index. Additionally, it is found that financial development and foreign direct investment enhance sustainable development. These findings underscore the critical role of targeted financial instruments and robust financial and investment environments in promoting sustainable development in the UAE.

The findings present significant policy recommendations. For instance, the positive impact of Green Sukuk on SDGs underscores the need for policymakers to actively promote the issuance of Green Sukuk. The government can introduce incentives such as tax breaks, subsidies, and streamlined regulatory processes to make Green Sukuk more attractive to both issuers and investors. By doing so, they can channel more funds into environmentally sustainable projects, thereby accelerating the country's progress toward its Sustainable Development Goals. Additionally, public awareness campaigns can be initiated to educate potential investors about the benefits and opportunities associated with Green Sukuk investments. Furthermore, there is a need to strengthen the financial sector in the UAE given the direct impact of financial development on SDGs. Policymakers should focus on improving access to credit, particularly for small- and medium-sized enterprises (SMEs) that are often key drivers of innovation and sustainable practices. Enhancing financial infrastructure, such as improving payment systems and increasing financial literacy among the population, can also contribute to a more robust financial sector. By fostering an environment that supports financial development, the UAE can ensure that financial resources are efficiently allocated toward sustainable and productive uses. Moreover, the UAE should continue to create a conducive environment for attracting FDI. This can be achieved by maintaining political and economic stability, simplifying business regulations, and offering investment incentives. Additionally, the government can focus on attracting FDI into sectors that directly contribute to sustainable development, such as renewable energy, sustainable agriculture, and green infrastructure. By doing so, the UAE can leverage foreign capital, technology, and expertise to further its sustainable development agenda.

This study, while providing valuable perspectives into the impact of Green Sukuk investment by Islamic banks on sustainable development in the UAE, has certain limitations. The dataset spans only 5 years, which may not capture long-term trends and fluctuations in Green Sukuk investments and their impacts. Additionally, this study focuses exclusively on the UAE, limiting the generalizability of the findings to other regions with different economic, social, and regulatory environments. Future research should consider a longer time frame and a broader geographic scope to enhance the robustness and applicability of the results. Furthermore, exploring the qualitative aspects of Green Sukuk investments, such as the specific projects financed and their direct environmental and social impacts, would provide a more comprehensive understanding of their role in sustainable development. Investigating the interplay between different types of sustainable financial instruments and their cumulative effects on development goals could also offer valuable insights for policymakers and investors.

References

Aassouli, D., Asutay, M., Mohieldin, M., & Nwokike, T. C. (2018). *Green Sukuk, energy poverty, and climate change: A roadmap for sub-Saharan Africa*. The World Bank Group.

Abdullah, N., & Nayan, M. A. (2020). Green Sukuk, financing the future to sustainable environment. *International Journal of Zakat and Islamic Philanthropy, 2*(1).

Abdul-Rahman, A., & Nor, S. M. (2016). Challenges of profit-and-loss sharing financing in Malaysian Islamic banking. *Geografia, 12*(2).

Abdullah, M. S., & Keshminder, J. S. (2022). What drives green sukuk? A leader's perspective. *Journal of Sustainable Finance & Investment, 12*(3), 985-1005.

Abubakar, L., & Handayani, T. (2020). Green Sukuk: sustainable financing instruments for infrastructure development in Indonesia. In *1st Borobudur International Symposium on Humanities, Economics and Social Sciences (BIS-HESS 2019)* (pp. 983–987). Atlantis Press.

Ahmed, H. (2017). Contribution of Islamic finance to the 2030 agenda for sustainable development. In *High-level conference on financing for development and the means of implementation of the 2030 agenda for sustainable development* , (pp. 1–53).

Ahmed, M. (2021). Islamic social finance: Case studies of funding models. In *Innovative humanitarian financing* (pp. 101–134). Springer. Retrieved from https://link.springer.com/chapter/10.1007/978-3-030-83209-4_5

Ahmed, H., Mohieldin, M. V., & Aboulmagd, F. (2015). *On the sustainable development goals and the role of Islamic finance*. World Bank Policy Research. Retrieved from https://papers.ssrn.com/sol3/papers.cfm?abstract_id=2606839.

Akin, T., & Mirakhor, A. (2016). Efficiency with rule-compliance: A contribution to the theory of the firm in Islamic economics. *Journal of Economics and Political Economy, 3*(3), 560–574. Retrieved from https://journals.econsciences.com/index.php/JEPE/article/view/864

Alam, N., Duygun, M., & Ariss, R. T. (2016a). Green Sukuk: An innovation in Islamic capital markets. In A. Dorsman, Ö. Arslan-Ayaydin, & M. B. Karan (Eds.), *Energy and finance: Sustainability in the energy industry* (pp. 167–185). Springer. Retrieved from https://link.springer.com/book/10.1007/978-3-319-32268-1

Alam, N., Duygun, M., & Ariss, R. T. (2016b). *Green sukuk: An innovation in Islamic capital markets. In energy and finance: Sustainability in the energy industry*. Springer International Publishing.

AlGhoul, R., & Esmail, E. (2024, April). *Emirates Islamic Issues $750 million Sustainability Sukuk*. Emirates News Agency-WAM. Retrieved from http://www.wam.ae

Ali, Q., Rusgianto, S., Parveen, S. Y., & Zin, R. M. (2023). An empirical study of the effects of green Sukuk spur on economic growth, social development, and financial performance in Indonesia. *Environment, Development and Sustainability*, 1–27. Retrieved from https://link.springer.com/article/10.1007/s10668-023-03520-6

Alkadi, R. S. (2024). Towards a sustainable future: a comprehensive review of Green Sukuk. *Review of Accounting and Finance*. Retrieved from https://www.emerald.com/insight/content/doi/10.1108/RAF-03-2023-0105/full/html#sec006

Alvin, R. C. (2024). *AD Ports-led consortium to develop green methanol plant in Egypt*. Retrieved from https://www.thenationalnews.com/business/energy/2024/05/28/ad-ports-led-consortium-to-develop-green-methanol-plant-in-egypt/

Araminta, D. V., & Timur, Y. P. (2022). The role of green Sukuk in realizing the sustainable development goals 2030 agenda. *Jurnal Ekonomi & Bisnis Islam, 8*(2).

Araminta, D. V., Qudziyah, Q., & Timur, Y. P. (2022). The role of green Sukuk in realizing the sustainable development goals 2030 agenda. *Journal of Islamic Economics and Business, 8*(2), 251–266. https://doi.org/10.20473/jebis.v8i2.37531

Aziz, Z. A. (2008). Enhancing interlinkages and opportunities: The role of Islamic finance. *The National Breau of Asian Research, 18*(4). Retrieved from https://www.nbr.org/publication/enhancing-interlinkages-and-opportunities-the-role-of-islamic-finance/

Barchiche, D. (2023). *A new momentum for the Sustainable Development Goals*. Retrieved from https://www.iddri.org/en/about-iddri/team/damien-barchiche

Bennett, M. S., & Iqbal, Z. (2013). How socially responsible investing can help bridge the gap between Islamic and conventional financial markets. *International Journal of Islamic and Middle Eastern Finance and Management, 6*(3), 211–225. Retrieved from https://www.emerald.com/insight/content/doi/10.1108/IMEFM-Aug-2012-0078/full/html

Biancone, P. P., Hassan, M. K., Uluyol, B., & Paltrinieri, A. (2022). Social impact management and finance and Islamic finance for sustainable development. https://doi.org/10.13140/RG.2.2.27057.45925.

Billah, M., Elsayed, A. H., & Hadhri, S. (2023). Asymmetric relationship between green bonds and Sukuk markets: The role of global risk factors. *Journal of International Financial Markets, Institutions and Money, 83*, 101–728. Retrieved from https://www.sciencedirect.com/science/article/pii/S1042443122002001

Cortallini, G., & Panetta, I. C. (2021). Green bond: A systematic literature review for future research agendas. *Journal of Risk Financial Management, 14*(12), 589. https://doi.org/10.3390/jrfm14120589

Ebrahim, R., Abdulla, Y., & Kumaraswamy, S. (2021). Islamic finance and sustainable development goals in Bahrain. In *International conference on sustainable Islamic business and finance* (pp. 8–13). IEEE.

El-Hawary, D., Grais, W., & Iqbal, Z. (2007). Diversity in the regulation of Islamic Finance Institutions. *The Quarterly Review of Economics and Finance, 46*, 778–800. Retrieved from https://doi.org/10.1016/j.qref.2006.08.010

Elkington, J. (1998). Accounting for the triple bottom line: Measuring business excellence, 18–22. https://doi.org/10.1108/eb025539

Gujarati, D. (2003). *Basic Econometrics*. 4th Edn. Singapura: McGraw-Hill.

GULF NEWS. (2024). *Gulf News*. Retrieved from Gulf News: http://www.gulfnews.com

Habib, A. (2017). Contribution of Islamic finance to the 2030 agenda for sustainable development. In *High-level conference on financing for development and the means of implementation of the 2030 agenda for sustainable development*. Durham University.

Habibullah, M., & Hassan, R. (2023). Developing a conceptual Shari'ah framework of Green Sukuk for Bangladesh: Its implications on social welfare by realizing SDGs. *Journal of Islamic Finance*. Retrieved from https://journals.iium.edu.my/iiibf-journal/index.php/jif/article/view/748

Harahap, B., Risfandy, T., & Futri, I. N. (2023). Islamic law, Islamic finance, and sustainable development goals: A systematic literature review. *Sustainability, 15*(8). https://doi.org/10.3390/su15086626

Ibrahim, A. J. (2023). *Renewable energy financing in Islamic finance: A community green Waqf financing model for rural electrification*. SSRN.

IFSB. (2020). *Islamic financial services industry stability report 2020*. IFSB.

Iqbal, Z., & Mirakhor, A. (2011). *An introduction to Islamic finance: Theory and practice* (Vol. 687). John Wiley & Sons.

Iqbal, Z., & Mirakhor, A. (2017). *Ethical dimensions of Islamic finance: Theory and practice*. Springer.

ISFI. (2023). *Islamic Sustainable Finance and Investment*. Retrieved from Islamic Sustainable Finance and Investment: https://islamicsustainable.com/case-study-the-uaes-first-aed-denominated-green-sukuk/

Islam, S. M. (2015). Islamic banking: A true alternative to interest-based conventional banking. *International Scholar Journal of Accounting and Finance, 1*(1), 33–40. Retrieved from https://papers.ssrn.com/sol3/papers.cfm?abstract_id=2699451

Kareem, M. (2023). *Islamic Sustainable Finance and Investment*. Retrieved from Islamic Sustainable Finance and Investment: https://islamicsustainable.com/case-study-the-uaes-first-aed-denominated-green-sukuk/

Khairunnisa, M. (2022). Integrating green sukuk and cash waqf linked sukuk, the blended Islamic finance of fiscal instrument in Indonesia: A proposed model for fighting climate change. *International Journal of Islamic Khazanah, 12*(2), 133–144.

Kumar, P.. (2023). *Arabian Gulf Business Insight*. Retrieved from https://www.agbi.com/finance/2023/11/pifs-saudi-real-estate-refinance-company-issues-933m-sukuk/

Liu, F. H., & Lai, K. P. (2021). Ecologies of green finance: Green sukuk and development of green Islamic finance in Malaysia. *Economy and Space, 53*(8), 1896–1914. Retrieved from https://journals.sagepub.com/doi/pdf/10.1177/0308518X211038349

Mahama, A. J., & Mumin, M. A. (2023). Renewable energy resilience: Probing the role of political stability in UAE's sustainable energy transition. In *Governance quality, fiscal policy, and the path to a low-carbon future: Perspectives from developing economies* (pp. 1–31). IGI Global. https://www.igi-global.com/chapter/renewable-energy-resilience/329077

Mardi, M., Osman, M. R., & Aziz, A. H. (2020). *Pioneering the green Sukuk: Three years on*. World Bank.

MOCCAE. (2022). *Ministry Of Climate Change and Environment*. Retrieved from Ministry Of Climate Change and Environment: http://www.moccae.gov.ae

Mol, A. P., & Sonnenfeld, D. A. (2014). *Ecological modernisation around the world: An introduction*. Taylor and Francis. https://doi.org/10.1080/09644010008414510.

Musari, K. (2022). A comparative study of Islamic fiscal instrument securitization from history to the modern age: Esham, Sukuk, and Cash Waqf Linked Sukuk (CWLS). In *Handbook of research on challenges in public economics in the era of globalization* (pp. 397–418). IGI Global.

Musari, K., & Hidayat, S. E. (2023). The role of green Sukuk in Maqasid Al-Shariah and SDGs: Evidence from Indonesia. In *Islamic finance, FinTech, and the road to sustainability* (pp. 181–203). Retrieved from https://link.springer.com/chapter/10.1007/978-3-031-13302-2_9

Obaidullah, M. (2017). *Managing climate change: The role of Islamic finance*. Islamic Research and Training Institute. https://doi.org/10.13140/RG.2.2.25535.51361

Ozanne, L. K., Phipps, M., Weaver, T., Carrington, M., Luchs, M., Catlin, J., et al. (2016). Managing the tensions at the intersection of the triple bottom line: A paradox theory approach to sustainability management. *Journal of Public Policy*. Retrieved from https://journals.sagepub.com/doi/abs/10.1509/jppm.15.143

Paleologu, C., Benesty, J., Stanciu, C., Anghel, C., & Stenţa, M. (2016). Robust regularization of the recursive least-squares algorithm. In *8th International Conference on Electronics, Computers and Artificial Intelligence (ECAI)* (pp. 1–4). IEEE.

Pirgaip, B., & Ayaydin, O. A. (2024). Exploring the greenium in the green Sukuk universe: Evidence from the primary market. *International Journal of Islamic and Middle Eastern Finance and Management*.

Rahman, A. F., & Riyadh, H. A. (2016). Islamic finance: Current, future trends and challenges. *Journal of Islamic Banking and Finance*. https://doi.org/10.15640/jibf.v4n2a4

Rahman, M., Isa, C. R., Dewandaru, G., Hanifa, M. H., Chowdhury, N. T., & Sarker, M. (2020). Socially responsible investment sukuk (Islamic bond) development in Malaysia. *Qualitative Research in Financial Markets*. Retrieved from https://www.emerald.com/insight/content/doi/10.1108/QRFM-09-2019-0117/full/html

Sadiq, R., & Mushtaq, A. (2015). The role of Islamic finance in sustainable development. *Journal of Islamic Thought and Civilization, 5*(1).

Sambidge, A. (2023). *Arabian Gulf Business Insight*. Retrieved from https://www.agbi.com/banking-finance/2023/11/strong-growth-in-uae-green-sukuk/

Security Commission Malaysia. (2019). Retrieved from Security Commission Malaysia: https://www.sc.com.my

Sharma, A. (2023). *The National*. Retrieved from The National: https://www.thenationalnews.com/business/banking/2023/07/20/fab-issues-first-ever-uae-dirham-denominated-green-sukuk/

Shibli, A. (2024). *Sustainable sukuk issuance on the rise*. The Banker. Retrieved from https://www.thebanker.com/Sustainable-sukuk-issuance-on-the-rise-1706690029.

Sisdianto, E., Robiansyah, A., Razimi, M. S. B. A., & Afifah, A. L. (2024). The role of green sukuk (Islamic Bonds/Shariah-Compliant Bonds) in enhancing Indonesia economic development strategy in the new normal era. *KnE Social Sciences*, 608-625.

Sun, A. Y., Painter, S. L., & Wittmeyer, G. W. (2006). A constrained robust least squares approach for contaminant release history identification. *Water Resources Research, 42*(4), 1–13.

The World Bank. (2020). *The World Bank*. Retrieved from The World Bank: https://www.worldbank.org/en/country/malaysia/publication/pioneering-the-green-sukuk-three-years-on

Uddin, M. A. (2015). *Principles of Islamic finance: Prohibition of Riba, Gharar and Maysir*. MPRA.

UNDP. (2021). *A New Report Reveals $50 Billion Potential in Islamic Finance for Climate Emergency and SDGs*. Retrieved from https://www.undp.org/

UNDP. (2024). *United Nations Development Programme*. Retrieved from United Nations Development Programme. https://procurement-notices.undp.org/view_negotiation.cfm?nego_id=17985

Utermarck, S., Sharif, M., & Kohli, S. (2024). *Guidance on Green, Social and Sustainability Sukuk*. Retrieved from https://www.icmagroup.org/assets/documents/Sustainable-finance/ICMA-IsDB-LSEG-Guidance-on-Green-Social-and-Sustainability-Sukuk-April-2024.pdf

Wahab, M. Z., & Mohamed Naim, A. (2019). Malaysian initiatives to support sustainable and responsible investment (SRI) especially through sukuk approach. *Journal of Emerging Economies and Islamic Research, 7*(3), 1–11.

WAM. (2022). *Khaleej Times*. Retrieved from Khaleej Times: https://www.khaleejtimes.com/uae/cop27-uaes-green-bonds-sukuk-market-valued-at-17-billion

Weber, H., & Weber, M. (2020a). When means of implementation meet Ecological Modernization Theory: A critical frame for thinking about the sustainable development goals initiative. *World Development*,

105–129. Retrieved from https://www.sciencedirect.com/science/article/abs/pii/S0305750X20302564.

Weber, H., & Weber, M. (2020b). When means of implementation meet ecological modernization theory: A critical frame for thinking about the Sustainable Development Goals initiative. *World Development,* 136. Retrieved from https://www.sciencedirect.com/science/article/abs/pii/S0305750X20302564

Yakubu, I. N. (2020). Institutional quality and foreign direct investment in Ghana: A bounds-testing cointegration approach. *Review of International Business and Strategy, 30*(1), 109–122.

Part II

Financial Inclusion and Economic Empowerment

Microfinance and Women's Economic Empowerment in Nigeria: Challenges and Policy Perspectives

6

Hauwah K. K. AbdulKareem and Khadijah Iddrisu

6.1 Introduction

Women's empowerment remains a highly debated topic globally within the development discourse. It is integral to the Sustainable Development Goals (SDGs), with SDG 5 specifically targeting gender equality and women's empowerment, while its significance is also embedded in the targets and indicators of more than half of the other goals. At the regional level, women's empowerment is also a key component of Agenda 2063, with Aspiration 6 explicitly dedicated to this cause.[1] The Nigerian government has also introduced the National Women's Economic Empowerment (WEE) Policy and Action Plan aimed at advancing women's empowerment.[2] Despite these policies, women in Nigeria are greatly disempowered in various economic spheres—labour force participation, wage, self-employment, earnings and profits, asset ownership, and agriculture. For example, only 55% of women of working age in Nigeria are employed, representing a 14% gap compared to their male counterparts. Furthermore, a mere 12% of women in the country receive wages, which is 22% lower than the proportion of men earning an income (World Bank, 2022). Women own just 16% of the total equipment value in female-operated businesses, often in lower-profit sectors, reducing profits by 46%. Men are five times more likely to own land, with women holding only 8% of individually owned land compared to 70% for men (World Bank, 2022). In agriculture, men use fertilizer and herbicides eight times and 50% more than women, respectively, which diminishes women's productivity by 6% and 18% (World Bank, 2022).

Nigeria also consistently ranks among the lowest-performing countries in terms of gender equality. In 2024, Nigeria is ranked 125th out of 144 countries regarding gender equality (World Economic Forum, 2024). The Social Institutions and Gender Index places Nigeria among the top 10% of countries with the highest levels of gender discrimination, rating "high" or "very high" across all evaluated categories.[3] Nigeria also falls into the group of countries with the highest gender inequality in human development outcomes. These gender disparities result in an economic loss of approximately US$9.3 billion, or 2.3% of

[1] https://au.int/en/articles/au-priorities-gender-equality-and-womens-empowerment-agenda-2063

[2] https://nigeria.iom.int/sites/g/files/tmzbdl1856/files/documents/2024-02/nigeria-national-wee-policy-and-action-plan-may-_230519_133503.pdf

[3] https://www.oecd.org/en/data/dashboards/social-institutions-gender-index.html

H. K. K. AbdulKareem (✉)
Department of Economics and Development Studies, Kwara State University, Malete, Nigeria
e-mail: hauwah.kamaluddeen@kwasu.edu.ng

K. Iddrisu
School of Business, Simon Diedong Dombo University of Business and Integrated Development Studies, Wa, Ghana
e-mail: khadijahiddris@iug.edu.gh

I. N. Yakubu (ed.), *Strategic Approaches to Banking Business and Sustainable Development Goals*, Sustainable Development Goals Series, https://doi.org/10.1007/978-3-031-80744-2_6

Nigeria's gross domestic product (GDP), potentially increasing to around US$23 billion when considering multiplier effects across various sectors (World Bank, 2022). Women make up approximately half (49.5%) of the Nigerian population[4]; therefore, their economic empowerment is crucial for Nigeria if the country must achieve its national development objectives and the SDGs. However, microfinance has been noted to be a key intervention that addresses the twin problems of poverty and disempowerment (Addai, 2017; Lamichhane, 2020).

Microfinance refers to the provision of financial services, including small loans, savings accounts, and insurance, to individuals or small businesses that lack access to traditional banking services (Addai, 2017). Microfinance can empower women by enhancing their access to and use of financial products and services (AbdulKareem, 2019). This, in turn, can increase their income levels, savings rates, and asset ownership and decrease their vulnerabilities. This chapter seeks to discuss how microfinance can empower women. It also seeks to understand the challenges financial institutions encounter in delivering microfinance services and the related policy implications.

The rest of this chapter is structured as follows: Section 6.2 discusses the concepts, barriers, and policies of women's empowerment in Nigeria. Section 6.3 explains microfinance in Nigeria and some of its institutions. Section 6.4 shows how microfinance empowers women, whereas Sect. 6.5 discusses the challenges and policy perspectives. This chapter ends with a concluding remark.

6.2 Women's Empowerment in Nigeria: Concepts, Policies, and Barriers

The term women's empowerment is usually employed to explain different concepts, from intervention programmes to the promotion of gender equality and advocacy for women's rights (Mayoux, 2000). Various organizations, authors, and feminist activities, however, conceptualize and discuss it in various ways. For example, Malhotra et al. (2002) developed four indicators of women's economic empowerment from their synthesis of the measurement framework of different authors. Income and other non-income variants, namely control over resources, contribution to household expenditure, and financial decision-making, are identified as the key indicators. According to World Vision Australia (2024), women's empowerment encompasses the enhancement of women's self-esteem, their autonomy in making decisions, and their capacity to drive social change for both themselves and others. This concept is intrinsically linked to female empowerment, which is not only a fundamental human right but also essential for fostering a more peaceful and prosperous world. The Economic and Social Commission for Western Asia (ESCWA) (2022) viewed women's empowerment as women recognizing and addressing gender-based power imbalances, thereby amplifying their voices to challenge inequalities in the home, workplace, and community. This process includes women taking charge of their lives by setting personal goals, acquiring skills, solving problems, and cultivating self-reliance. The European Institute for Gender Equality (2024) also explained that women's empowerment comprises five key elements: a woman's sense of self-worth, her right to make and shape her own choices, her access to opportunities and resources, her power to govern her own life both within and beyond the home, and her ability to drive social change towards a more equitable social and economic order at both national and international levels.

In Nigeria, women's empowerment involves increasing access to education, improving healthcare services, enhancing economic opportunities through entrepreneurship and employment, and advocating for legal reforms to protect women's rights.[5] Programmes aimed at empowering women often focus on providing skills training, providing access to microfinance, and creating

[4] https://data.worldbank.org/indicator/SP.POP.TOTL.FE.ZS?locations=NG

[5] https://www.globalgiving.org/projects/empower-nigerian-young-women/

platforms for women's participation in decision-making processes at both community and national levels. The empowerment of women in Nigeria is seen as essential for the country's overall development, as it contributes to economic growth, reduces poverty, and promotes a more just and equitable society. However, achieving true empowerment requires sustained efforts from the government, civil society, and international organizations to address the deep-rooted barriers that women face in various aspects of life.

6.2.1 Women's Empowerment Policies in Nigeria

There are various policies, including the global (see Box 6.1), regional (see Box 6.2), and several national policies in Nigeria. Some of these national policies are the Better Life Programme for Rural Women, the National Gender Policy, the National Women's Economic Empowerment (WEE) Policy and Action Plan, the Violence Against Persons Prohibition (VAPP) Act 2015, and the National Action Plan (NAP) on Women, Peace, and Security (see Table 6.1). Early policy is the Better Life Programme aimed to boost rural women's productivity and living standards by promoting education and business skills.[6] Its goals included raising awareness of women's rights, fostering collective action, and improving life standards through projects like women's cooperatives, small-scale industries, and multi-purpose centres for training. The programme also emphasized healthcare and local crafts. It had a hierarchical structure with local committees, state chapters led by the governor's spouse, and national coordination by the president's spouse. Despite some criticisms, this structure ensured effective access to funding. Financing came from self-help organizations, loans, state and local grants, and international donors. The next policy is the National Gender Policy of Nigeria, introduced in 2006 and revised in 2021, which addresses gender inequality and promotes women's empowerment.[7] Its key objectives include eliminating discriminatory practices, ensuring equal opportunities in education, employment, politics, and health, and addressing the feminization of poverty. The policy emphasizes improving women's health, particularly in reproductive and maternal care, and advocates for increased female participation in politics. It calls for legal reforms, gender-responsive budgeting, and the use of media to challenge stereotypes, aiming for broad societal change and institutional strengthening to achieve gender equality.

Box 6.1 Sustainable Development Goal 5—Gender Equality

Gender equality is a critical objective aiming to achieve full gender parity and empower all women and girls by 2030. It addresses the need to end all forms of discrimination, violence, and harmful practices, such as child marriage and female genital mutilation, that disproportionately affect women and girls. Additionally, it seeks to ensure that women and girls have equal opportunities for leadership roles and decision-making in all sectors of society and access to education, healthcare, and economic resources. However, progress towards gender equality is significantly off track. While harmful practices like child marriage and female genital mutilation are decreasing, the rate of change is insufficient to keep pace with population growth. Achieving gender parity in managerial positions remains a distant goal, with current projections estimating it could take 176 years. Many women still lack control over their sexual and reproductive health, and violence against women continues to be pervasive.

The urgency of addressing these issues is underscored by recent data:

(continued)

[6]https://books.openedition.org/ifra/2867?lang=en

[7]https://www.wrapanigeria.org/revised-2021-2026-national-gender-policy/

Box 6.1: (continued)

- *Target 5.1*: In 2022, 55% of 120 surveyed countries lacked nondiscrimination laws, and half had no quotas for women in national parliaments.
- *Target 5.3*: Around 640 million women were married before 18, and over 230 million have undergone female genital mutilation as of 2024.
- *Target 5.4*: Women spend about 2.5 times more hours in unpaid domestic and care work than men.
- *Target 5.5*: As of January 2024, women held 26.9% of seats in national parliaments and 27.5% of managerial positions, reflecting slow progress.
- *Target 5.a*: Legal protections for women's land rights are minimal in 58% of reporting countries.
- *Target 5.b*: Women were 8% less likely to own a mobile phone than men in 2023.
- *Target 5.c*: Only 26% of countries globally have comprehensive systems for tracking gender equality.

Source: UN (2023)

Box 6.2 Africa Agenda 2063 Aspiration 6—Gender Equality and Women's Empowerment

Aspiration 6 of Agenda 2063 envisions an Africa driven by its people, particularly women and youth, with a focus on inclusivity and active citizen participation in decision-making. The African Union (AU) emphasizes the importance of gender equality as a fundamental human right and a crucial element for regional integration, economic growth, and social development. To support this, the AU developed the Gender Equality and Women's Empowerment (GEWE) strategy, which focuses on six key pillars: women's economic empowerment, social justice and rights, leadership and governance, gender management systems, women's roles in peace and security, and access to media and information and communication technologies (ICTs). The AU also enforces the African Charter on Human and Peoples' Rights on the Rights of Women in Africa, requiring states to combat discrimination against women. The Women, Gender and Youth Directorate (WGDD) leads and coordinates the AU's efforts to achieve gender equality and uphold the Solemn Declaration on Gender Equality in Africa (SDGEA).

Source: AU (2021)

Another important policy is the National Women's Economic Empowerment (WEE) Policy and Action Plan, which is a Nigerian government initiative to enhance women's economic participation and independence.[8] It seeks to overcome barriers to women's access to economic resources like credit and land and improve their skills through education and training. The policy supports the growth of women-owned businesses by providing access to markets and technology and advocates for gender-responsive policies. It also addresses gender-based violence and discrimination, recognizing these as significant obstacles to women's economic empowerment. The policy aims to create a more inclusive and equitable economic environment for women.

The next women's empowerment policy is the Violence Against Persons Prohibition (VAPP) Act, enacted in Nigeria in 2015, which is a comprehensive law aimed at eliminating all forms of violence, particularly against women and children. It broadens the definition of violence to include physical, psychological, sexual, emotional, and economic abuse, as well as harmful traditional practices like female genital

[8] https://www.policyvault.africa/policy/national-womens-economic-empowerment-wee-policy-and-action-plan/#:~:text=The%20main%20goal%20of%20the,girls%20in%20Nigeria"s%20dynamic%20economy

Table 6.1 National women's empowerment policies in Nigeria

Period	Policy	Objective
1987	Better Life Programme for Rural Women	To enhance rural women's productivity and living conditions by advancing their education and business acumen
2006 (2021)	National Gender Policy	To eliminate gender-based discrimination and promote gender equality in Nigeria
2015	Violence Against Persons Prohibition Act	To provide comprehensive legal protection against all forms of violence, including domestic violence, sexual harassment, and harmful traditional practices
2017–2020	National Action Plan on Women, Peace, and Security	To align with the United Nations Security Council Resolution 1325 and focus on the inclusion of women in peacebuilding and conflict resolution
2022	National Women's Economic Empowerment (WEE) Policy and Action Plan	To specifically target the economic empowerment of women by improving their access to financial resources, entrepreneurship opportunities, and skills development

Source: Authors (2024)

mutilation and forced marriage. The act criminalizes domestic violence and spousal rape and provides legal protection and support for victims, including shelters and compensation. It also mandates the establishment of specialized courts for enforcement and encourages the reporting of violence to ensure accountability. Last but not least is the National Action Plan on Women, Peace, and Security (NAP) in Nigeria, developed to implement United Nations Security Council Resolution 1325, which ensures women's active involvement in peace and security efforts. It promotes women's participation in peacebuilding and conflict resolution, protects women and girls from violence, and advances women's rights in conflict settings. The NAP also integrates gender perspectives in peace initiatives and includes strategies for the rehabilitation and reintegration of women and girls affected by conflicts, providing them with access to education, healthcare, and livelihoods. The plan aims to build a more inclusive and sustainable peace in Nigeria.

6.2.2 Barriers to Women's Empowerment in Nigeria

Though policies exist to promote women's empowerment in Nigeria, there are barriers that hinder some Nigerian women from receiving empowerment. Some of these barriers include the following.

6.2.2.1 Cultural and Social Norms

In many parts of Nigeria, traditional beliefs and practices enforce gender norms that position women in subordinate roles both at home and in society. These cultural expectations often confine women to domestic duties, while men are seen as the primary decision-makers and breadwinners. Consequently, women face significant barriers to accessing education, employment, and financial resources. These norms also limit women's participation in decision-making processes, further entrenching their economic and social dependence on men. Gender-based violence, including domestic abuse and harmful practices like female genital mutilation, exacerbates these challenges by undermining women's physical and mental health and diminishing their confidence. Discrimination, including legal inequalities and workplace bias, reinforces women's marginalization, making it even more difficult for them to overcome these restrictive norms. This combination of cultural practices, violence, and discrimination creates a cycle of disadvantage that perpetuates women's limited access to resources and opportunities, maintaining their subordinate status in society.

6.2.2.2 Educational Disparities

In Nigeria, gender disparities in education and training significantly impact women's job opportunities and earning potential. Despite progress in recent years, many girls in rural and underserved areas still face obstacles to accessing quality education, including cultural biases that prioritize boys' education, early marriage, and financial constraints that force families to withdraw girls from school. This educational gap results in fewer women acquiring the necessary skills and qualifications to compete in the labour market, limiting their access to formal employment and high-paying jobs. Additionally, women often lack access to vocational training and higher education opportunities that could prepare them for careers in growing sectors such as technology, finance, and engineering. As a result, many women are confined to low-paying, informal, or part-time jobs with limited career advancement prospects. This disparity in education and training perpetuates economic inequality, reinforcing the cycle of poverty and restricting women's ability to contribute fully to Nigeria's economic development. Addressing these educational gaps is essential for achieving gender equality and empowering women to reach their full economic potential.

6.2.2.3 Political and Legal

In Nigeria, the underrepresentation of women in political and decision-making positions significantly hinders their ability to influence policies that directly impact their lives. Despite women making up nearly half of the population, they hold less than 10% of elected positions in national and state assemblies. This imbalance means that policies and laws often fail to reflect the specific needs and concerns of women, such as gender-based violence, maternal health, and economic empowerment. A practical example is the Gender and Equal Opportunities Bill, which has faced multiple rejections in the Nigerian National Assembly. This bill, aimed at promoting gender equality and protecting women's rights, has struggled to gain traction due to the male-dominated legislature, where many lawmakers hold traditional views that resist changes to the status quo. Additionally, the lack of legal protections, such as inadequate enforcement of laws against domestic violence and discrimination in the workplace, further marginalizes women. For instance, the Violence Against Persons Prohibition (VAPP) Act, though enacted, has not been effectively implemented across all states, leaving many women without recourse when facing violence. This lack of representation and legal protection perpetuates systemic gender inequalities, limiting women's ability to advocate for and secure their rights in Nigeria.

6.2.2.4 Low Financial Services

In Nigeria, women often encounter significant obstacles in accessing essential economic resources such as credit, land, and capital, which are crucial for starting or expanding businesses. Cultural norms and legal restrictions frequently hinder women's ownership of land and property, making it difficult for them to secure loans or collateral needed for business ventures. Financial institutions may also view women as higher-risk borrowers due to these constraints, leading to stricter lending criteria or outright denial of credit. Moreover, women's limited access to formal education and financial literacy further exacerbates these challenges, reducing their ability to navigate complex financial systems or understand the terms of credit agreements. This exclusion from financial resources not only stifles entrepreneurial potential but also limits women's ability to achieve financial independence and contribute to economic growth. As a result, many women are confined to small-scale, low-profit businesses, often in the informal sector, where they face additional vulnerabilities such as inadequate legal protections and lower returns on investment. These barriers perpetuate a cycle of economic dependency and inequality, making it difficult for women to break free from poverty and fully participate in the economic development of Nigeria. Addressing these challenges is crucial for unlocking the economic potential of women and achieving inclusive growth in the country.

6.3 Microfinance in Nigeria: Institutions and Functions

Microfinance provides financial services to poor and low-income households lacking access to formal institutions, often described as "*banking for the poor*" (Olonade et al., 2021). These programmes offer loans, savings, and other financial services to support small businesses in underserved communities. Rooted in traditional community financing, microfinance has spread globally, including in Africa, Latin America, and Asia. The movement gained momentum in the early 1980s in countries like Bangladesh and Bolivia, attracting the interest of multilateral donors and private sector bankers over the past 20 years (Olonade et al., 2021). Microfinance bridges finance and development assistance, aiming to empower the economically disadvantaged.

Microfinance in Nigeria has deep cultural roots, with traditional practices of providing financial services dating back centuries. These services primarily target low-income earners in rural and urban areas, facilitated through informal microfinance institutions (MFIs) such as self-help groups (SHGs) and rotating savings and credit associations (ROSCAs). Additionally, savings collectors and cooperative societies play a role in delivering microfinance services. However, these informal institutions often face limited outreach due to a scarcity of loanable funds. To address these limitations, the Nigerian government has historically implemented several publicly financed microfinance and rural credit programmes aimed at the poor. Notable initiatives include the rural banking programme, sectoral credit allocations, a concessionary interest rate, and the Agricultural Credit Guarantee Scheme (ACGS). Other institutional arrangements, such as the Nigerian Agricultural and Cooperative Bank Limited (NACB), the National Directorate of Employment (NDE), the Nigerian Agricultural Insurance Corporation (NAIC), the Peoples Bank of Nigeria (PBN), community banks (CBs), and the Family Economic Advancement Programme (FEAP), were established to enhance financial services in rural areas. In 2000, the government consolidated NACB, PBN, and FEAP into the Nigerian Agricultural Cooperative and Rural Development Bank Limited (NACRDB) to improve finance provision to the agricultural sector. Additionally, the National Poverty Eradication Programme (NAPEP) was created with a mandate to alleviate poverty through financial services.

Despite these efforts, government-sponsored microfinance services often relied on a traditional supply-led, subsidized credit approach, focusing on the agricultural sector and non-farm activities like trading, tailoring, weaving, blacksmithing, agro-processing, and transportation. Although these programmes led to increased credit disbursement and gains in agricultural production and other activities, the benefits were often short-lived due to the unsustainable nature of the programmes. Since the 1980s, non-governmental organizations (NGOs) have emerged as key players in Nigeria's microfinance sector, particularly in supporting micro- and rural entrepreneurs. Unlike the earlier supply-led approach, these NGOs have adopted a demand-driven strategy, addressing the needs of low-income groups and the poor who are underserved by the formal financial sector. The rise of NGOs in microfinance has also been driven by declining support from development partners. These NGOs typically operate as charity organizations, capital lending institutions, or credit-only membership-based institutions, registered under the Trusteeship Act. They fund their activities through grants, fees, interest on loans, and member contributions. Despite their growing presence, NGOs face challenges in expanding their outreach, primarily due to the unsustainable nature of their funding sources.

In December 2005, the Central Bank of Nigeria (CBN) introduced a Microfinance Policy Framework to improve financial access for micro-entrepreneurs and low-income households, aiming to drive economic growth by expanding their financial services. This policy emerged alongside a significant increase in the minimum capital requirement for deposit money banks (DMBs), from NGN 2.5 billion to NGN 25 billion, highlighting the need for robust microfinance regulation. The revised National Microfinance Policy Framework of April 2011 builds on the 2005 policy, targeting economically active poor and low-income households by providing affordable credit and innovative savings products. With over 800 microfinance banks operating across Nigeria, the policy addresses the large unserved market segment and aims to enhance financial inclusion.

In summary, Nigeria's microfinance landscape has evolved from traditional, community-based practices to more formalized, government-sponsored initiatives and NGO-driven approaches. While these efforts have made significant strides in improving access to financial services for low-income earners, the sector still faces challenges related to sustainability and outreach. Addressing these challenges is crucial for ensuring that microfinance continues to play a vital role in alleviating poverty and promoting economic development in Nigeria.

6.4 Enhancing Women's Empowerment Through Microfinance in Nigeria

Enhancing women's empowerment through microfinance in Nigeria involves using financial services as a tool to increase women's economic independence, social status, and decision-making power within their communities. Some of the ways through which microfinance empowers women include providing access to credit facility, partnering with development partners to support women, providing financial literacy and education, promoting the adoption of financial technology, and enhancing financial inclusion.

6.4.1 Access to Credit Facility

Box 6.3 ASA Nigeria Provides Loans to Women

ASA Nigeria is a for-profit microfinance bank that accepts deposits and provides small, socially responsible loans to low-income entrepreneurs, with a focus on women. The institution has developed specialized financial facilities to support female entrepreneurs, including the small business loan. This loan is designed to empower women, facilitate the expansion of small businesses, and help underprivileged individuals move away from reliance on moneylenders. By promoting self-employment and self-reliance through income-generating activities, the small business loan aims to enhance women's economic contributions. Key features of the small business loan include loan types with amounts ranging from NGN 50,000 to 200,000 for the first loan and NGN 20,000 to 50,000 for follow-on loans, no processing fees, a repayment term of 24 weeks (6 months), a one-week processing time for the first loan, a one-week grace period, an age limit of 21 to 55 years, a maximum loan limit of NGN 500,000, and no collateral requirement. This loan product supports the development of underprivileged communities by fostering self-reliance and encouraging business growth among women.

Source: ASA Nigeria (2024)

MFIs empower women through the provision of credit facilities. This enables women to start or expand small businesses, increasing their income and financial stability. With access to capital, women can invest in various entrepreneurial activities such as trading, agriculture, tailoring, or small-scale manufacturing. This economic empowerment allows women to contribute to household income, reduce poverty, and improve the living standards of their families.

Box 6.4 Letshego Microfinance Bank named as AFAWA Bank of the Year

Letshego MFB Nigeria initially secured a \$3.5 million guarantee line, which was leveraged to extend an additional \$1.5 million in loans. Over the course of a year, the bank disbursed approximately \$5 million in total, with \$2 million specifically allocated to women-led small- and medium-sized enterprises (SMEs). During this period, Letshego MFB Nigeria allocated 63% of its 5115 loans to women-led micro- and small enterprises, benefiting 3241 women entrepreneurs. The proportion of loans directed towards women increased from 25% in 2020 to 51% in 2022. This achievement earned Letshego MFB an award from the African Development Bank's Affirmative Finance Action for Women in Africa (AFAWA) programme.

Source: Marsh (2024)

For example, ASA Nigeria provides loans to women on favourable terms such as no collateral and no processing fees (see Box 6.3). Additionally, Letshego MFB increased its loan allocation to women, which contributed to its recognition as the Affirmative Finance Action for Women in Africa (AFAWA) Bank of the Year (see Box 6.4).

6.4.2 Partnering with Development Partners to Support Women

Box 6.5 Diamond Bank Partnered with Women's World Banking to Enhance Financial Inclusion for Women

Diamond Bank has partnered with Women's World Banking to enhance financial inclusion for women by focusing on expanding access to financial services and supporting women entrepreneurs. The collaboration involved developing and tailoring financial products, such as savings accounts, loans, and insurance, specifically designed to meet women's needs. It will also offer training and support to improve financial literacy and business skills among women entrepreneurs, thereby enhancing their access to capital. Furthermore, the partnership aims to increase credit accessibility for women-owned businesses, particularly in underserved communities, and advocate for policies that promote women's economic empowerment. By exploring technological innovations, the collaboration seeks to improve outreach and offer advanced financial solutions. This initiative supports broader objectives of fostering economic growth and gender equality within the financial sector.

Source: Oji (2017)

MFIs are increasingly partnering with development partners to bolster support for women entrepreneurs. These collaborations aim to address the specific challenges faced by women-led businesses, particularly in accessing finance. Development partners, such as international organizations and financial development institutions, often provide guarantees, funding, and technical assistance to mitigate the risks associated with lending to women entrepreneurs. For example, some MFIs such as Diamond Bank partnered with Women's World Banking (WWB) to enhance financial inclusion for women (see Box 6.5).

6.4.3 Financial Literacy and Education

Box 6.6 Accion Microfinance Bank Provided Financial Literacy and Education to Women

Accion Microfinance Bank recently hosted a one-day digital skills boot camp for female entrepreneurs, emphasizing the importance of social media and digital tools for business growth. The event featured keynote speeches by the chief digital officer, who discussed strategies for digital marketing, e-commerce, and operational efficiency. The corporate branding and digital communications manager provided guidance on social media marketing and content creation. Participants received certificates upon completion, highlighting Accion MFB's commitment to enhancing financial inclusion and empowering women through innovative solutions and digital literacy.

Source: Aspyee Admin (2024)

MFIs offer financial literacy programmes to help women develop essential skills for effective money management. These programmes cover budgeting, saving, credit management, and investment strategies, equipping women with the knowledge to make informed financial decisions. Participants learn to create budgets, track expenses, manage debt, build credit, and plan for future needs like retirement or emergencies. The programmes also provide practical training on using financial products such as savings accounts, loans, and insurance. This education boosts women's confidence in handling finances, leading to proactive decision-making in personal and business contexts.

By empowering women with financial knowledge, MFIs enhance their economic stability and independence, enabling them to contribute more effectively to their households and communities and promoting broader social and economic development. For example, Accion Microfinance Bank educated women on innovative business and financial strategies (see Box 6.6).

6.4.4 Promote the Adoption of Financial Technology Among Women

Some MFIs in Nigeria empower women by promoting fintech adoption through various strategies. They provide digital financial literacy programmes to teach women about mobile banking, digital payments, and online savings. MFIs also offer user-friendly digital products, like mobile wallets and online loans, and partner with fintech companies to integrate advanced technologies. Incentives such as reduced fees encourage fintech use, while community outreach and dedicated support help women in rural areas transition to digital services. Awareness campaigns further promote fintech benefits, enhancing women's financial literacy, access to services, and economic opportunities, leading to greater financial independence and empowerment. For instance, some MFIs such as Accion Microfinance Bank and LAPO Microfinance Bank aid women in the adoption of financial technology (fintech) as shown in Boxes 6.3–6.7.

Box 6.7 LAPO Microfinance Bank Promotes Financial Inclusion Among Women in Nigeria

LAPO Microfinance Bank addresses the financial challenges faced by poor women in Nigeria, who often rely on high-interest moneylenders due to limited access to affordable credit. As a Scale2Save partner, LAPO aims to empower women through accessible financial services and savings promotion. Their "Supporting Female Entrepreneurs (SUFEN)" loan and the Scale2Save-supported "My Pikin & I" savings account are key initiatives. The savings account allows women to start saving with just 200 Naira and offers incentives such as free insurance and education scholarships based on savings milestones. With over 4.5 million customers, 73% of whom are women, and a workforce that is 57% female, LAPO's approach includes deploying numerous agents and branches to ensure accessibility. This innovative product design, low entry requirements, and targeted benefits effectively support women's financial inclusion and empowerment, reflecting a positive shift in savings habits and economic stability.

Source: WSBI and ESBG (2024)

6.4.5 Financial Inclusion

MFIs play a crucial role in facilitating women's access to formal banking services by offering tailored financial products and services designed to meet their unique needs. By bridging the gap between informal financial practices and formal banking systems, MFIs enable women to open and manage savings accounts, which helps in fostering a savings culture among them. These institutions often encourage women to develop disciplined saving habits through various incentives and educational programmes. In addition to savings, MFIs provide specialized financial products aimed at encouraging investments. These products may include microloans for small businesses, investment savings plans, and tailored credit facilities that support entrepreneurial ventures. By offering these financial products, MFIs empower women to invest in their businesses, expand their economic activities, and achieve financial stability. The focus on both saving and investing ensures that women not only build a secure financial foundation but also have the means to grow their assets and contribute to their economic development. Consequently, LAPO MFI promotes financial inclusion among women through education, encouraging savings and helping the businesses of women (see Box 6.7).

6.5 Challenges and Policy Perspectives in the Microfinance–Women's Empowerment Nexus

MFIs not only empower women through their business model but also encounter some challenges. Therefore, this section discusses the challenges and provides some policy perspectives to minimize such challenges.

6.5.1 Challenges

While MFIs encounter various challenges, this chapter focuses on key issues, including the high cost of delivering financial services to women, difficulties in effective outreach and support, regulatory constraints, and information asymmetry.

6.5.1.1 High Cost of Providing Financial Services to Women

The high cost of providing financial services to women impacts the sustainability of MFIs. Key factors include substantial operational costs for infrastructure in remote areas, such as setting up branches or mobile units, and high per-transaction costs for processing small loans. Limited economies of scale occur when serving small, dispersed client bases, increasing per-client costs. Regulatory compliance adds further expense, particularly for adhering to anti-money laundering and reporting standards. Additionally, higher credit risks and default rates in low-income areas exacerbate financial strain. These challenges require MFIs to adopt efficient solutions to maintain sustainability. For example, a similar study conducted in Ghana revealed that MFIs face the challenges of the high cost of extending financial services to marginalized populations such as women (see Ebenezer et al., 2020).

6.5.1.2 Effective Outreach and Support Challenges

Ensuring effective outreach and support for women in rural regions poses significant challenges, particularly due to limited access to technology and financial services. For instance, in remote areas of Nigeria, women may have to travel long distances to access banking services, as there are few local branches or ATMs. In such cases, MFIs often deploy mobile banking units or community agents to reach these women, but these solutions can be costly and logistically complex. Additionally, limited Internet connectivity in rural areas can hinder the adoption of digital financial tools, requiring MFIs to invest in alternative methods such as radio broadcasts or community meetings to provide financial education and support.

6.5.1.3 Regulatory Constraints

Regulatory constraints significantly challenge MFIs in empowering women. Stringent requirements and bureaucratic procedures, such as high capital demands and extensive documentation, can limit MFIs' flexibility and effectiveness, particularly in underserved areas. These constraints can prevent MFIs from offering affordable financial products tailored to women's needs. Additionally, compliance costs and complex regulations can divert resources from programme development and client support, impeding initiatives like financial literacy programmes. Restrictive regulations may also stifle innovation, delaying the introduction of technologies like mobile banking that could enhance financial inclusion. Overall, these challenges impact MFIs' ability to provide essential financial tools and support for women. For instance, the minimum capital requirement for national microfinance banks is set at NGN 5 billion, which can be prohibitively high for smaller MFIs operating in underserved areas.[9] Also, a newly licensed National Microfinance Bank (MFB) is limited to operating a maximum of ten branches at the outset, regardless of its potential capacity to open additional branches. This restriction can limit the growth and outreach of MFIs, especially in underserved areas.

6.5.1.4 Information Asymmetry

Information asymmetry poses a challenge for MFIs in empowering women by creating gaps in knowledge and access. For instance, in Nigeria, women in rural areas may lack detailed information about financial products, leading to underutilization of services like microloans or savings accounts. MFIs might also struggle to accurately assess the creditworthiness of women borrowers due to incomplete or unreliable data, increasing the risk of default. Additionally, limited financial literacy among women can hinder their ability to make informed decisions about managing loans and savings, impacting their economic empowerment.

6.5.2 Policy Perspectives

6.5.2.1 Regulatory Reform

Simplifying regulatory requirements and reducing compliance costs can significantly improve the flexibility and effectiveness of MFIs in serving women. By streamlining regulations, MFIs can more easily navigate legal and bureaucratic hurdles, allowing them to focus on developing and offering tailored financial products that meet the specific needs of women. Simplified procedures can also reduce operational expenses, making it feasible for MFIs to innovate and introduce new services. This increased flexibility enables MFIs to adapt their offerings more readily, fostering greater financial inclusion and support for women, particularly in underserved and rural areas. For example, the CBN should revise the limit on new MFBs operating a maximum of ten branches. Increasing this limit and promoting agency banking could enhance outreach and financial access, particularly in underserved areas, by leveraging local businesses and community centres to support women.

6.5.2.2 Enhanced Financial Literacy Programmes

Expanding financial education initiatives to reach more women, particularly in rural areas, is crucial for addressing information asymmetry. Training programmes should be designed to provide practical financial management skills, such as budgeting, saving, and investment strategies, tailored to the specific needs of women in these communities. Additionally, these programmes should emphasize the benefits of digital financial tools, including mobile banking, online payments, and digital savings platforms. By offering hands-on training and real-world applications, these initiatives can empower women with the knowledge and skills necessary to make informed financial decisions, enhance their economic opportunities, and improve overall financial inclusion.

6.5.2.3 Incentives for Innovation

Providing incentives for MFIs and fintech companies to innovate can significantly improve financial access for women. By offering financial grants, tax breaks, or subsidies, policymakers can encourage these institutions to develop and deploy cutting-edge solutions tailored to women's needs. Support for technology adoption is essential; this includes facilitating the creation of user-friendly digital products such as mobile banking apps, online loan platforms, and digital savings tools. These innovations can make financial services more accessible and practical for women, especially those in underserved or rural areas. Enhanced product design and technologi-

[9] https://adeolaoyinlade.com/en/how-to-establish-a-microfinance-bank-in-nigeria/

cal advancement help bridge the gap in financial inclusion, ultimately empowering women economically.

6.5.2.4 Targeted Support for Rural Areas

Implementing policies that address the high operational costs of serving rural areas is crucial for ensuring that women in these regions have access to financial services. This can involve providing support for infrastructure development, such as building and maintaining branch offices or mobile banking units, which are essential for reaching remote communities. Additionally, offering subsidies or financial assistance to offset operational expenses can help reduce the burden on microfinance institutions (MFIs) and make it financially viable for them to operate in underserved areas. These measures can help bridge the access gap, enabling women in rural areas to benefit from essential financial services and opportunities.

6.6 Conclusion

The global focus on women's empowerment has led to concerted efforts by both academics and policymakers to address gender disparities. Despite these efforts, Nigeria continues to grapple with significant gender discrimination, leaving many women disempowered. However, the rise of MFIs in Nigeria presents a potential avenue for fostering women's empowerment. This chapter explored how MFIs can contribute to this cause by first defining women's empowerment in the Nigerian context and highlighting relevant policies and barriers. This chapter then provided an overview of microfinance in Nigeria and examined how MFIs can drive women's empowerment.

In Nigeria, key policies aimed at enhancing women's empowerment include the Better Life Programme for Rural Women, the National Gender Policy, the National Women's Economic Empowerment Policy and Action Plan, Violence Against Persons Prohibition Act 2015, and National Action Plan on Women, Peace, and Security. However, despite these policies, barriers such as cultural and social norms, educational disparities, and limited access to financial services persist. This chapter identified that microfinance can mitigate some of these barriers by providing access to finance, partnering with development organizations to support women, offering financial literacy and education, and promoting financial inclusion through technology adoption. This chapter also addressed the challenges MFIs face in this empowerment process, including the high cost of delivering financial services to women, difficulties in effective outreach and support, regulatory constraints, and information asymmetry. It concluded by offering four key policy recommendations: regulatory reform to ease the operational burden on MFIs, enhanced financial literacy programmes to empower women, incentives for innovation in financial services, and targeted support to ensure rural women have better access to financial resources.

The policy implication of this analysis is the need for a coordinated approach that not only strengthens existing policies but also encourages the development of an enabling environment for MFIs to operate more effectively. This could involve revising regulatory frameworks to reduce operational constraints, expanding financial education initiatives, and providing targeted support to overcome the unique challenges faced by women in rural and underserved areas.

References

AbdulKareem, H. K. K. (2019). Financial Inclusion: Effectiveness of Microfinance Banks in Promoting Access and Usage of Financial Services. *International Journal of Research in Management, Economics and Commerce, 9*(6), 30–40.

Addai, B. (2017). Women Empowerment Through Microfinance: Empirical Evidence from Ghana. *Journal of Finance and Accounting, 5*(1), 1. https://doi.org/10.11648/j.jfa.20170501.11

ASA Nigeria. (2024). Small business loan. https://nigeria.asa-international.com/loans/small-business-loan/

Aspyee Admin. (2024, July 17). Accion MfB empowers female entrepreneurs with digital skills. https://aspyee.org/news/accion-mfb-empowers-female-entrepreneurs-digital-skills

AU. (2021, January 29). Gender Equality & Development. https://au.int/en/gender-equality-development#:~:text=Aspiration%206%20of%20Agenda%202063,are%20actively%20involved%20in%20decision

Ebenezer, O.-Y., Gideon, O.-Y., & Uwineza, D. (2020). Microfinance and Women's Empowerment in Ghana: Challenges and Prospects. *International Journal of Academic Research in Business and Social Sciences, 10*(10). https://doi.org/10.6007/IJARBSS/v10-i10/7945

ESCWA. (2022). Women's empowerment. https://www.unescwa.org/sd-glossary/women%E2%80%99s-empowerment

European Institute for Gender Equality. (2024). Empowerment of women. https://eige.europa.eu/publications-resources/thesaurus/terms/1246?language_content_entity=en

Lamichhane, D. B. (2020). Microfinance for women empowerment: a review of best practices. *Interdisciplinary Journal of Management and Social Sciences, 1*, 1–25.

Malhotra, A., Schuler, S. R., & Boender, C. (2002). Measuring women's empowerment as a variable in international development.

Mayoux, L. (2000). Micro-finance and the empowerment of women: A review of the key issues.

Oji, H. (2017). Diamond bank partners with women world Banking for financial inclusion. https://guardian.ng/business-services/diamond-bank-partners-wwb-mtn-others-to-deepen-womens-financial-inclusion/

Olonade, O. Y., Oyibode, B. O., Idowu, B. O., George, T. O., Iwelumor, O. S., Ozoya, M. I., Egharevba, M. E., & Adetunde, C. O. (2021). Understanding gender issues in Nigeria: The imperative for sustainable development. *Heliyon, 7*(7), e07622. https://doi.org/10.1016/j.heliyon.2021.e07622

UN. (2023). Achieve gender equality and empower all women and girls. Retrieved August 26, 2024, from https://sdgs.un.org/goals/goal5

Van Marsh, A.. (2024, May). Letshego Microfinance Bank Nigeria named Affirmative Finance Action for Women in Africa initiative's first "AFAWA Bank of the Year" at African Business Awards. https://am.afdb.org/en/press-releases/letshego-microfinance-bank-nigeria-named-affirmative-finance-action-women-africa#:~:text=Letshego%20MFB%20Nigeria%20said%20that,2020%20to%2051%25%20in%202022

World Bank. (2022). Closing gaps, increasing opportunities: A diagnostic on women's economic empowerment in Nigeria.

World Economic Forum. (2024). Global Gender Gap 2024: Insight Report. https://www3.weforum.org/docs/WEF_GGGR_2024.pdf

World Vision Australia. (2024). What is women's empowerment?

WSBI, & ESBG. (2024). Microfinance bank empowers Nigerian women through financial inclusion. https://www.wsbi-esbg.org/microfinance-bank-empowers-nigerian-women-through-financial-inclusion/

7 The Impact of Financial Inclusion on Economic Growth: Does Institutional Quality Matter?

Ibrahim Nandom Yakubu, Abdul Razak Yussif, and Mohammed Abdul-Rahaman

7.1 Introduction

Financial inclusion has emerged as a critical component in the discourse on development policies, especially within the context of emerging markets, where its significance is underscored by its alignment with seven of the 17 Sustainable Development Goals (SDGs) (Klapper et al., 2016). The essence of financial inclusion lies in facilitating access to and the utilization of formal financial services for individuals and entities traditionally underserved or excluded from the formal banking sector. This inclusive approach targets various demographics, including low-income earners, smallholder farmers, small businesses, and underbanked firms (Ofosu-Mensah Ababio et al., 2021).

In the past decade, substantial efforts have been directed toward enhancing accessibility to financial services, marked by notable progress. However, a significant segment of the global population remains without access to formal financial services, highlighting persistent challenges and the need for further advancements (Klapper et al., 2016; Demirguc-Kunt et al., 2018). The evolving landscape of financial inclusion reflects ongoing endeavors to bridge the gap and bring financial services to the forefront of development initiatives. While strides have been made, the recognition of the existing disparities emphasizes the continuous and dynamic nature of the financial inclusion agenda on the global stage.

Most African nations have experienced notable advancements in financial inclusion in recent years, with mobile money services playing a pivotal role in delivering financial services (Balele, 2019). Financial technology has played a significant role in reducing geographic access restrictions to financial services, particularly benefiting rural communities. Additionally, the development of financial technology has streamlined non-paperwork processes, such as opening a bank account, associated with the traditional brick-and-mortar banking system. Despite the adoption of financial technology in developing nations, including those in Africa and other regions, substantial barriers to financial inclusion persist. These obstacles include high lending rates, low savings rates, and financial illiteracy. Nevertheless, it is maintained that financial inclusion enhances economic prosperity in underdeveloped nations (Shetty, 2008).

Given its purported connection to economic growth, financial inclusion has been a focal point

I. N. Yakubu
School of Business, University for Development Studies, Tamale, Ghana

A. R. Yussif (✉)
Tamale Technical University, Tamale, Ghana
e-mail: yarazak@tatu.edu.gh

M. Abdul-Rahaman
Department of Accounting, S.D Dombo University of Business and Integrated Development Studies, Wa, Ghana

I. N. Yakubu (ed.), *Strategic Approaches to Banking Business and Sustainable Development Goals*, Sustainable Development Goals Series, https://doi.org/10.1007/978-3-031-80744-2_7

for researchers and policymakers since the release of the Maya Declaration on financial inclusion for the unbanked in 2011. However, conflicting empirical data in the literature persists regarding the relationship between financial inclusion and economic growth. Numerous empirical studies, including those by Inoue and Hamori (2016), Sharma (2016), Kim et al. (2018), Sethi and Acharya (2018), Balele (2019), Yakubu and Bunyaminu (2021), Ozturk and Ullah (2022), Biswas (2023) among others, have examined the impact of financial inclusion on economic growth. A fundamental drawback in most earlier studies lies in how they measure financial inclusion, often relying on single measures like bank accounts per thousand adults, number of ATMs, borrowers, commercial bank depositors, and banking outlets. However, employing single measures of financial inclusion may lead to biased and misleading findings (Sarma, 2008). Hence, this study takes a unique approach by using a financial inclusion index that captures both the demand and access to financial services. Also, by introducing institutional quality as a moderating factor that is virtually ignored in prior studies, the study adds novelty to the literature on the link between financial inclusion and economic growth. It is worth noting that Ifediora et al. (2022) have looked at the nexus between financial inclusion and economic growth in Sub-Saharan Africa (SSA). However, the authors dwelled on a sample of 22 countries with data spanning from 2012 to 2018. This study however differs from the work of Ifediora et al. (2022) by employing more countries and a larger sample period. Also, examining how financial inclusion and institutional quality interactively influence growth further distinguishes this study from Ifediora et al. (2022) research.

Consequently, our study contributes significantly to both literature and policy. For literature, the study adds to the very scanty empirical research on the link between financial inclusion, institutional quality, and economic growth, particularly in the context of developing economies. Second, to the best of the researchers' knowledge, this study is the first to investigate how quality institutions moderate the impact of financial inclusion on economic growth in Sub-Saharan Africa, employing an institutional quality index. For policymakers, examining how financial inclusion and institutional quality affect economic growth will help in policy formulation, especially on how to increase access to financial services to achieve the desired growth in Sub-Saharan Africa.

The rest of the study is organized as follows: Sect. 7.2 reviews the literature on the effect of financial inclusion and institutional quality on economic growth. Section 7.3 presents the research methodology. While Sect. 7.4 analyses the data and discusses the findings, Sects. 7.5 and 7.6 present the conclusion and recommendations, respectively.

7.2 Literature Review

7.2.1 Theoretical Background

This study is grounded in Schumpeter's (1911) finance-led hypothesis, which posits that economic growth and development hinge on identifying and fully utilizing production factors to innovate and enhance output, necessitating access to financial resources. Schumpeter asserts that entrepreneurs with inventiveness, imagination, and foresight require financial support to translate their ideas into reality. According to this theory, a robust financial system serves as a conduit, making financial resources available to the most effective users and productive sectors, thereby propelling economic growth. The finance-led theory, outlined in this study, underscores the pivotal role played by financial institutions in stimulating innovation and fostering growth.

The fundamental proposition of this theory is that financial institutions play a pivotal role in driving innovation and fostering growth. Consequently, a critical component of any expansion process involves mobilizing factors of production, with financial capital being a fundamental element. The Schumpeterian model of economic growth revolves around inventions and innovations, wherein credit assumes a significant role by enabling entrepreneurs to exert control over other production elements. Schumpeter emphasizes that the technical conditions of the economy, influenced significantly by credit creation and supported by an expansion of bank credit, are crucial for economic growth. However, a limitation to his argument is that while bank credit may be advantageous for short-term industrial development, it may not suffice for long-term development. To secure long-term financing, alternative funding sources, such as the issue of shares, need to be considered.

7.2.2 Empirical Literature

Given the objectives of the study, this section is divided into two sub-sections. That is, the researchers look at the empirical literature on the nexus between financial inclusion and economic growth and the literature on the impact of institutional quality on economic growth.

7.2.2.1 The Effect of Financial Inclusion on Economic Growth

Several studies have examined the link between financial inclusion and economic growth in the literature. For instance, Okoye et al. (2017) investigated the role of financial inclusion as a strategy for enhancing economic growth and development in Nigeria, utilizing data from the Central Bank of Nigeria and other relevant sources. The findings from multiple regression analysis revealed that indicators such as broad money to Gross Domestic Product (GDP), credit to the private sector to GDP, loan to deposit ratios, and liquidity ratios exert a significant effect on economic growth. In a study focused on the relationship between financial inclusion and economic growth in Organization of Islamic Cooperation (OIC) countries, Kim et al. (2018) established panel data for 55 OIC countries. Dynamic panel estimations from their study indicated a positive effect of financial inclusion on economic growth. Sethi and Acharya (2018) conducted a comprehensive analysis spanning the years 2004–2010, examining the relationship between financial inclusion and economic growth across both developing and developed nations. Employing various econometric techniques, including fixed effect and random effects models, their empirical results consistently demonstrated a positive long-term correlation between financial inclusion and economic growth. Furthermore, utilizing the Johansen cointegration test, the findings indicated that, while total loans and advances (a measure of financial inclusion) significantly impact economic growth, total deposits mobilized, the number of bank branches, and investments do not affect growth. Nwafor and Yomi (2018) utilized data from 2001 to 2016 to investigate the relationship between financial inclusion and economic growth in Nigeria. Using the two-stage least squares regression method, the results established a significant effect of financial inclusion on economic growth. Suidarma (2019) examined the nexus between financial inclusion and economic growth in ASEAN for the period 2008–2015 using the Vector Error Correction Model (VECM). The findings indicated that financial inclusion significantly contributed to the economic development of the ASEAN region. Peterand and Okpebru (2020) researched the impact of financial inclusion on economic growth in Nigeria from 2000 to 2018. The study employed the autoregressive distribution lag, and the findings showed that loans to small and medium enterprises (LSME) and rural bank deposits (RBD) as measures of financial inclusion have a significant effect on economic growth. Iramayasari and Adry (2020) explored the impact of financial inclusion, as measured by the number of ATMs and bank branches, on financial stability and economic growth in ASEAN countries. This study used panel data from 2004 to 2017 consisting of six countries in ASEAN. The results showed that financial inclusion has a significant effect on economic growth in ASEAN. Using the extended technique of moments and the Dumitrescu-Hurlin causality test, Adedokun and Ağa (2023) empirically evaluated the relationship between financial inclusion and economic growth in Sub-Saharan African (SSA) nations for the period 2004–2017. Principal component analysis was used to create a composite financial inclusion index based on the multidimensionality of the financial inclusion metrics. The findings demonstrated that financial inclusion significantly boosts economic growth in SSA nations. Van et al. (2021) probed the relationship between financial inclusion and economic growth in emerging markets. The study adopted the panel econometric technique to estimate the impact of financial inclusion on economic growth. The study found a positive relationship between financial inclusion and economic growth, especially in countries with low incomes and a lower degree of financial inclusion. Yakubu and Bunyaminu (2021) investigated the influence of financial inclusion on economic growth in the Economic Community of West African States (ECOWAS) nations using the pooled estimated generalized least squares (EGLS) for the period 2010–2017. The findings showed that financial inclusion has a positive and significant impact on economic growth, both directly and indirectly through financial openness. Younas et al. (2022) examined the impact of financial inclusion and the size of the shadow economy on the economic growth of developing economies over the period 2008–2017. The study applied a panel ordinary least square (OLS) fixed effect, a two-step difference generalized method of moments (GMM), and a panel Granger causality approach. The findings showed that financial inclusion has a significant positive impact on economic growth. Ozturk and Ullah (2022) investigated the impact of digital financial inclusion on economic growth and environmental sustainability in OBRI countries from 2007 to 2019. The study applied the generalized method of moments (GMM), two-stage least squares (2SLS), and pooled ordinary least squares (OLS) techniques. The empirical results demonstrated that digital financial inclusion spurs economic growth. In a recent investigation, Biswas (2023) demonstrated a direct influence of financial inclusion on economic growth within South Asian Economies. Similarly, Anthony-Orji et al. (2023) uncovered a positive correlation between financial inclusion and economic growth in the context of Nigeria.

7.2.2.2 The Impact of Institutional Quality on Economic Growth

The impact of institutional quality on economic growth has been extensively documented in the literature. In West Africa, for instance, Iheonu et al. (2017) investigated the effect of institutional quality on economic performance from 1996 to 2015. The study employed various analytical techniques, including the fixed effect model, the random effect model, and the panel two-stage least squares technique. The findings revealed that all institutional quality measures considered had a positive and significant impact on economic performance. Nguyen et al. (2018), using the System Generalized Method of Moments (SGMM) estimators, explored the effect of institutional quality on economic growth in 29 emerging economies from 2002 to 2015. The study concluded that institutional quality has a robust positive impact on economic growth. Similarly, Hayat (2019) analyzed the contribution of institutional quality to economic growth in a sample of 104 countries. Employing the GMM estimation method, the author found that higher institutional quality and FDI inflows both lead to faster economic growth. Abubakar (2020), using annual time-series data spanning 1979–2018, determined the relationship between institutional quality, contract-intensive money, and an effective governance index to economic growth in Nigeria. The study indicated that while the effective governance index had a positive but statistically insignificant impact on the economy, institutional quality (contract-intensive money) had a positive and statistically significant impact. Asghar et al. (2020) examined the effect of institutional quality on economic growth in 13 developing economies in Asia using panel data from 1990 to 2013. Employing the panel ARDL technique, the results demonstrated that institutional quality has a positive effect on economic growth. Canh et al. (2021) explored the impact of institutional quality and economic integration on the shadow economy. The study, using the panel-corrected standard errors (PCSE) estimator and the dynamic fixed effects autoregressive distributed lag (DFE ARDL) estimator, analyzed a global sample of 112 countries between 2005 and

2015. The results suggested that trade openness had a smaller negative impact on the shadow economy than institutional quality and FDI inflows. Singh and Pradhan (2022) studied the link between institutional quality and economic performance in South Asia from 2002 to 2016. The findings indicated that institutional quality has a favorable long-term impact on economic growth. Ahmed et al. (2022), using the dynamic OLS (DOLS) and completely modified OLS (FMOLS) techniques, analyzed the long-term effects of institutional quality and financial development on green growth. The findings highlighted that the long-term promotion of green economic growth is being driven by institutional quality and financial development. In another context, Islam and Mustafa Shindaini (2022) examined the impact of institutional quality (INQ) and human capital creation (HCC) on the economic growth (EG) linkage in Bangladesh using an ARDL approach. The study employed time-series annual data over the period 1990–2019 and constructed an institutional quality index. The study found that institutional quality affects economic growth positively in the long run. Pradhan et al. (2023) disclosed that institutional quality has the potential to stimulate economic growth in lower-income nations in the long run.

7.2.3 Gaps in the Literature

The empirical literature review highlights several notable limitations that are pervasive in the existing studies on the relationship between financial inclusion and institutional quality. First, a common drawback is the reliance on single measures to represent the complex constructs of financial inclusion and institutional quality. This oversimplification may lead to an incomplete understanding of the multifaceted nature of these variables and hinder the accuracy of the findings. Second, a substantial portion of the literature employs traditional panel methods, which, while valuable, may not capture the nuanced dynamics and temporal aspects of the relationship under investigation. The limitation of solely relying on these conventional techniques raises questions about the robustness and comprehensiveness of the analyses conducted in previous studies. Moreover, the existing body of empirical literature tends to analyze financial inclusion and institutional quality in isolation, overlooking the potential interactive effects between these two crucial factors. This literature lapse may result in an incomplete understanding of their combined influence on economic growth. To address these shortcomings, this study adopts a more comprehensive approach. It utilizes an index of financial inclusion, incorporating multiple dimensions for a more holistic representation. Additionally, the study employs the Generalized Method of Moments (GMM) technique, recognized for its superiority in handling endogeneity issues and providing more efficient estimates compared to the traditional regression methods commonly employed in the literature. Also, the study investigates the interactive impact of financial inclusion and institutional quality on economic growth, which is largely overlooked in extant studies. By addressing these methodological gaps, the study aims to contribute to a more nuanced and robust understanding of the intricate relationship between financial inclusion, institutional quality, and their impact on economic growth.

7.3 Research Methodology

7.3.1 Data and Sources

The study employs data from 33 countries in Sub-Saharan Africa. The data covers the period 2000–2017. The sample countries and the study period are based on the complete availability of data for the variables of interest. For the variables of the study, economic growth serves as the dependent variable whereas financial development and institutional quality are the main explanatory variables. The study controls for the effect of financial openness, bank stability, and inflation. The data for economic growth and inflation are obtained from the World

Development Indicators (WDI) of the World Bank. Financial inclusion indices are sourced from the Financial Access Survey of the International Monetary Fund (IMF). The Worldwide Governance Indicators (WGI) provide the data for institutional quality. Bank stability is gleaned from the Global Financial Development Database (GFDD) of the World Bank.

7.3.2 Description of Variables

7.3.2.1 Dependent Variable

Economic growth is the dependent variable measured by Gross Domestic Product (GDP) growth (annual percentage).

7.3.2.2 Independent Variables

In measuring financial inclusion, the study employs principal component analysis (PCA), following the empirical works of Yakubu and Musah (2024) and Bashiru et al. (2023), to construct a financial inclusion index using various dimensions of financial inclusion. These sub-dimensions are "ATMs per hundred thousand adults, bank branches per hundred thousand adults, bank accounts per thousand adults, borrowers from commercial banks per hundred adults, depositors with commercial banks per thousand adults, outstanding loans with commercial banks, and outstanding deposits with commercial banks." With the PCA method, *j*th factor index can be expressed as:

$$\boldsymbol{FINDEX}_j = W_{J1}X_1 + W_{J2}X_2 + \ldots\ldots + W_{JP}X_P \qquad (7.1)$$

$FINDEX_j$ is the index of financial inclusion. W_j is the weight on the factor score coefficient. X is the original figure of the various components and the number of variables in the equation is represented by P.

In line with the work of Yakubu (2020), institutional quality is measured by employing an index of institutional quality. This index is created using sub-measures of institutional quality, which include voice and accountability, political stability, government effectiveness, regulatory quality, rule of law, and control of corruption.

For the control variables, the KAOPEN index by Chinn and Ito is used as a proxy for financial openness following the work of Yakubu and Bunyaminu (2021). Bank stability is gauged using the z-score of banks. This proxy captures the default possibilities in the banking sector. Consumer price (annual %) is employed as a measure of inflation. In Table 7.1, the variable descriptions are summarized.

7.3.3 Model Specification

The basic empirical model for the study is specified as follows:

$$Y_{it} = \alpha + \beta' X_{it} + \varepsilon_{it} \qquad (7.2)$$

From the equation, the subscript i indicates the cross-sectional dimension, and t signifies the

Table 7.1 Description of variables

Variables	Acronym	Definition	Data source
Dependent variable			
Economic growth	ECG	GDP growth (annual percentage)	WDI
Independent variables			
Financial inclusion	FINDEX	Financial inclusion index	IMF
Financial openness	FOP	KAOPEN index by Chinn and Ito	KAOPEN index of financial openness
Bank stability	STA	Bank Z-score	GFDD
Inflation	INF	Consumer price (annual %)	WDI

Source: Authors' Own

time period. The coefficients of the independent factors are represented by β and the error term is ε. Y is economic growth, serving as the dependent variable and the explanatory factors are connoted by X.

Given the objectives of this study, which are to examine the direct effect of financial inclusion and institutional quality on economic growth and how they interactively affect economic growth, Eq. (7.2) is re-stated as follows.

$$\mathrm{ECG}_{it} = \alpha_0 + \beta_1 \mathrm{FINDEX}_{it} + \beta_2 \mathrm{FOP}_{it} + \beta_3 \mathrm{STA}_{it} + \beta_4 \mathrm{INF}_{it} + \varepsilon_{it} \quad (7.3)$$

$$\mathrm{ECG}_{it} = \alpha_0 + \beta_1 \mathrm{INSQ}_{it} + \beta_2 \mathrm{FOP}_{it} + \beta_3 \mathrm{STA}_{it} + \beta_4 \mathrm{INF}_{it} + \varepsilon_{it} \quad (7.4)$$

$$\mathrm{ECG}_{it} = \alpha_0 + \beta_1 \mathrm{FINDEX}_{it} + \beta_2 \mathrm{INSQ}_{it} + \beta_3 \mathrm{FINDEX}_{it}{}^{*}\mathrm{INSQ}_{it} + \beta_3 \mathrm{FOP}_{it} + \beta_4 \mathrm{STA}_{it} + \beta_5 \mathrm{INF}_{it} + \varepsilon_{it} \quad (7.5)$$

Equation (7.3) examines the effect of financial inclusion on economic growth alongside the control factors. Equation (7.4) looks at the impact of institutional quality on economic growth together with the control factors and Eq. (7.5) explores the interactive effect of financial inclusion on economic growth.

The acronyms ECG, FINDEX, INSQ, FOP, STA, and INF are economic growth, financial inclusion, institutional quality, financial openness, bank stability, and inflation, respectively.

7.3.4 Estimation Technique

The generalized method of moments (GMM) approach developed by Arellano and Bond (1991) is used as an analytical technique in this study. Compared to other panel estimation techniques such as the fixed effects, random effects, and the least squares methods, the GMM estimator eliminates unobserved country-specific effects in panel data. The GMM method also addresses the endogeneity problem, in which the explanatory factors may be correlated with the error term.

7.4 Results and Discussion

7.4.1 Descriptive Statistics

In Table 7.2, the descriptive statistics of the variables are shown. The mean of economic growth is 4.6329% with maximum and minimum values of 37.9987% and −36.3920%, respectively. The average value of the financial inclusion index is −0.0959 with a standard deviation of 1.9404. The institutional quality index has a mean of 0.0655 and ranges from −4.2229 to 5.4779. The mean of financial openness is −0.6096. The average values of bank stability and inflation are 11.2521% and 6.3377%, respectively. The variables are positively skewed except economic growth, which is skewed to the left. Aside from institutional quality, the Kurtosis value shows that the

Table 7.2 Descriptive statistics

	ECG	FINDEX	INSQ	FOP	STA	INF
Mean	4.6329	−0.0959	0.0655	−0.6096	11.2521	6.3377
Maximum	37.9987	10.6417	5.4779	2.3467	44.4128	37.1422
Minimum	−36.3920	−1.3978	−4.2229	−1.9166	2.1599	−9.6162
Std. dev.	4.7762	1.9404	2.1662	1.2652	5.9691	6.7604
Skewness	−0.4176	2.7958	0.2946	1.5200	1.5940	1.8382
Kurtosis	21.2516	12.1249	2.8112	3.8164	7.6776	7.8474
Jarque-Bera	7900.3990	2710.5580	9.0625	234.5012	758.3588	875.9764
Probability	0.0000	0.0000	0.0108	0.0000	0.0000	0.0000
Observations	594	594	594	594	568	594

Source: Authors' Own

Table 7.3 Correlation analysis

	FINDEX	INSQ	FOP	STA	INF
FINDEX	1.0000				
INSQ	0.4560	1.0000			
FOP	0.4177	0.4048	1.0000		
STA	0.1344	0.0457	0.0324	1.0000	
INF	−0.0258	−0.0287	0.1427	0.0120	1.0000
VIF	1.4051	1.3641	1.3476	1.0196	1.0352
Tolerance	0.7117	0.733 1	0.7421	0.9808	0.9660

Source: Authors' Own

variables have leptokurtic-shaped distributions. Also, the probability values of the Jarque-Bera test indicate that the distributions of the variables are not normal.

7.4.2 Correlation Analysis

The results of the correlation analysis are presented in Table 7.3. The results clearly indicate that the variables have weak correlation coefficients. According to Kennedy (2003), a high correlation, and, thus, multicollinearity, is evident when the correlation coefficient exceeds 0.80. Also, the results of the variance inflation factor (VIF) and the tolerance values are further evidence that the variables are free from multicollinearity. Gujarati posits that there is a multicollinearity problem when the VIF values exceed 10 and the tolerance (1/VIF) is less than 0.10.

7.4.3 Regression Results

The main findings of the study are discussed in this section based on the GMM estimation method. The study examines the direct effect of financial inclusion and institutional quality on economic growth and how financial inclusion and institutional quality interactively drive economic growth.

7.4.3.1 The Impact of Financial Inclusion on Economic Growth

In Table 7.4, the GMM estimates on the direct effect of financial inclusion on economic growth are presented. The results evidence that financial inclusion exerts a direct significant negative effect on economic growth. This suggests that an increase in financial inclusion does not enhance economic growth in SSA. This negative effect can be attributed to the underdeveloped nature of the banking sector in the region, where it is still difficult for rural residents and the most disadvantaged individuals to receive financial services. As a result, the initiatives to increase financial inclusion are insufficient to sustain the needed growth. The result is in sharp contrast with most prior studies (e.g., Adedokun and Aga, 2023; Van et al., 2021; Yakubu & Bunyaminu, 2021; Younas et al., 2022; Ozturk & Ullah, 2022). For the con-

Table 7.4 GMM results on the direct effect of financial inclusion on growth

Variables	GMM estimates
$\text{Lag}_{t\text{-}1}$	−0.0691***
	(0.0065)
FINDEX	−1.466570***
	(0.1428)
INSQ	
FOP	0.5308
	(0.3735)
STA	−0.1151**
	(0.0410)
INF	−0.1740***
	(0.0084)
Diagnostics	
Hansen J test	30.5908
Prob. Hansen J	0.3355
AR(2)	−1.8478
Prob. AR(2)	0.0646
Number of countries	33

Source: Authors' Own
Note: *** and ** indicate significance levels at 1% and 5% respectively

trol variables, the findings indicate that financial openness has a positive insignificant impact on economic growth. It is noted that there is a negative significant relationship between banking sector stability and economic growth. The implication is that in a highly stable banking sector, bank operations are restricted to safeguard banks' effectiveness. One of these limitations is the adoption of rigid criteria for credit access. This impacts businesses' ability to obtain additional finance to fund their operations, which may have an impact on domestic output. Contrary to this finding, Bayar et al. (2021) and Sunney (2020) find that bank stability significantly enhances economic growth. The disparity in results might be attributed to the study period and analytical approach used by the researchers. The findings show that inflation has a strong negative influence on economic growth, depicting that growth is stifled when inflation rises. Given the surrounding uncertainties, high inflation discourages investors from committing their investable capital to the local market. Also, rising inflation can reduce consumers' purchasing power and make it more difficult for them to acquire essential goods and services, stifling economic growth. The negative link between inflation and economic growth agrees with most earlier findings (Adaramola and Dada, 2020; Yakubu et al., 2021; Tien, 2021; Junejo et al., 2021).

7.4.3.2 The Impact of Institutional Quality on Economic Growth

The direct impact of institutional quality on economic growth is shown in Table 7.5. The study finds that institutional quality has a positive significant influence on economic growth. This demonstrates that improving democratic and political institutions promotes economic growth in SSA considerably. This is due to the fact that high-quality institutions boost investor confidence in the country. As a result, investors are willing to commit their capital to feasible initiatives that boost domestic output. This finding agrees with some earlier studies (see Hayat, 2019; Asghar et al., 2020; Ahmed et al., 2022; Islam & Mustafa Shindaini, 2022). The results further evidence that financial openness, bank stability, and inflation have a significant negative impact on economic growth.

Table 7.5 GMM results on the direct effect of institutional quality on growth

Variables	GMM estimates
Lag_{t-1}	−0.0192**
	(0.0077)
FINDEX	
INSQ	1.5446***
	(0.1624)
FOP	−1.1474***
	(0.3220)
STA	−0.1197**
	(0.0475)
INF	−0.1280***
	(0.0106)
Diagnostics	
Hansen J test	28.8373
Prob. Hansen J	0.4208
AR(2)	−1.4162
Prob. AR(2)	0.1567
Number of countries	33

Source: Authors' Own

7.4.3.3 The Interactive Effect of Financial Inclusion and Institutional Quality on Economic Growth

The results of the interactive effect of financial inclusion and institutional quality are depicted in Table 7.6. It can be observed that financial inclusion and institutional quality maintain their respective coefficient signs. That is, while financial inclusion negatively affects economic growth, institutional quality drives economic growth positively. It is found that the interactive impact of financial inclusion and institutional quality is positive and significant. This implies that for financial inclusion to contribute significantly to enhancing economic growth, it must be coupled with strong institutions. Quality institutions and good governance can create a stable and predictable business environment, which can attract investment and encourage economic activity including the boom in the banking sector. Bank stability and inflation also maintain their negative coefficient signs, suggesting that in all cases, the stability of the banking sector and the

Table 7.6 GMM results on the interactive effect of financial inclusion and institutional quality on growth

Variables	GMM estimates
Lag_{t-1}	−0.0712***
	(0.0080)
FINDEX	−2.3335***
	(0.2070)
INSQ	2.21999***
	(0.2173)
FINDEX * INSQ	0.5615***
	(0.0903)
FOP	0.1977
	(0.3408)
STA	−0.1533***
	(0.0455)
INF	−0.1919***
	(0.0188)
Diagnostics	
Hansen J test	28.2435
Prob. Hansen J	0.3466
AR(2)	−1.7322
Prob. AR(2)	0.0832
Number of countries	33

Source: Authors' Own

high level of inflation impede economic growth in Sub-Saharan Africa.

7.4.4 Model Diagnostics

From the model diagnostics in Tables 7.4, 7.5, and 7.6, the Hansen J tests suggest that the null hypothesis, which specifies that the over-identification restrictions are valid, cannot be rejected in all cases, implying that the instruments in the study are apt. Also at a 5% significance level, the AR(2) test shows that there is no second-order autocorrelation.

7.5 Conclusion

This study empirically examines the impact of financial inclusion on economic growth and how institutional quality moderates the financial inclusion-economic growth relationship. To achieve the objectives of the study, the researchers employ data covering from 2000 to 2017 for 33 countries in Sub-Saharan Africa. The system generalized method of moments (GMM) technique is applied as an analytical technique. The GMM method is superior to other panel techniques given that it deals with the issue of endogeneity and eliminates biases in model estimates. The empirical results show that financial inclusion has a negative significant impact on economic growth while the direct effect of institutional quality on economic growth is positive. The findings indicate that the interactive effect of financial inclusion and institutional quality on economic growth is positive. Furthermore, for the control factors included in the study, the results reveal that bank stability and inflation significantly and negatively affect growth. In conclusion, the findings posit that an increase in financial inclusion does not necessarily translate to economic growth. To achieve the desired growth via inclusive finance, efforts to promote financial inclusion must move in tandem with strong institutions.

7.6 Recommendations

Given that financial inclusion negatively drives growth when examined directly, policymakers need to take steps to mitigate this negative effect. In doing so, there is a need to implement proper regulations and oversight. This can help prevent over-indebtedness and increased vulnerability to financial fraud. It is also important to encourage responsible lending and borrowing. This can include measures to ensure that borrowers have the ability to repay loans and that lenders are not engaging in predatory lending practices. It is also important to assess the impact of policies and programs aimed at increasing financial inclusion and making adjustments as necessary to ensure that they are sustainable and geared toward promoting economic growth.

To strengthen the impact of institutional quality and its moderating effect on economic growth, there is a need to enhance the rule of law by promoting independent and effective judiciaries and providing secure property rights to individuals and businesses, including financial institutions. It is also imperative to encourage private sector development by reducing bureaucratic regulations and providing access to finance.

For bank stability to enhance economic growth, bank regulations and supervisory frameworks must be strengthened to ensure that banks are well-capitalized and able to withstand financial shocks. To curb the adverse effect of inflation on economic growth, there is a need to implement monetary policies such as increasing interest rates or reducing the money supply to reduce inflationary pressures. Adopting fiscal policies such as reducing government spending may also help control rising inflation levels.

The study dwells on panel data from Sub-Saharan African countries and the results cannot be generalized for the entire African continent. The researchers, therefore, suggest that future research may consider a broader sample consisting of African countries to provide more comprehensive evidence. A comparative analysis between Africa and other continents such as Asia may offer an interesting picture of the role of financial inclusion in promoting growth. In addition to the composite index of financial inclusion, future research may investigate how the individual dimensions of financial inclusion affect growth. Further studies may also consider different econometric techniques to check for the robustness of findings.

References

Abubakar, S. (2020). Institutional quality and economic growth: Evidence from Nigeria. *African Journal of Economic Review, 8*(1), 48–64.

Adaramola, A. O., & Dada, O. (2020). Impact of inflation on economic growth: Evidence from Nigeria. *Investment Management & Financial Innovations, 17*(2), 1–13.

Adedokun, M. W., & Ağa, M. (2023). Financial inclusion: A pathway to economic growth in sub-Saharan African economies. *International Journal of Finance & Economics, 28*(3), 2712–2728.

Ahmed, F., Kousar, S., Pervaiz, A., & Shabbir, A. (2022). Do institutional quality and financial development affect sustainable economic growth? Evidence from South Asian countries. *Borsa Istanbul Review, 22*(1), 189–196.

Anthony-Orji, O. I., Orji, A., Ogbuabor, J. E., & Uka, L. C. (2023). Money matters a lot: Empirical analysis of financial development, financial inclusion and economic growth in Nigeria. *International Journal of Economic Policy in Emerging Economies, 17*(1), 100–117.

Arellano, M., & Bond, S. (1991). Some tests of specification for panel data: Monte Carlo evidence and an application to employment equations. *The Review of Economic Studies, 58*(2), 277–297.

Asghar, N., Qureshi, D. S., & Nadeem, M. (2020). Institutional quality and economic growth: Panel ARDL analysis for selected developing economies of Asia. *South Asian Studies, 30*(2), 381–404.

Balele, N. P. (2019). The impact of financial inclusion on economic growth in sub-Saharan Africa. *Journal of Applied Economics and Business, 7*(4), 51–68.

Bashiru, S., Bunyaminu, A., Yakubu, I. N., & Al-Faryan, M. A. S. (2023). Drivers of financial inclusion: Insights from sub-Saharan Africa. *Economies, 11*(5), 146.

Bayar, Y., Borozan, D., & Gavriletea, M. D. (2021). Banking sector stability and economic growth in post-transition European Union countries. *International Journal of Finance & Economics, 26*(1), 949–961.

Biswas, G. K. (2023). Financial inclusion and its impact on economic growth: An empirical evidence from south Asian countries. *European Journal of Business and Management Research, 8*(4), 163–167.

Canh, P. N., Schinckus, C., & Dinh Thanh, S. (2021). What are the drivers of shadow economy? A further evidence of economic integration and institutional quality. *The Journal of International Trade & Economic Development, 30*(1), 47–67.

Demirguc-Kunt, A., Klapper, L., Singer, D., & Ansar, S. (2018). *The global Findex database 2017: Measuring financial inclusion and the fintech revolution*. World Bank Publications.

Hayat, A. (2019). Foreign direct investments, institutional quality, and economic growth. *The Journal of International Trade & Economic Development, 28*(5), 561–579.

Ifediora, C., Offor, K. O., Eze, E. F., Takon, S. M., Ageme, A. E., Ibe, G. I., & Onwumere, J. U. (2022). Financial inclusion and its impact on economic growth: Empirical evidence from sub-Saharan Africa. *Cogent Economics & Finance, 10*(1), 2060551.

Iheonu, C., Ihedimma, G., & Onwuanaku, C. (2017). *Institutional quality and economic performance in West Africa*. MPRA Papers.

Inoue, T., & Hamori, S. (2016). Financial access and economic growth: Evidence from sub-Saharan Africa. *Emerging Markets Finance and Trade, 52*(3), 743–753.

Iramayasari, I., & Adry, M. R. (2020). Pengaruh Inklusi Keuangan Terhadap Stabilitas Sistem Keuangan dan pertumbuhan Ekonomi di ASEAN. *Jurnal Kajian Ekonomi dan Pembangunan, 2*(2), 29–38.

Islam, M. S., & Mustafa Shindaini, A. J. (2022). Impact of institutional quality and human capital creation on economic growth in Bangladesh: Evidence from an ARDL approach. *International Journal of Social Economics, 49*(12), 1787–1802.

Junejo, I., Faiz, M., Qazi, N., & Tipu, A. A. K. (2021). Impact of inflation, imports and tax revenue on economic growth of Pakistan: An empirical study from 1990–2020. *International Research Journal of Management and Social Sciences, 2*(2), 87–95.

Kennedy, P. (2003). *A guide to econometrics*. Cambridge, Massachusetts: MIT press.

Kim, D. W., Yu, J. S., & Hassan, M. K. (2018). Financial inclusion and economic growth in OIC countries. *Research in International Business and Finance, 43*, 1–14.

Klapper, L., El-Zoghbi, M., & Hess, J. (2016). Achieving the sustainable development goals. *The role of financial inclusion*. Available online: http://www.ccgap.org

Nguyen, C. P., Su, T. D., & Nguyen, T. V. H. (2018). Institutional quality and economic growth: The case of emerging economies. *Theoretical Economics Letters, 8*(11), 1943–1956.

Nwafor, M. C., & Yomi, A. I. (2018). The nexus between financial inclusion and economic growth: Evidence from Nigeria. *International Journal of Research and Innovation in Social Sciences, 2*(4), 143–149.

Ofosu-Mensah Ababio, J., Attah-Botchwey, E., Osei-Assibey, E., & Barnor, C. (2021). Financial inclusion and human development in frontier countries. *International Journal of Finance & Economics, 26*(1), 42–59.

Okoye, L. U., Adetiloye, K. A., Erin, O., & Modebe, N. J. (2017). Financial inclusion as a strategy for enhanced economic growth and development. *Journal of Internet Banking and Commerce, 22*(S8), 1–12.

Ozturk, I., & Ullah, S. (2022). Does digital financial inclusion matter for economic growth and environmental sustainability in OBRI economies? An empirical analysis. *Resources, Conservation and Recycling, 185*, 106489.

Pradhan, R. P., Nair, M. S., Arvin, M. B., & Hall, J. H. (2023). Institutional quality, financial development and sustainable economic growth among lower income countries. *Natural Resources Forum, 47*(3), 435–483.

Peterand, E., & Okpebru, O. O. (2020). Effects of financial inclusion on economic growth in Nigeria. *IOSR Journal of Research & Method in Education (IOSR-JRME), 10*(1), 44–49.

Sarma, M. (2008). *Index of financial inclusion, Indian Council for Research on international economic relations (ICRIER)*. Working Paper.

Schumpeter, J. A. (1911). *The theory of economic development*. Cambridge: Harvard University Press.

Sethi, D., & Acharya, D. (2018). Financial inclusion and economic growth linkage: Some cross-country evidence. *Journal of Financial Economic Policy, 10*(3), 369–385.

Sharma, D. (2016). Nexus between financial inclusion and economic growth: Evidence from the emerging Indian economy. *Journal of Financial Economic Policy, 8*(1), 13–36.

Shetty, N. K. (2008). The microfinance promise in financial inclusion and welfare of the poor: Evidence from India. *The IUP Journal of Applied Economics, 8*(5), 1–21.

Singh, B. P., & Pradhan, K. C. (2022). Institutional quality and economic performance in South Asia. *Journal of Public Affairs, 22*(1), e2401.

Suidarma, I. M. (2019). The nexus between financial inclusion and economic growth in ASEAN. *JEJAK, 12*(2), 267–281.

Sunney, I. A. (2020). *Bank stability and economic growth in Sub-Saharan Africa*. MPhil Thesis, University of Cape Coast, Ghana.

Tien, N. H. (2021). Relationship between inflation and economic growth in Vietnam. *Turkish Journal of Computer and Mathematics Education (TURCOMAT), 12*(14), 5134–5139.

Van, L. T. H., Vo, A. T., Nguyen, N. T., & Vo, D. H. (2021). Financial inclusion and economic growth: International evidence. *Emerging Markets Finance and Trade, 57*(1), 239–263.

Yakubu, I. N. (2020). Institutional quality and foreign direct investment in Ghana: A bounds-testing cointegration approach. *Review of International Business and Strategy, 30*(1), 109–122.

Yakubu, I. N., & Bunyaminu, A. (2021). Financial inclusion and economic growth in West Africa: The moderating effect of financial openness. *Empirical Economics Letters, 20*(2), 155–164.

Yakubu, I. N., & Musah, A. (2024). The nexus between financial inclusion and bank profitability: A dynamic panel approach. *Journal of Sustainable Finance & Investment, 14*(2), 430–443.

Yakubu, I. N., Abokor, A. H., & Gedik Balay, I. (2021). Re-examining the impact of financial intermediation on economic growth: Evidence from Turkey. *Journal of Economics and Development, 23*(2), 116–127.

Younas, Z. I., Qureshi, A., & Al-Faryan, M. A. S. (2022). Financial inclusion, the shadow economy and economic growth in developing economies. *Structural Change and Economic Dynamics, 62*, 613–621.

8 The Interplay Between Financial Inclusion and the Sustainable Development Goals Index in ASEAN-5 Countries

Abdul-Latif Mohammed, Ibrahim Nandom Yakubu, and Alhassan Bunyaminu

8.1 Introduction

The achievement of the Sustainable Development Goals (SDGs) relies heavily on the importance of financial inclusion (Yap et al., 2023, 2024). Financial inclusion, which involves using formal financial services, became popular in the early 2000s as it was recognized that poverty and financial exclusion are closely linked (Babajide et al., 2015; Erlando et al., 2020). Thus, financial inclusion is a focus for researchers and policymakers, with studies highlighting its benefits in developing countries. The increasing body of empirical evidence emphasizes the positive outcomes related to enhancing financial inclusion (Allen et al., 2016; Demirguc-Kunt et al., 2018). In essence, financial inclusion pertains to the provision and utilization of formal financial services among disadvantaged and low-income populations, ensuring their active involvement in economic activities (Banerjee et al., 2021; Demirguc-Kunt et al., 2018). Studies (e.g., Ellili, 2024; Razzaq et al., 2024) highlight the benefits of financial inclusion such as enabling access to financial products and services for all. This includes access to various financial services such as banking, savings, insurance, and credit.

Despite these benefits, around 1.4 billion people are excluded from formal financial systems due to their involvement in the informal sector (Yap et al., 2023). This limits their access to affordable financial services (Achugamonu et al., 2020). In effect, financial exclusion limits resources and perpetuates poverty and inequality, hindering sustainable development (World Bank, 2022; Yap et al., 2023). Thus, addressing financial inclusion is crucial for achieving the SDGs (Kara et al., 2021; Yap et al., 2023). This is because financial inclusion and sustainable development yield beneficial outcomes for society and the environment with the international development community focusing on these two agendas (Yap et al., 2023). But financial inclusion and sustainable development have traditionally been studied independently, without acknowledging their potential relationship (Ozili, 2022; Yap et al., 2023). This situation misreports the relationship between the two agendas since the UN's 2030 agenda positions financial inclusion as a tool for achieving seven SDGs: 1 (No Poverty), 2 (No Hunger), 3 (Good health and Well-being), 5 (Gender Equality), 8 (Decent Work and Economic Growth), 9 (Industry, Innovation and Infrastructure), and 10 (Reduced Inequalities) (Ohiomu & Ogbeide-Osaretin, 2019; Salampasis & Mention, 2018; Yap et al., 2023).

A.-L. Mohammed (✉)
Newcastle Business School, University of Newcastle, Newcastle, NSW, Australia
e-mail: AbdulLatif.mohammed@uon.edu.au

I. N. Yakubu
School of Business, University for Development Studies, Tamale, Ghana

A. Bunyaminu
Faculty of Accounting and Finance, University of Professional Studies, Accra, Ghana

I. N. Yakubu (ed.), *Strategic Approaches to Banking Business and Sustainable Development Goals*, Sustainable Development Goals Series, https://doi.org/10.1007/978-3-031-80744-2_8

Accordingly, this study is motivated to examine the relationship between financial inclusion and the Sustainable Development Index in Asian countries, specifically in the ASEAN-5 economies (Indonesia, Malaysia, the Philippines, Thailand, and Vietnam). The sustainable development index is the most commonly used tool for measuring sustainable development (Yap et al., 2023). Scholars (e.g., Yap et al., 2023) report that prior research only examined certain indicators of sustainable development, such as poverty (Burgess & Pande, 2005; Inoue & Hamori, 2012), inequality (De Haan & Sturm, 2017; Demir et al., 2022), and growth (Levine, 2005). However, very few studies analyze finance-related aspects of the SGDs and their impact on sustainable development through financial inclusion (Yap et al., 2023).

This study considers this background to provide insights into the link between financial inclusion and the SDG Index from Asian countries, which are less focused in the literature (Hussain et al., 2024). The study focuses on Asia because of its unique characteristics, such as a low level of financial inclusion and an underdeveloped financial system. Asian nations are projected to experience significantly higher macroeconomic growth rates (Hussain et al., 2024). Recent statistics indicate that emerging Asian countries are projected to witness a substantial increase in economic development, with a growth rate of 6.3% predicted over the next 2 years. This forecast indicates positive progress in these regions, suggesting potential opportunities for growth and investment (Hasan et al., 2021; Le et al., 2019). Additionally, researchers observe that Asia's strong economic growth is hindered by over one billion people still lacking access to formal financial services, requiring policymakers' attention for inclusive growth (Bhardwaj et al., 2018). Narrowing the focus to the ASEAN-5, these nations have made significant strides in promoting financial inclusion through digital innovations and policies designed to expand access to financial services (Ong et al., 2023). Despite these advancements, they continue to face unique challenges, such as large informal economies and geographical barriers. Analyzing financial inclusion in this region provides valuable insights that can guide global strategies for achieving inclusive and sustainable growth in other developing economies.

Consequently, this study makes several key contributions to the existing literature on financial inclusion and sustainable development. First, it provides a comprehensive analysis of the relationship between financial inclusion and the Sustainable Development Index within the specific context of ASEAN-5 countries. While previous research has largely focused on the economic benefits of financial inclusion (see Yakubu & Bunyaminu, 2021; Azimi, 2022; Alosheibat et al., 2023; Ofoeda et al., 2024; Chinoda & Kapingura, 2024, Yakubu et al., 2024), this study extends the discourse by linking financial inclusion to broader sustainability outcomes, thus addressing a critical gap in the literature. Second, the study employs a robust methodological framework, utilizing the Sustainable Development Index as a holistic measure of progress toward the SDGs. This approach allows for a comprehensive understanding of how financial inclusion interacts with various dimensions of sustainable development, including social, economic, and environmental factors. By doing so, the study offers a more integrated perspective on the role of financial inclusion in promoting sustainable growth. Moreover, the study will contribute to the global discourse on sustainable development by highlighting the importance of inclusive financial systems in achieving the SDGs. Through its regional focus, this study seeks to provide a valuable addition to the ongoing efforts to create a more inclusive and sustainable global economy.

The structure of this study is as follows. In Sect. 8.2, the literature on financial inclusion and its connection to the SDGs index is thoroughly reviewed. The data and methodology are discussed in Sect. 8.3, while the empirical results are presented and discussed in Sect. 8.4. The study is concluded in Sect. 8.5.

8.2 Literature Review

Achieving SDGs requires global cooperation from governments, the private sector, civil society organizations, the education sector, and the media. Within this cooperation framework, the financial system is a critical mechanism that grants households access to formal financial services such as savings accounts, consumer credit, and mortgages (Kara et al., 2021). Sustainability studies emphasize the importance of financial inclusion. For example, using diverse channels for financial inclusion promotes sustainable development, improves earning potential, empowers women entrepreneurs, reduces transaction costs, fosters good health, accelerates fund accumulation, and encourages the use of digital technologies (Yap et al., 2023; Yikun et al., 2023).

Considering this, financial inclusion has a positive impact on social and economic outcomes at both the household and national levels (Chowdhury & Chowdhury, 2024), as it facilitates economic development by broadening access to financial services (Chowdhury & Chowdhury, 2024; Law et al., 2014). Studies such as Swamy (2014) discovered that female-headed households are particularly impacted by financial inclusion, potentially resulting in income inequality and perpetuating cycles of poverty (Aghion & Bolton, 1997). Moreover, the availability of savings instruments through an inclusive financial system enhances savings, investment, consumption, and female empowerment (Ashraf et al., 2010), ultimately benefiting impoverished individuals and enhancing the quality of financial services (Chowdhury & Chowdhury, 2024). For example, Demirguc-Kunt and Klapper (2012) identified barriers to financial inclusion and highlighted the benefits of efficient financial systems for the poor and disadvantaged.

Furthermore, Sarma and Pais (2011) found a connection between financial inclusion and human development in a country, and Prina (2015) found a positive relationship between financial inclusion and improved health, as it enables individuals to handle medical expenses during a health crisis. In the Asia region, scholars (e.g., Gupta et al., 2014; Kuri & Laha, 2011) employed financial inclusion indices encompassing variables such as the penetration, availability, and usage of banking services. Their analysis demonstrated a positive association between human development and financial inclusion within India. While the advantages of financial inclusion have been documented in these studies, other research has identified potential disadvantages. For example, Arora (2012) found that low literacy and awareness hinder financial inclusion, leading to low human development in South Asia.

These studies provide insights into the relationship between financial inclusion and the SDGs. Researchers suggest financial inclusion aids sustainable development goals (Chitimira & Warikandwa, 2023; Yap et al., 2023), such as improving female entrepreneurship and decreasing transaction costs. The World Bank highlights how financial inclusion enables certain SDGs (Chitimira & Warikandwa, 2023). For example, improved access to financial services benefits people globally, especially in remote areas. Financial inclusion promotes health under SDG 3. Financial inclusion motivates women entrepreneurs to pursue new opportunities, which helps to promote SDG 5 (gender equality) (Chitimira & Warikandwa, 2023).

Although studies demonstrate the connection between financial inclusion and various aspects of development, research on the correlation between financial inclusion and sustainable development is scarce (Ozili, 2022). By examining global data on the sustainable development index, this study contributes to the existing literature by exploring how financial inclusion and sustainable development are related in the Asia region. This region shares numerous political, economic, and cultural similarities (Chowdhury & Chowdhury, 2024). The next discusses the study methodology.

8.3 Methodology

8.3.1 Data

The data for the study spans from 2015 to 2021, which aligns with the adoption and implementation phase of the SDGs. The data are sourced from the SDG Transformation Center and the World Bank's World Development Indicators, ensuring a comprehensive and reliable dataset for the analysis.

8.3.2 Description of Variables

The primary dependent variable in this study is the Sustainable Development Goals Index (SDG Index), which measures the overall progress of each country toward achieving sustainable development goals. The SDG Index serves as a composite indicator, capturing the multi-dimensional aspects of sustainable development, including economic, social, and environmental progress.

The independent variable of interest is financial inclusion, which has been operationalized through the construction of a Financial Inclusion Index. In line with the work of Yakubu and Musah (2024), the index was developed using Principal Components Analysis (PCA), a statistical technique that reduces the dimensionality of data while retaining most of the variability present in the original dataset (Bashiru et al., 2023a). The Financial Inclusion Index is constructed from four key sub-indicators that reflect the availability and accessibility of financial services: automated teller machines (ATMs) per 100,000 adults, borrowers from commercial banks per 1000 adults, depositors with commercial banks per 1000 adults, and commercial bank branches per 100,000 adults. These sub-indicators were selected based on their relevance to the availability and accessibility dimensions of financial inclusion. The PCA was applied to aggregate these sub-indicators into a single financial inclusion index, which serves as the main independent variable in the study. Thus, the financial inclusion index (FI) is a weighted sum of the selected principal components:

$$\text{FINC} = \sum_{j=1}^{k} \alpha_j \text{PC}_j \tag{8.1}$$

where α_j is the weights derived from the eigenvalues and k is the number of retained principal components.

The study controls for financial development, economic growth, and education. Table 8.1 provides a summary of the variables, their definitions, and the respective data sources.

8.3.3 Model Specification

The empirical model for this study is specified as follows:

$$Y_{it} = \alpha + \beta' X_{it} + \varepsilon_{it} \tag{8.2}$$

In this model, Y_{it} represents the dependent variable, which in this study is the SDG Index for country i at time t. The term α is the constant term, while $\beta' X_{it}$ captures the relationship between the independent variables and the dependent variable. Here, X_{it} represents the vector of independent variables, which includes the financial inclusion index and the control variables (financial development, economic growth, and education) for each country i at time t. Finally, ε_{it} is the error term that accounts for unobserved factors affecting the dependent variable.

8.3.4 Analytical Technique

The study employs the Pooled Estimated Generalized Least Squares (EGLS) method with cross-section weights as the primary analytical technique. The EGLS method is particularly suitable for handling panel data with cross-sectional heteroscedasticity and autocorrelation, which are common issues in multi-country studies like this one. By applying cross-section weights, the EGLS method adjusts for the varying variances across different countries, providing more efficient and unbiased estimates compared to Ordinary Least Squares (OLS).

The choice of EGLS is justified by its ability to account for heterogeneity across the ASEAN-5

Table 8.1 Description of variables

Variable	Abbreviation	Measurement	Data Source
Dependent variable			
Sustainable Development Goals Index	SDG Index	Composite indicator	SDG Transformation Center
Independent variable			
Financial Inclusion Index	FINC	Constructed using PCA from sub-indicators below:	World Development Indicators
Automated teller machines (ATMs)	ATMS	ATMs per 100,000 adults	World Development Indicators
Borrowers from commercial banks	BOR	Borrowers per 1000 adults	World Development Indicators
Depositors with commercial banks	DEP	Depositors per 1000 adults	World Development Indicators
Commercial bank branches	BRA	Branches per 100,000 adults	World Development Indicators
Control variables			
Financial development	FIND	Domestic credit to private sector by banks (% of GDP)	World Development Indicators
Economic growth	ECG	GDP growth (annual %)	World Development Indicators
Education	EDU	Secondary school enrollment (% gross)	World Development Indicators

Source: Authors

countries, where economic conditions and institutional quality may differ significantly. This technique ensures that the estimates reflect the true relationships between financial inclusion, the SDG Index, and other control variables, without being biased by cross-country differences in variability (Wooldridge, 2010; Baltagi, 2008).

8.4 Results and Discussion

8.4.1 Descriptive Statistics

The descriptive statistics for the variables in this study are presented in Table 8.2, summarizing key measures such as the mean, median, maximum, minimum, and standard deviation for each variable. The SDG Index has a mean value of 66.39, with a relatively narrow range between the maximum (68.95) and minimum (63.07) values, indicating moderate variability across the ASEAN-5 countries. The financial inclusion variables, including Automated Teller Machines (ATMS), Borrowers from Commercial Banks (BOR), Depositors with Commercial Banks (DEP), and Commercial Bank Branches (BRA), exhibit varying degrees of dispersion, with BOR showing the highest standard deviation, suggesting significant differences in borrowing rates across the region. The Financial Inclusion Index (FINC) is centered around zero, as expected from a PCA-derived index. The control variables also display distinct ranges and variability, reflecting the economic diversity within ASEAN-5.

8.4.2 Correlation and Multicollinearity Analysis

Table 8.3 presents the correlation analysis between the variables and examines multicollinearity through Variance Inflation Factor (VIF) and Tolerance values. The correlation coefficients between the variables indicate the strength and direction of their linear relationships. Financial inclusion (FINC) shows a moderate positive correlation with financial development (FIND) at 0.3820, while economic growth (ECG) and education (EDU) exhibit negative correlations with both FINC and FIND. The VIF values for all

Table 8.2 Descriptive statistics

	SDG Index	ATMS	BOR	BRA	DEP	FINC	FIND	ECG	EDU
Mean	66.3925	76.5805	568.5565	9.9045	1387.3490	−9.77E-16	120.7273	2.3562	101.4252
Median	66.8700	58.4150	350.1350	10.1550	1274.5100	−0.7393	120.8303	3.4761	105.2796
Maximum	68.9500	117.7900	1165.390	12.5300	2404.8700	2.9399	133.7857	5.8127	130.9335
Minimum	63.0700	50.2400	309.4100	7.0200	705.2400	−2.01117	111.2873	−6.0500	78.5831
Std. dev.	1.8144	29.2510	363.3467	1.5691	643.2888	1.6987	6.0427	3.4318	16.7205

Source: Authors' Computation

Table 8.3 Correlation analysis

	FINC	FIND	ECG	EDU	VIF	Tolerance
FINC	1.0000				1.2851	0.7781
FIND	0.3820	1.0000			3.8481	0.2599
ECG	−0.1524	−0.5949	1.0000		2.3113	0.4327
EDU	−0.0657	−0.5426	−0.0404	1.0000	2.1667	0.4615

Source: Authors' Computation

Table 8.4 Pooled EGLS estimation results

Variables	Pooled EGLS estimates
FINC	3.2034***
	(0.6122)
FIND	−0.1446***
	(0.0287)
ECG	−0.1395***
	(0.0316)
EDU	−0.0543**
	(0.0247)
C	89.6775***
	(2.9527)
Diagnostics	
R^2	0.9707
Adj. R^2	0.9572
F-statistic	71.7432
Prob. (F-statistic)	[0.0000]
Wald test, χ^2	465.1702
Prob. (χ^2)	[0.0000]

Source: Authors' Computation

Notes: ** and *** denote significance at 5% and 1% respectively. Values in () and [] are standard errors and p–values, respectively

variables are below the commonly accepted threshold of 10, suggesting that multicollinearity is not a significant concern in this model (Bashiru et al., 2023b). The Tolerance values, all above 0.2, further confirm the absence of problematic multicollinearity, ensuring the reliability of the regression coefficients in the analysis.

8.4.3 Regression Results

The Pooled EGLS estimation results in Table 8.4 provide insightful findings on the relationship between the sustainable development index and the explanatory variables, including financial inclusion, financial development, economic growth, and education within the ASEAN-5 countries.

Financial inclusion exhibits a positive and highly significant impact on the SDG Index. This strong positive relationship indicates that improvements in financial inclusion are crucial for advancing the Sustainable Development Goals in the ASEAN-5 region. The significance of this result underscores the importance of policies and initiatives that enhance access to financial services, which can drive broader development outcomes.

Conversely, financial development is found to have a negative and significant effect on the sustainable development index, with a coefficient of −0.1446, also significant at the 1% level. This negative relationship suggests that increased financial development, as measured by domestic credit to the private sector, may not necessarily lead to improved SDG outcomes. This could be due to the complexities of financial development, where factors such as financial instability or inequality may counteract the potential benefits. Economic growth similarly shows a negative relationship with the SDG Index, with a coefficient of −0.1395, significant at the 1% level. This finding indicates that economic growth alone does not guarantee progress toward the SDGs, highlighting the need for growth strategies that are inclusive and sustainable. The negative impact suggests that without a focus on equity and sustainability, economic expansion may fall short of contributing positively to SDG achievement. Education is also negatively associated with the SDG Index, with a coefficient of −0.0543, significant at the 5% level. Although education is a critical factor in sustainable development, this result suggests that higher levels of secondary school enrollment may not directly translate into immediate improvements in SDG outcomes. The complexity of education's impact on development likely requires a longer-term perspective to fully realize its benefits.

8.5 Conclusion and Recommendation

This study aimed to explore the impact of financial inclusion on sustainable development in the ASEAN-5 countries, utilizing data from 2015 to 2021, a period coinciding with the adoption of the SDGs. Employing the Pooled Estimated Generalized Least Squares (EGLS) technique, the analysis revealed that financial inclusion has a significant positive effect on the SDG Index, highlighting its critical role in advancing sustainable development. Conversely, financial development, economic growth, and education showed negative impacts on the SDG Index, suggesting that while these factors are important, their benefits are not straightforward and may require more nuanced approaches to fully align with the SDGs.

Based on the study's findings, several policy recommendations are proposed to enhance sustainable development through financial inclusion in the ASEAN-5 countries. First, policymakers should prioritize the expansion of financial inclusion by implementing supportive regulations and developing innovative financial products that cater to underserved populations. This includes enhancing access to banking services through digital platforms and financial literacy programs. Continuous monitoring and evaluation of financial inclusion initiatives are crucial. Governments should establish mechanisms for tracking progress and assessing the effectiveness of financial inclusion policies. This will help in making data-driven adjustments and ensuring that financial inclusion efforts contribute effectively to achieving the SDGs.

Second, it is essential to address the negative impacts associated with financial development, economic growth, and education on the SDG Index. Policymakers should adopt a more integrated approach that balances financial growth with sustainable practices and educational advancements. This might involve strengthening institutional frameworks to ensure that financial development contributes positively to sustainable goals and aligns with broader economic policies.

For further research, it is recommended to explore the longitudinal effects of financial inclusion on sustainable development across different regions within the ASEAN-5 countries, including a more granular analysis of sub-national variations. Future studies could also investigate the impact of specific financial inclusion interventions, such as digital financial services or microfinance programs, on achieving particular SDGs. Additionally, incorporating qualitative approaches, such as case studies or interviews, could provide deeper insights into the challenges and success factors influencing financial inclusion. Expanding the scope to include a broader set of countries or regions could also enhance the generalizability of the findings and offer comparative perspectives on financial inclusion's role in sustainable development.

References

Achugamonu, U. B., Adetiloye, K. A., Adegbite, E. O., Babajide, A. A., & Akintola, F. A. (2020). Financial exclusion of bankable adults: Implication on financial inclusive growth among twenty-seven SSA countries. *Cogent Social Sciences, 6*(1), 1730046.

Aghion, P., & Bolton, P. (1997). A theory of trickle-down growth and development. *The Review of Economic Studies, 64*(2), 151–172.

Allen, F., Demirguc-Kunt, A., Klapper, L., & Peria, M. S. M. (2016). The foundations of financial inclusion: Understanding ownership and use of formal accounts. *Journal of Financial Intermediation, 27*, 1–30.

Alosheibat, S. D., Banikhalid, H. H., & Alawneh, A. (2023). Impact of financial inclusion on economic growth during the period 2011–2021: A cross-sectional study of lower-middle-income countries. *International Journal of Economics and Financial Issues, 13*(6), 148–154.

Arora, R. U. (2012). Financial inclusion and human capital in developing Asia: The Australian connection. *Third World Quarterly, 33*(1), 177–197.

Ashraf, N., Karlan, D., & Yin, W. (2010). Female empowerment: Impact of a commitment savings product in The Philippines. *World Development, 38*(3), 333–344.

Azimi, M. N. (2022). New insights into the impact of financial inclusion on economic growth: A global perspective. *PLoS One, 17*(11), e0277730.

Babajide, A. A., Adegboye, F. B., & Omankhanlen, A. E. (2015). Financial inclusion and economic growth in Nigeria. *International Journal of Economics and Financial Issues, 5*(3), 629–637.

Baltagi, B. H. (2008). *Econometric analysis of panel data*. John Wiley & Sons.

Banerjee, R., Donato, R., & Maruta, A. A. (2021). The effects of financial inclusion on development outcomes: New insights from ASEAN and East Asian countries. In *Financial inclusion in Asia and Beyond* (pp. 54–97). Routledge.

Bank, W. (2022). *Financial inclusion*. https://www.worldbank.org/en/topic/financialinclusion/overview

Bashiru, S., Bunyaminu, A., Yakubu, I. N., & Al-Faryan, M. A. S. (2023a). Drivers of financial inclusion: Insights from Sub-Saharan Africa. *Economies, 11*(5), 146.

Bashiru, S., Bunyaminu, A., Yakubu, I. N., Al-Faryan, M. A. S., Mac Carthy, J., & Nyamekye, K. A. (2023b). Quantifying the link between financial development and bank profitability. *Theoretical Economics Letters, 13*(2), 284–296.

Bhardwaj, M., Hedrick-Wong, Y., & Howard, T. (2018). *Financial inclusion for Asia's unbanked*. World Bank Blogs.

Burgess, R., & Pande, R. (2005). Do rural banks matter? Evidence from the Indian social banking experiment. *American Economic Review, 95*(3), 780–795.

Chinoda, T., & Kapingura, F. M. (2024). Digital financial inclusion and economic growth in Sub-Saharan Africa: The role of institutions and governance. *African Journal of Economic and Management Studies, 15*(1), 15–30.

Chitimira, H., & Warikandwa, T. V. (2023). Financial inclusion as an enabler of United Nations sustainable development goals in the twenty-first century: An introduction. In H. Chitimira & T. V. Warikandwa (Eds.), *Financial inclusion and digital transformation regulatory practices in selected SADC countries: South Africa, Namibia, Botswana and Zimbabwe* (pp. 1–22). Springer International Publishing. https://doi.org/10.1007/978-3-031-23863-5_1

Chowdhury, E. K., & Chowdhury, R. (2024). Role of financial inclusion in human development: Evidence from Bangladesh, India and Pakistan. *Journal of the Knowledge Economy, 15*(1), 3329–3354. https://doi.org/10.1007/s13132-023-01366-x

De Haan, J., & Sturm, J.-E. (2017). Finance and income inequality: A review and new evidence. *European Journal of Political Economy, 50*, 171–195.

Demir, A., Pesqué-Cela, V., Altunbas, Y., & Murinde, V. (2022). Fintech, financial inclusion and income inequality: A quantile regression approach. *The European Journal of Finance, 28*(1), 86–107.

Demirguc-Kunt, A., & Klapper, L. (2012). Measuring financial inclusion: The global findex database. *The World Bank Development Policy Research Working Paper*(6025).

Demirguc-Kunt, A., Klapper, L., Singer, D., & Ansar, S. (2018). *The Global Findex Database 2017: Measuring financial inclusion and the fintech revolution*. World Bank Publications.

Ellili, N. O. D. (2024). Financial inclusion and sustainable development: A review and research agenda. *Journal of Financial Services Marketing, 29*, 1345. https://doi.org/10.1057/s41264-024-00269-5

Erlando, A., Riyanto, F. D., & Masakazu, S. (2020). Financial inclusion, economic growth, and poverty alleviation: Evidence from eastern Indonesia. *Heliyon, 6*(10), e05235.

Gupta, A., Chotia, V., & Rao, N. M. (2014). Financial inclusion and human development: A state-wise analysis from India. *International Journal of Economics, Commerce and Management, 2*(5), 1–23.

Hasan, M., Le, T., & Hoque, A. (2021). How does financial literacy impact on inclusive finance? *Financial Innovation, 7*(1), 40.

Hussain, S., Rehman, A., Ullah, S., Waheed, A., & Hassan, S. (2024). Financial inclusion and economic growth: Comparative panel evidence from developed and developing Asian countries. *SAGE Open, 14*(1), 21582440241232585. https://doi.org/10.1177/21582440241232585

Inoue, T., & Hamori, S. (2012). How has financial deepening affected poverty reduction in India? Empirical analysis using state-level panel data. *Applied Financial Economics, 22*(5), 395–408.

Kara, A., Zhou, H., & Zhou, Y. (2021). Achieving the United Nations' sustainable development goals through financial inclusion: A systematic literature review of access to finance across the globe. *International Review of Financial Analysis, 77*, 101833. https://doi.org/10.1016/j.irfa.2021.101833

Kuri, P. K., & Laha, A. (2011). Financial inclusion and human development in India: An inter-state analysis. *Indian Journal of Human Development, 5*(1), 61–77.

Law, S. H., Azman-Saini, W., & Tan, H. B. (2014). Economic globalization and financial development in East Asia: A panel cointegration and causality analysis. *Emerging Markets Finance and Trade, 50*(1), 210–225.

Le, T.-H., Chuc, A. T., & Taghizadeh-Hesary, F. (2019). Financial inclusion and its impact on financial efficiency and sustainability: Empirical evidence from Asia. *Borsa Istanbul Review, 19*(4), 310–322.

Levine, R. (2005). Finance and growth: Theory and evidence. *Handbook of Economic Growth, 1*, 865–934.

Ofoeda, I., Amoah, L., Anarfo, E. B., & Abor, J. Y. (2024). Financial inclusion and economic growth: What roles do institutions and financial regulation play? *International Journal of Finance & Economics, 29*(1), 832–848.

Ohiomu, S., & Ogbeide-Osaretin, E. N. (2019). Financial inclusion and gender inequality reduction: Evidence from sub-Saharan Africa. *The Indian Economic Journal, 67*(3–4), 367–372.

Ong, H. B., Wasiuzzaman, S., Chong, L. L., & Choon, S. W. (2023). Digitalisation and financial inclusion of lower middle-income ASEAN. *Heliyon, 9*(2), e13347.

Ozili, P. K. (2022). Financial inclusion and sustainable development: An empirical association. *Journal of Money and Business, 2*(2), 186–198.

Prina, S. (2015). Banking the poor via savings accounts: Evidence from a field experiment. *Journal of Development Economics, 115*, 16–31.

Razzaq, A., Qin, S., Zhou, Y., Mahmood, I., & Alnafissa, M. (2024). Determinants of financial inclusion gaps in Pakistan and implications for achieving SDGs. *Scientific Reports, 14*(1), 13667.

Salampasis, D., & Mention, A.-L. (2018). FinTech: Harnessing innovation for financial inclusion. In *Handbook of blockchain, digital finance, and inclusion, volume 2* (pp. 451–461). Elsevier.

Sarma, M., & Pais, J. (2011). Financial inclusion and development. *Journal of International Development, 23*(5), 613–628.

Swamy, V. (2014). Financial inclusion, gender dimension, and economic impact on poor households. *World Development, 56*, 1–15.

Wooldridge, J. M. (2010). *Econometric analysis of cross section and panel data*. MIT press.

Yakubu, I. N., & Bunyaminu, A. (2021). Financial inclusion and economic growth in West Africa: The moderating effect of financial openness. *Empirical Economics Letters, 20*(2), 155–164.

Yakubu, I. N., & Musah, A. (2024). The nexus between financial inclusion and bank profitability: A dynamic panel approach. *Journal of Sustainable Finance & Investment, 14*(2), 430–443.

Yakubu, I. N., Salifu, M., & Saani, M. R. (2024). Financial inclusion, corruption control, and inclusive growth: Exploring the dynamic linkages. In *Examining corruption and the sustainable development goals* (pp. 169–180). IGI Global.

Yap, S., Shan Lee, H., & Xin Liew, P. (2023). The role of financial inclusion in achieving financerelated sustainable development goals (SDGs): A cross-country analysis. *Economic research-Ekonomska istraživanja, 36*(3).

Yap, S., Lee, H. S., & Liew, P. X. (2024). The roles of insurance and banking services on financial inclusion. *SAGE Open, 14*(2), 21582440241252268.

Yikun, Z., Leong, L. W., Abu-Rumman, A., Shraah, A. A., & Hishan, S. S. (2023). Green growth, governance, and green technology innovation. How effective towards SDGs in G7 countries? *Economic research-Ekonomska istraživanja, 36*(2).

Part III

Ethical and Inclusive Banking for Sustainable Development

Leveraging Central Bank Digital Currency for Sustainable Development in Emerging Economies: Challenges and Strategic Approaches

9

Ibrahim Nandom Yakubu, Mubarik Abdul Mumin, and Ibrahim Osman Adam

9.1 Introduction

Central bank digital currency (CBDC) has emerged as a revolutionary concept in the realm of digital finance, attracting significant attention from policymakers, economists, and technologists alike. CBDCs represent a digital form of a country's fiat currency, issued and regulated by the central bank (Avgouleas & Blair, 2024; Clarke & Hrnjic, 2024; Lannquist & Tan, 2023). Unlike cryptocurrencies such as Bitcoin, CBDCs are centralized and backed by the full faith and credit of the issuing government. The concept has gained traction in recent years, with over 100 countries exploring or actively developing CBDCs as of 2023 (Allen et al., 2022). This surge in interest is driven by various factors, including the digitalization of economies, the decline in cash usage, and the potential to enhance financial inclusion and monetary policy transmission.

As the world grapples with these technological advancements in finance, it simultaneously faces pressing global challenges addressed by the Sustainable Development Goals (SDGs). Adopted by all United Nations Member States in 2015, the SDGs represent a universal call to action to end poverty, protect the planet, and ensure prosperity for all by 2030 (Halisçelik & Soytas, 2019; Sachs, 2015). The 17 interconnected goals address global challenges such as poverty, inequality, climate change, environmental degradation, peace, and justice. As we approach the midpoint of the SDG timeline, progress has been made in some areas, but significant challenges remain, particularly in developing economies. The COVID-19 pandemic has further exacerbated these challenges, reversing years of progress in some cases (United Nations Department of Economic and Social Affairs, 2016). In this context, innovative financial technologies, including CBDCs, are being explored as potential tools to accelerate progress toward the SDGs.

Developing economies, characterized by lower levels of industrialization, income per capita, and human development indices, face unique challenges in achieving the SDGs. These countries often struggle with limited access to financial services, high levels of poverty and inequality, and inadequate infrastructure (Dahlman & Mealy, 2016; Jaiyesimi, 2016). According to the World Bank, about 1.4 billion adults remain unbanked globally, with the majority residing in developing countries (Demirgüç-Kunt et al., 2022). Furthermore, these economies are often more vulnerable to economic shocks, climate change

I. N. Yakubu · M. A. Mumin (✉) · I. O. Adam
School of Business, University for Development Studies, Tamale, Ghana
e-mail: mamubarik@uds.edu.gh

I. N. Yakubu (ed.), *Strategic Approaches to Banking Business and Sustainable Development Goals*, Sustainable Development Goals Series, https://doi.org/10.1007/978-3-031-80744-2_9

impacts, and global crises. The digital divide, in terms of access to both technology and digital literacy, presents another significant hurdle. However, developing economies also offer fertile ground for innovative solutions, as they can potentially leapfrog traditional financial systems and adopt cutting-edge technologies like CBDCs more readily than developed economies with entrenched financial infrastructures.

This research explores the potential role of Central Bank Digital Currencies in accelerating the achievement of Sustainable Development Goals in developing economies. By examining the intersection of digital finance innovation and sustainable development, we aim to assess how CBDCs can address specific challenges faced by developing countries, such as financial exclusion, inefficient government services, and limited economic opportunities. This study will investigate both the opportunities and the risks associated with CBDC implementation in the context of SDGs, providing insights for policymakers, central banks, and development practitioners. Ultimately, this research seeks to contribute to the growing body of knowledge on how financial technologies can be leveraged to create more inclusive, sustainable, and resilient economies in the developing world.

The rest of this chapter is structured as follows: Sect. 9.2 highlights the understanding of Central Bank Digital Currencies (CBDCs). Section 9.3 examines the Sustainable Development Goals (SDGs) in context. Section 9.4 discusses the potential impact of CBDCs on the SDGs. Section 9.5 reviews case studies on CBDCs. Section 9.6 elaborates on challenges and key considerations associated with CBDCs, while Sect. 9.7 offers significant recommendations for their implementation. Finally, Sect. 9.8 concludes this study.

9.2 Understanding Central Bank Digital Currency (CBDC)

Central bank digital currency (CBDC) represents a groundbreaking form of digital money issued by a country's central bank, designed to serve as a complement or alternative to traditional physical currency and commercial bank deposits. The Bank for International Settlements (BIS) defines CBDC as "a digital payment instrument, denominated in the national unit of account, that is a direct liability of the central bank" (Kosse & Mattei, 2023). This definition encapsulates several key characteristics that distinguish CBDCs from other forms of digital money. Unlike cryptocurrencies or stablecoins issued by private entities, CBDCs are direct liabilities of the central bank, backed by its credibility and the nation's economic strength (Auer et al., 2023). They exist purely in digital form, without physical representation, enabling instant transfers and programmability. In many proposed implementations, CBDCs would have legal tender status, meaning they must be accepted for payments of debts and taxes (European Central Bank, 2023a). Furthermore, CBDCs are designed to be interoperable with existing payment systems and financial infrastructure, and their digital nature allows for the implementation of smart contracts and conditional payments, enabling new functionalities in the financial system (Berg, 2022; Sethaput & Innet, 2023).

CBDCs are generally categorized into two main types: wholesale and retail, each serving different purposes and user groups. Wholesale CBDCs are designed for use by financial institutions and are not accessible to the general public. They are primarily intended to enhance the efficiency and security of interbank settlements and large-value transfers (Secretariat, 2022; Lannquist & Tan, 2023; Bindseil et al., 2021). Key features of wholesale CBDCs include restricted access to authorized financial institutions, potential for 24/7 settlement of high-value transactions, improved liquidity management for banks, and enhanced cross-border payment capabilities. The Bank of France, for instance, has conducted several experiments with wholesale CBDCs, demonstrating their potential to streamline cross-border and cross-currency transactions (Jahan et al., 2022).

In contrast, retail CBDCs, also known as general purpose CBDCs, are designed for use by the general public and businesses for everyday transactions. They are intended to complement or potentially replace physical cash and provide an alternative to commercial bank deposits (Schumacher, 2024; Bindseil et al., 2021). Key features of retail CBDCs include universal accessibility for all citizens and businesses, potential for offline functionality to ensure inclusivity, privacy considerations balanced with regulatory compliance, and integration with existing payment infrastructures. China's digital yuan (e-CNY) is a prominent example of a retail CBDC in the advanced stages of testing, with millions of users participating in pilot programs across multiple cities (Lannquist & Tan, 2023).

The global landscape of CBDC initiatives is rapidly evolving, with countries at various stages of research, development, and implementation. As of 2024, over 100 countries, representing more than 95% of the global GDP, are exploring CBDCs (Brokke & Engen, 2019). Many central banks, including the US Federal Reserve and the European Central Bank, are in the research phase, examining the potential benefits, risks, and technical challenges associated with CBDCs. Several countries have moved beyond research to launch pilot programs. Notable examples include China's e-CNY, which has been tested in multiple cities with millions of users; the Eastern Caribbean Central Bank's (ECCB) DCash, launched in multiple Caribbean nations; and Sweden's e-krona pilot, testing various technical solutions and use cases.

As of 2024, the Bahamas remains one of the few countries to have fully launched a CBDC, with its Sand Dollar in circulation since October 2020 (Colebrook, 2022). Nigeria's eNaira, launched in October 2021, is another example of a fully implemented CBDC. Recognizing the potential impact of CBDCs on cross-border transactions, several international initiatives are underway. The BIS Innovation Hub is coordinating multiple projects, including Project Dunbar and Project mBridge, which explore multi-CBDC arrangements for international settlements (Themistocleous et al., 2023).

The landscape of CBDC initiatives is characterized by diversity in approach, with each country tailoring its CBDC strategy to its specific economic context, policy objectives, and technological capabilities. As more countries progress from research to pilot programs and full implementation, the global financial system is poised for potentially significant transformation, with implications for monetary policy, financial stability, and the future of money itself.

9.3 Sustainable Development Goals in Context

9.3.1 Brief Overview of the 17 SDGs

The Sustainable Development Goals (SDGs), adopted by the United Nations General Assembly in 2015, represent a global commitment to create a more equitable, sustainable, and prosperous world for all. These 17 interconnected goals form the core of the 2030 Agenda for Sustainable Development, encompassing a wide range of social, economic, and environmental issues. From eradicating poverty and hunger to ensuring quality education, gender equality, and climate action, the SDGs are designed to be universal, integrated, and indivisible, recognizing the interconnected nature of global challenges (Weiland et al., 2021).

As we approach the midpoint of this ambitious agenda, progress has been mixed. The United Nations' latest progress report highlights both achievements and setbacks. Significant strides have been made in areas such as maternal and child health, access to electricity, and marine protection. However, progress in many areas has slowed or reversed due to the COVID-19 pandemic, conflicts, and climate change (United Nations, 2024). For instance, global extreme poverty rose in 2020 for the first time in over 20 years, with an estimated 119–124 million people pushed back into extreme poverty in 2020 alone (Millard & Fucci, 2023). This setback underscores the fragility of development gains and the need for sustained, innovative approaches to achieving the SDGs.

9.3.2 Specific Challenges in Achieving SDGs in Developing Economies

Developing economies face particularly acute challenges in achieving the SDGs, often grappling with a complex web of interconnected obstacles. Resource constraints stand out as a primary hurdle, with many developing countries lacking the financial capacity to invest adequately in critical areas such as healthcare, education, and infrastructure (Sachs et al., 2024). The United Nations Conference on Trade and Development (UNCTAD) estimates an annual investment gap of $2.5 trillion in developing countries for achieving the SDGs (UNCTAD, 2023). This financial shortfall is exacerbated by institutional weaknesses, including weak governance structures, corruption, and inefficient public services, which hinder progress across multiple SDGs. The World Bank's Worldwide Governance Indicators consistently show that low-income countries score lower on measures of government effectiveness, regulatory quality, and control of corruption (Nikolic, 2023).

Economic vulnerabilities further compound these challenges, with developing economies often more susceptible to economic shocks, commodity price fluctuations, and debt crises. The COVID-19 pandemic has starkly illustrated this vulnerability, with the International Monetary Fund (IMF) projecting slower economic recovery for emerging markets and developing economies compared to advanced economies (Ruch, 2020). Climate change adds another layer of complexity, disproportionately affecting developing countries, particularly Small Island Developing States and least developed countries. The Global Climate Risk Index reveals that eight of the ten countries most affected by extreme weather events between 2000 and 2019 were developing countries (Germanwatch, 2024).

The digital divide presents yet another significant barrier to progress on several SDGs, including education, innovation, and economic growth. The International Telecommunication Union (ITU) reports that only 27% of people in least developed countries used the Internet in 2022, compared to 90% in developed countries (ITU, 2023). This disparity in access to digital technologies hampers the ability of developing economies to leverage technological advancements for sustainable development. Additionally, demographic pressures, particularly rapid population growth in many developing countries, strain resources and service delivery systems. The UN projects that more than half of the global population growth between now and 2050 will occur in sub-Saharan Africa (United Nations Population Division, 2023), highlighting the need for sustainable development strategies that can keep pace with demographic changes.

9.3.3 The Role of Financial Technology in Sustainable Development

In the face of these challenges, financial technology (fintech) has emerged as a powerful tool for advancing sustainable development, particularly in developing economies. Fintech solutions, such as mobile money and digital banking platforms, have dramatically expanded access to financial services, directly contributing to SDG 1 (No Poverty) and SDG 10 (Reduced Inequalities). For instance, M-Pesa in Kenya has significantly increased financial inclusion, with 96% of households outside Nairobi having at least one M-Pesa account (FSD Kenya, 2023). Beyond financial inclusion, blockchain- and AI-powered fintech solutions are enhancing transparency and efficiency in resource allocation, potentially addressing corruption and improving public service delivery (SDG 16). The World Food Programme's Building Blocks project, which uses blockchain to deliver cash assistance more efficiently to refugees in Jordan, exemplifies this potential (World Food Programme, 2024).

Fintech is also facilitating the growth of sustainable finance by enabling green bonds, impact investing, and environmental, social, and governance (ESG)-focused financial products. For instance, crowdfunding and peer-to-peer lending platforms enabled by fintech are opening new avenues for funding SDG-related projects. The United Nations Development Programme's (UNDP) Digital X Scale Accelerator has leveraged such platforms to fund innovative solutions for the SDGs in multiple developing countries (Ingram et al., 2022).

The vast amounts of data generated by fintech are also being leveraged for development planning and impact assessment. The Global Partnership for Sustainable Development Data highlights how fintech-generated data is being used to track progress on the SDGs and inform policy decisions (Honcharenko et al., 2021). Additionally, digital platforms and mobile apps are making financial education more accessible, contributing to SDG 4 (quality education) and enhancing overall financial literacy. The Organisation for Economic Co-operation and Development (OECD) reports that fintech-based financial education initiatives have reached millions of people in developing countries (Ariana et al., 2024; Fauziyah et al., 2024).

As we consider the potential of Central Bank Digital Currencies (CBDCs) in achieving the SDGs, it's important to view them within this broader context of fintech's role in sustainable development. CBDCs could potentially amplify many of these positive impacts while addressing some of the limitations of current fintech solutions, particularly in terms of financial stability, monetary policy effectiveness, and regulatory oversight. The intersection of CBDCs, sustainable development, and the specific challenges faced by developing economies presents a rich area for exploration, offering the potential to accelerate progress toward the SDGs in ways that were previously unimaginable.

9.4 Potential Impacts of CBDC on SDGs in Developing Economies

Central Bank Digital Currencies (CBDCs) have the potential to significantly impact the achievement of Sustainable Development Goals (SDGs) in developing economies. This section explores the various ways in which CBDCs could contribute to specific SDGs, focusing on financial inclusion, economic growth, gender equality, improved governance, and environmental sustainability.

9.4.1 Financial Inclusion (SDG 1: No Poverty, SDG 10: Reduced Inequalities)

CBDCs have the potential to dramatically improve financial inclusion, contributing directly to SDG 1 (No Poverty) and SDG 10 (Reduced Inequalities). By providing a digital means of payment and store of value, CBDCs can offer unbanked and underbanked populations a secure, low-cost entry point into the formal financial system. The World Bank estimates that 1.4 billion adults remain unbanked globally, with the majority in developing countries (Ozili, 2023). CBDCs can significantly reduce the cost of financial transactions by leveraging digital technology. The Bank for International Settlements (BIS) suggests that CBDCs could lower the cost of remittances, which averaged 6.3% of the transfer amount in 2023, well above the SDG target of 3% (Westermeier, 2018). Furthermore, CBDCs could enhance financial resilience by facilitating more efficient distribution of social benefits and emergency aid. During the COVID-19 pandemic, countries with advanced digital payment systems were able to disburse financial aid more quickly and efficiently (Bidder et al., 2024). In the realm of cross-border transactions, multi-CBDC arrangements could streamline processes, potentially reducing costs and increasing speed. The BIS's Project mBridge demonstrated the potential for CBDCs to reduce cross-border transaction costs by up to 50% (Sethaput & Innet, 2023).

9.4.2 Economic Growth and Job Creation (SDG 8: Decent Work and Economic Growth)

CBDCs could contribute to economic growth and job creation in several ways. By reducing transaction costs and increasing the speed of payments, CBDCs could enhance overall economic productivity. The European Central Bank estimates that the introduction of a digital euro could increase the European Union's GDP by up to 0.5% over a decade (Brokke & Engen, 2019).

CBDCs could also help bring informal economic activities into the formal sector, potentially increasing tax revenues and improving worker protections. The International Labour Organization estimates that the informal economy in developing countries accounts for about 70% of total employment (Tan, 2023). The programmability feature of CBDCs could spur innovation in financial services and beyond. For example, smart contracts enabled by CBDCs could automate various business processes, potentially creating new jobs in the digital economy (World Economic Forum, 2023). Additionally, the digital footprint created by CBDC transactions could help individuals and small businesses build credit histories, potentially improving their access to loans and other financial services (Wenker, 2022).

9.4.3 Gender Equality in Financial Access (SDG 5: Gender Equality)

CBDCs could play a significant role in promoting gender equality in financial access. In some cultures, where women face restrictions on mobility or interactions with male bank staff, CBDCs could provide a means of accessing financial services from home. The Global System for Mobile Communications Association (GSMA) reports that the gender gap in mobile Internet use in low- and middle-income countries narrowed to 16% in 2023, suggesting that digital solutions can help bridge gender divides (GSMA, 2024). CBDCs could provide women with greater control over their finances, potentially increasing their economic participation and decision-making power within households. A study by UN Women found that digital financial services can have this empowering effect (UN Women, 2023). Moreover, the programmability of CBDCs could enable the creation of financial products tailored to women's needs, such as savings accounts with specific goals or micro-insurance products (Alliance for Financial Inclusion, 2024).

9.4.4 Improved Government Services and Reduced Corruption (SDG 16: Peace, Justice, and Strong Institutions)

CBDCs have the potential to enhance governance and reduce corruption. The digital nature of CBDCs allows for better tracking of financial flows, potentially reducing corruption and illicit financial activities. It is estimated that corruption costs developing countries $1.26 trillion per year (Transparency International, 2019). CBDCs could streamline the delivery of government services, including social benefits, reducing leakages, and ensuring that funds reach intended beneficiaries. The Indian government's direct benefit transfer system, which leverages digital payments, saved an estimated $33 billion over 6 years by reducing fraud and inefficiencies. Furthermore, CBDCs could improve tax collection by providing a more comprehensive view of economic activities. The International Monetary Fund (2024) suggests that the digitalization of tax systems in developing countries could increase tax revenues.

9.4.5 Environmental Sustainability Through Digital Transactions (SDG 13: Climate Action)

While the environmental impact of CBDCs is a subject of ongoing research, they could potentially contribute to environmental sustainability. CBDCs could reduce the need for physical cash, potentially decreasing the environmental impact of cash production and transportation. The European Central Bank estimates that producing and distributing euro banknotes generates about 500,000 tons of CO_2 equivalent per year (ECB, 2023b). Unlike some cryptocurrencies that use energy-intensive consensus mechanisms, CBDCs can be designed to be more energy-efficient. The People's Bank of China (PBOC) claims that its digital yuan uses a fraction of the energy consumed by major cryptocurrencies (Hamilton, 2024). Additionally, CBDCs could potentially facilitate the implementation and tracking of green finance initiatives. The Bank of England is exploring how CBDCs could support the transition to a net-zero economy by enabling more efficient allocation of capital to sustainable projects (Tan, 2023).

While these potential impacts are promising, it's important to note that the actual effects of CBDCs will depend on their specific design, implementation, and the broader policy environment in which they operate. As developing economies consider CBDC initiatives, careful consideration must be given to potential risks and challenges, including privacy concerns, cybersecurity risks, and the need for digital literacy and infrastructure development.

9.5 Case Studies

As Central Bank Digital Currencies (CBDCs) gain traction globally, several developing countries have launched pilot programs or full implementations. This section examines key case studies, analyzing their approaches, outcomes, and the lessons learned.

9.5.1 Analysis of CBDC Pilots or Implementations in Developing Countries

While China is not typically classified as a developing country, its CBDC implementation offers valuable insights for developing economies. The People's Bank of China (PBOC) began piloting the e-CNY in 2020 and has since expanded to multiple cities. The e-CNY operates on a two-tier system where the PBOC issues e-CNY to authorized operators (commercial banks), who then distribute it to the public. It features controllable anonymity, allowing small transactions to be anonymous while larger ones require identification. Notably, the e-CNY can be used without an Internet connection. As of December 2023, the e-CNY pilot had over 300 million individual users and 5.6 million merchant accounts, with transaction volumes exceeding 1.8 trillion yuan (approximately $250 billion) (Mu, 2023). The e-CNY has been used for various purposes, including retail payments, government subsidies, and transportation fees.

Nigeria launched the eNaira in October 2021, becoming the first African country to introduce a CBDC. The eNaira features a tiered Know Your Customer (KYC) system with different wallet tiers having varying transaction limits based on the level of user identification. It integrates with existing banking apps, allowing users to access eNaira through their regular banking applications. The eNaira also offers Unstructured Supplementary Service Data (USSD) functionality, enabling access via feature phones, not just smartphones. As of January 2024, about 13 million eNaira wallets had been created, with transaction volumes reaching 10 billion naira (approximately $22 million) (Ree, 2023). However, adoption has been slower than anticipated, with challenges including low public awareness and technical issues.

The Eastern Caribbean Central Bank (ECCB) launched DCash in March 2021, making it the first currency union to issue a CBDC. DCash is implemented across multiple islands in the Eastern Caribbean, in collaboration with Bitt, a fintech company. It is designed with a focus on disaster recovery, facilitating quick fund transfers in case of natural disasters. However, the system faced a significant outage in January 2022, highlighting the importance of robust technical infrastructure.

The Bahamas launched the Sand Dollar in October 2020, becoming one of the first countries globally to fully implement a CBDC. The Sand Dollar is designed with a focus on financial inclusion, aiming to serve the unbanked population across the country's many islands. It offers interoperability across different wallet providers and supports offline functionality, allowing transactions without Internet connectivity. By the end of 2023, approximately 20% of the Bahamian population had adopted the Sand Dollar, with transaction volumes steadily increasing (Central Bank of The Bahamas, 2024). The CBDC has been particularly useful in facilitating payments on smaller, less-connected islands.

9.5.2 Lessons Learned and Best Practices

The experiences of these early CBDC adopters offer valuable lessons and best practices for other developing countries considering CBDC implementation. Nigeria's experience with the eNaira highlights the importance of prioritizing user experience and education. CBDCs should be easily accessible and understandable to the general public, with significant resources devoted to public awareness and user-friendly interfaces. The DCash outage underscores the need for ensuring robust technical infrastructure. Adequate investment in resilient and well-tested systems is crucial for CBDC success.

Successful implementations like China's e-CNY and the Bahamas' Sand Dollar demonstrate the importance of focusing on specific use cases, such as government payments or serving unbanked populations. The public–private partnership seen in the DCash implementation shows that collaboration with the private sector can leverage existing expertise and infrastructure effectively. The approaches to tiered KYC and privacy in China's and Nigeria's CBDCs demonstrate the importance of balancing regulatory requirements with user privacy concerns.

Ensuring interoperability, as seen in Nigeria and the Bahamas, is crucial for widespread adoption. The ability to integrate with existing financial systems can significantly boost CBDC uptake. Given infrastructure challenges in many developing countries, offline capability, as implemented in China and the Bahamas, is essential for inclusivity. As digital systems, CBDCs are vulnerable to cyber threats, making robust security measures crucial. A Bank for International Settlements study emphasizes the importance of addressing cybersecurity concerns in CBDC design and implementation (BIS, 2023a).

Environmental considerations should not be overlooked. A United Nations Environment Programme report highlights that the energy efficiency of CBDC systems, compared to some cryptocurrencies, should be a key consideration in their design and deployment (Suratini et al., 2024). While most current implementations focus on domestic use, an International Monetary Fund working paper suggests that the potential for cross-border transactions should be considered in the design phase of CBDCs (Ginneken, 2019).

These case studies and lessons learned provide valuable insights for other developing countries considering CBDC implementation. However, it is important to note that each country's unique economic, social, and technological context will necessitate a tailored approach to CBDC design and deployment. As more countries progress with their CBDC initiatives, continued research and knowledge sharing will be crucial in refining best practices and maximizing the potential of CBDCs to contribute to sustainable development in the developing world.

9.6 Challenges and Considerations

While Central Bank Digital Currencies (CBDCs) offer significant potential benefits for developing economies, their implementation also presents several challenges and important considerations. This section explores these issues in detail, focusing on technological infrastructure, regulatory frameworks, privacy and security concerns, the digital skills gap, and potential impacts on traditional banking systems.

9.6.1 Technological Infrastructure Requirements

Implementing a CBDC requires robust technological infrastructure, which can be a significant challenge for many developing economies. The Bank for International Settlements (BIS) emphasizes that a reliable and widespread telecommunications network is crucial for CBDC adoption (BIS, 2023b). However, many developing countries still struggle with limited Internet connectivity, especially in rural areas. The International Telecommunication Union (ITU) reports that as of 2023, only 36% of individuals in least developed countries used the Internet, compared to 92% in developed countries (ITU, 2023).

Beyond basic connectivity, CBDC systems require significant computational power and data storage capabilities. Cloud computing solutions might offer a way forward, but they come with their own set of challenges related to data sovereignty and security. The World Bank's Digital Economy for Africa initiative highlights the need for substantial investments in digital infrastructure to support financial innovation in developing economies (Resilient Digital Africa, 2024).

Moreover, to ensure financial inclusion, CBDC systems must be designed to work on a variety of devices, including feature phones, and potentially in offline modes. The GSMA's Mobile Economy Report 2024 indicates that while smartphone adoption is growing in developing countries, feature phones still account for a significant portion of mobile devices in use (Joiner et al., 2024).

9.6.2 Regulatory and Legal Frameworks

The introduction of CBDCs necessitates the development of comprehensive regulatory and legal frameworks. Many developing countries may need to update their existing financial regulations or create entirely new ones to accommodate CBDCs. The International Monetary Fund (IMF) emphasizes the need for clear legal foundations for CBDCs, including defining their legal tender status and the central bank's authority to issue digital currency (Bossu et al., 2020).

Regulatory challenges also extend to cross-border transactions. As CBDCs have the potential to facilitate easier cross-border payments, international cooperation and regulatory harmonization become crucial. The Financial Stability Board (FSB) has called for enhanced cross-border cooperation in CBDC regulation to address potential risks to global financial stability (FSB, 2024).

Another key regulatory consideration is the prevention of money laundering and terrorist financing. The Financial Action Task Force (FATF) has issued guidelines for a risk-based approach to virtual assets, including CBDCs, emphasizing the need for robust Know Your Customer (KYC) and anti-money laundering (AML) procedures (FATF, 2021).

9.6.3 Privacy and Security Concerns

Privacy and security are paramount concerns in CBDC implementation. Unlike physical cash, which offers anonymity, digital transactions can potentially be tracked, raising privacy concerns. The European Central Bank's report on the digital euro emphasizes the need to balance transaction privacy with the Anti-Money Laundering and Combating the Financing of Terrorism (AML/CFT) requirements (Thompsett, 2024).

Cybersecurity is another critical concern. As digital systems, CBDCs are vulnerable to cyberattacks, hacking, and fraud. The cost of cybercrime is projected to reach $10.5 trillion annually by 2025, according to Cybersecurity Ventures (2021). Developing economies, which may have less advanced cybersecurity infrastructure, could be particularly vulnerable.

Moreover, the centralized nature of CBDCs raises concerns about potential government surveillance and control. Human Rights Watch has warned about the potential misuse of CBDCs for financial surveillance and control, particularly in countries with weak democratic institutions (Jiang, 2023).

9.6.4 Financial Literacy and Digital Skills Gap

The successful implementation of CBDCs requires a population with adequate financial literacy and digital skills. However, many developing countries face significant challenges in this area. The OECD/INFE 2023 International Survey of Adult Financial Literacy reveals that financial literacy levels in developing countries are generally lower than those in developed countries (OECD, 2023).

The digital skills gap compounds this challenge. The United Nations Educational, Scientific and Cultural Organization (UNESCO) Global Education Monitoring Report 2024 highlights the persistent digital divide, with many adults in developing countries lacking basic digital skills (UNESCO, 2024). This gap could lead to the exclusion of significant portions of the population from CBDC-based financial systems, potentially exacerbating existing inequalities.

Addressing this challenge requires comprehensive education and training programs. The G20 Global Partnership for Financial Inclusion emphasizes the need for digital financial literacy initiatives to support the adoption of digital financial services, including CBDCs (G20 GPFI, 2023).

9.6.5 Potential Negative Impacts on Traditional Banking Systems

The introduction of CBDCs could have significant impacts on traditional banking systems, particularly in developing economies where banking sectors may be less robust. One key concern is the potential for bank disintermediation. If individuals and businesses can hold CBDCs directly with the central bank, they might reduce their deposits in commercial banks, potentially affecting banks' ability to lend and their overall stability.

The Bank for International Settlements warns that in times of financial stress, the ease of converting bank deposits to CBDC could accelerate bank runs (BIS, 2023a). This risk could be particularly acute in developing economies with less stable banking sectors.

Furthermore, CBDCs could impact banks' business models and profitability (Auer et al., 2024). A report by McKinsey & Company suggests that banks might need to offer new, value-added services to remain competitive in a CBDC environment. This could be challenging for banks in developing economies with limited resources for innovation.

The potential impact on monetary policy transmission is another consideration. While CBDCs could enhance the effectiveness of monetary policy by providing central banks with more direct tools, they could also complicate policy implementation, particularly in developing economies with less sophisticated financial markets (Yang & Zhou, 2022).

In conclusion, while CBDCs offer significant potential benefits for developing economies, their implementation comes with substantial challenges. Addressing these challenges requires careful planning, significant investment in infrastructure and education, and the development of robust regulatory frameworks. Moreover, it necessitates a balanced approach that leverages the benefits of CBDCs while mitigating potential risks to financial stability and individual privacy. As developing economies consider CBDC implementation, these challenges and considerations should be at the forefront of policy discussions and planning processes.

9.7 Recommendations for Implementation

The implementation of Central Bank Digital Currencies (CBDCs) in developing economies requires careful planning and strategic execution. This section provides recommendations for policymakers, central banks, and other stakeholders involved in CBDC implementation, drawing on lessons from early adopters and insights from international organizations.

9.7.1 Policy Considerations for Central Banks and Governments

Central banks and governments must carefully consider several key policy aspects when implementing CBDCs. Firstly, it's crucial to clearly define objectives and use cases for the CBDC, aligning them with broader national development goals. The Bank for International Settlements (BIS) emphasizes that CBDCs should be designed to serve specific policy objectives, such as enhancing financial inclusion or improving payment system efficiency (Lovegrove, 2023). For instance, Nigeria's eNaira was explicitly designed to increase financial inclusion and facilitate remittances. Adopting a phased approach is also recommended, as implementing a CBDC is a complex process that benefits from gradual rollout. The International Monetary Fund (IMF) suggests starting with pilot programs in limited geographical areas or specific use cases before full-scale implementation. China's approach with the e-CNY, involving multiple pilot phases across different cities, provides a good example of this strategy.

Ensuring interoperability is another critical consideration. CBDCs should be designed to be interoperable with existing payment systems and potentially with other countries' CBDCs. The G20's Roadmap for Enhancing Cross-border Payments emphasizes the importance of interoperability in improving the efficiency of cross-border transactions (Bank for International Settlements, 2022a). While promoting innovation, central banks must also balance this with maintaining financial stability. The European Central Bank's report on the digital euro highlights the need to design CBDCs in a way that prevents excessive shifts from bank deposits to CBDC holdings (ECB, 2023b). Lastly, developing comprehensive legal and regulatory frameworks is essential. Governments need to establish clear legal foundations for CBDCs, including defining their status as legal tender and the central bank's authority to issue digital currency. The World Bank's guide on CBDC implementation stresses the importance of updating existing financial regulations to accommodate CBDCs.

9.7.2 Strategies for Overcoming Challenges

Implementing CBDCs in developing economies comes with unique challenges that require strategic solutions. To address technological infrastructure challenges, governments should prioritize investments in digital infrastructure. The United Nations Conference on Trade and Development (UNCTAD) recommends that developing countries increase their efforts in information and communication technology (ICT) infrastructure development (United Nations, 2021). Public–private partnerships can be an effective way to finance these investments. Balancing financial inclusion with regulatory compliance can be achieved through implementing tiered Know Your Customer (KYC) requirements and wallet limits. The Financial Action Task Force (FATF) guidelines support this approach, allowing for simplified due diligence for lower-risk, small-value accounts (FATF, 2024).

To ensure inclusivity, especially in areas with limited connectivity, CBDCs should have offline functionality. The Bank of Ghana's planned CBDC, the e-Cedi, is being designed with offline capabilities to serve rural areas with limited Internet access. Given the critical nature of CBDC systems, prioritizing cybersecurity is essential. The International Telecommunication Union (ITU) provides guidelines for developing countries on strengthening their national cybersecurity capabilities, which can be applied to CBDC systems (ITU, 2018). Addressing the digital skills gap requires implementing comprehensive digital and financial literacy programs. UNESCO's Strategy on Technological Innovation in Education provides a framework for integrating digital skills into educational curricula.

9.7.3 Collaboration with International Organizations and the Private Sector

Collaboration with international organizations and the private sector is crucial for successful CBDC implementation in developing economies. Engaging with international organizations can provide access to valuable expertise and resources. The IMF and World Bank offer technical assistance programs for countries exploring CBDCs, while the BIS Innovation Hub provides a platform for central banks to collaborate on financial technology projects, including CBDCs.

Fostering public–private partnerships can bring technological expertise and innovative solutions to CBDC projects. The World Economic Forum's CBDC Policy-Maker Toolkit emphasizes the importance of engaging with the private sector throughout the CBDC development process (World Economic Forum, 2020).

Participating in international CBDC initiatives can provide valuable insights and promote interoperability. For example, Project Dunbar, a multi-CBDC initiative involving the central banks of Australia, Malaysia, Singapore, and South Africa, explores the use of CBDCs for international settlements (BIS, 2022a, b). Collaboration with fintech companies can help in developing user-friendly CBDC wallets and innovative use cases. The Eastern Caribbean Central Bank's collaboration with the fintech company Bitt in developing DCash demonstrates the potential of such partnerships. Lastly, engaging in knowledge sharing is crucial. Developing countries should actively participate in international forums and knowledge-sharing platforms focused on CBDCs. The CBDC Think Tank, an initiative supported by the Bill & Melinda Gates Foundation, provides a platform for policymakers from developing countries to exchange experiences and best practices in CBDC implementation.

In a nutshell, the successful implementation of CBDCs in developing economies requires a multi-faceted approach that combines clear policy objectives, strategic planning, and collaboration with various stakeholders. By carefully considering these recommendations and adapting them to their specific contexts, developing countries can harness the potential of CBDCs to promote financial inclusion, enhance payment system efficiency, and support broader economic development goals. However, it is crucial to maintain a flexible approach, continuously learning from experiences and adjusting strategies as the CBDC landscape evolves.

9.8 Conclusion

In conclusion, Central Bank Digital Currencies represent a powerful tool with the potential to significantly contribute to the achievement of multiple Sustainable Development Goals in developing economies. Their ability to enhance financial inclusion, promote economic growth, improve government services, and potentially support environmental sustainability aligns closely with several SDG targets. However, it is crucial to recognize that CBDCs are not a panacea for development challenges. Their effectiveness in supporting SDGs will depend on thoughtful design, careful implementation, and integration with broader development strategies. As noted by the United Nations Development Programme (UNDP), "CBDCs should be viewed as part of a holistic approach to digital financial transformation that puts people and sustainability at the center." Moreover, the potential of CBDCs must be balanced against risks and challenges, including privacy concerns, cybersecurity threats, and potential disruptions to existing financial systems. Addressing these challenges will require ongoing research, international cooperation, and adaptive policymaking. While CBDCs offer significant promise for supporting sustainable development in emerging economies, their success will ultimately depend on how well they are designed and implemented to meet the specific needs and contexts of each country. As developing economies continue to explore and implement CBDCs, it will be crucial to maintain a focus on inclusivity, sustainability, and the broader goals of the SDG agenda. The journey of CBDCs in developing economies is just beginning, and their full potential to support sustainable development remains to be realized. As we move forward, continuous learning, adaptation, and collaboration will be key to harnessing the power of this innovative financial technology for the benefit of all.

References

Allen, F., Gu, X., & Jagtiani, J. (2022). Fintech, cryptocurrencies, and CBDC: Financial structural transformation in China. *Journal of International Money and Finance, 124*, 102625.

Alliance for Financial Inclusion. (2024). Digital Financial Services for Women: Policy Framework.

AFI Global. Available at: https://www.afi-global.org/wp-content/uploads/2019/10/AFI_DFS_Women-FI_AW_25.09.19_digital.pdf

Ariana, I., Wiksuana, I., Candraningrat, I., & Baskara, I. (2024). The effects of financial literacy and digital literacy on financial resilience: Serial mediation roles of financial inclusion and financial decisions. *Uncertain Supply Chain Management, 12*(2), 999–1014.

Auer, R., Böhme, R., Clark, J., & Demirag, D. (2023). Mapping the privacy landscape for central bank digital currencies. *Communications of the ACM, 66*(3), 46–53.

Auer, S., Branzoli, N., Ferrero, G., Ilari, A., Palazzo, F., & Rainone, E. (2024). *CBDC and the banking system.* Jahrbücher für Nationalökonomie und Statistik. Available at: https://www.degruyter.com/document/doi/10.1515/jbnst-2024-0006/html

Avgouleas, E., & Blair, W. (2024). A critical evaluation of central bank digital currencies (CBDCs): Payments' final frontier? *Capital Markets Law Journal, 19*(2), 103–112.

Bank for International Settlements. (2022a). *Options for access to and interoperability of CBDCs for cross-border payments.* Available at: https://www.bis.org/publ/othp52.pdf

Bank for International Settlements. (2022b). *Project Dunbar – International settlements using multi-CBDCs.* BIS Innovation Hub. Available at: https://www.bis.org/publ/othp47.pdf

Bank for International Settlements. (2023a). *Central bank digital currency (CBDC) information security and operational risks to central banks.* Available: https://www.bis.org/publ/othp81.pdf

Bank for International Settlements. (2023b). *Lessons learnt on CBDCs Report submitted to the G20 Finance Ministers and Central Bank Governors.* Available at: https://www.bis.org/publ/othp73.pdf

Berg, C. (2022). *Interoperability as a critical design choice for central bank digital currencies.* Available at SSRN 4205405.

Bidder, R. M., Jackson, T. P., & Rottner, M. (2024). *CBDC and banks: Disintermediating fast and slow.* Available at: https://www.bundesbank.de/resource/blob/931090/be2be8b2c5324245e4147d6306689312/mL/2024-04-29-dkp-15-data.pdf

Bindseil, U., Panetta, F., & Terol, I. (2021). Central Bank digital currency: Functional scope, pricing and controls. *ECB Occasional Paper*, (2021/286).

Bossu, W., Itatani, M., Margulis, C., Rossi, A., Weenink, H., & Yoshinaga, A. (2020). *Legal aspects of central bank digital currency: Central bank and monetary law considerations.* Available at: https://www.imf.org/en/Publications/WP/Issues/2020/11/20/Legal-Aspects-of-Central-Bank-Digital-Currency-Central-Bank-and-Monetary-Law-Considerations-49827

Brokke, O. G. J., & Engen, N. E. (2019). *Central bank digital currency (CBDC): An explorative study on its impact and implications for monetary policy and the banking sector.* Master's thesis. Norwegian School of Economics.

Clarke, G., & Hrnjic, E. (2024). Central bank digital currencies in emerging economies: Practical considerations. In *The Elgar companion to decentralized finance, digital assets, and blockchain technologies* (pp. 96–112). Edward Elgar Publishing.

Central Bank of The Bahamas. (2024). Public Update on The Bahamas Digital Currency SandDollar. Available at: https://www.centralbankbahamas.com/news/press-releases/press-release-public-update-on-the-bahamas-digital-currencysanddollar-1

Colebrook, C. R. (2022). *The digitalization of money in The Bahamas.* Central Bank of the Bahamas.

Cybersecurity Ventures. (2021). *Cybercrime facts and statistics.* Available at: https://cybersecurityventures.com/wp-content/uploads/2021/01/Cyberwarfare-2021-Report.pdf

Dahlman, C., & Mealy, S. (2016). Obstacles to achieving the sustainable development goals: Emerging global challenges and the performance of the least developed countries. In *Achieving the Istanbul Programme of Action by 2020* (p. 40).

Demirgüç-Kunt, A., Klapper, L., Singer, D., & Ansar, S. (2022). *The Global Findex Database 2021: Financial inclusion, digital payments, and resilience in the age of COVID-19.* World Bank Publications.

European Central Bank. (2023a). *ECB welcomes European Commission legislative proposals on digital euro and cash.* Available at: https://www.ecb.europa.eu/press/pr/date/2023/html/ecb.pr230628~e76738d851.en.html

European Central Bank. (2023b). *Report on a digital euro.* Available at: https://www.ecb.europa.eu/pub/pdf/other/Report_on_a_digital_euro~4d7268b458.en.pdf

FATF. (2021). *Updated guidance for a risk-based approach to virtual assets and virtual asset service providers.* Available at: https://www.fatf-gafi.org/content/dam/fatf-gafi/guidance/Updated-Guidance-VA-VASP.pdf.coredownload.inline.pdf

Fauziyah, A., Budiman, A., Sugiharti, H., Sartika, S. H., Khairunnisa, S., & Ruhayati, S. A. (2024, June). Digital financial literacy and digital financial inclusion: Efforts to overcome challenges after the Covid-19 pandemic and increase the independence of financial management. In *8th global conference on business, management, and entrepreneurship (GCBME 2023)* (pp. 115–121). Atlantis Press.

Financial Action Task Force. (2024). *Opportunities and challenges of new technologies for AML/CFT.* FATF Report. Available at: https://www.fatf-gafi.org/en/publications/Digitaltransformation/Opportunities-challenges-new-technologies-for-aml-cft.html

Financial Stability Board. (2024). *FSB consults on recommendations related to data flows and regulation and supervision in cross-border payments.* Available at: https://www.fsb.org/2024/07/fsb-consults-on-recommendations-related-to-data-flows-and-regulation-and-supervision-in-cross-border-payments/

FSD Kenya. (2023). FinAccess Household Survey. Available at: https://www.fsdkenya.org/research-and-publications/finaccess/

G20 Global Partnership for Financial Inclusion. (2023). *Digital financial inclusion: Emerging policy approaches*. G20 GPFI. Available at: https://www.gpfi.org/sites/gpfi/files/documents/Digital%20Financial%20Inclusion-CompleteReport-Final-A4.pdf

GSMA. (2024). Mobile Gender Gap Report. Available at: https://www.gsma.com/r/gender-gap/

Germanwatch. (2024). Global Climate Risk Index 2021. Available at: https://www.germanwatch.org/en/19777

Ginneken, C. L. (2019). *Settlement of cross-border transactions through central bank digital currency (CBDC): Analysis from a risk management perspective*. Master's thesis,. University of Twente.

Halisçelik, E., & Soytas, M. A. (2019). Sustainable development from millennium 2015 to sustainable development goals 2030. *Sustainable Development, 27*(4), 545–572.

Hamilton, C. (2024). Money is morphing-cryptocurrency can morph to be an environmentally and financially sustainable alternative to traditional banking. *DePaul Business & Commercial Law Journal, 22*(1), 5.

Honcharenko, I., Berezina, O., Berezhna, L., & Zhuk, V. (2021). Global partnership for sustainable development: Challenges and perspectives for Ukraine. In *SHS web of conferences* (Vol. 92, p. 09005). EDP Sciences.

Ingram, G., McArthur, J. W., & Vora, P. (2022). *How can digital public technologies accelerate progress on the sustainable development goals*. Brookings. Last modified May, 25.

International Monetary Fund. (2024). *IMF country report*. Available at: https://www.elibrary.imf.org/view/journals/002/2024/223/002.2024.issue-223-en.xml?cid=551877-com-dsp-crossref

International Telecommunication Union. (2018). *Guide to Developing A National Cybersecurity Strategy*. Available at: https://www.itu.int/dms_pub/itu-d/opb/str/D-STR-CYB_GUIDE.01-2018-PDF-E.pdf

International Telecommunication Union. (2023). *Measuring digital development*. Available at: https://www.itu.int/itu-d/reports/statistics/2023/10/10/ff23-foreword/

Jahan, S., Loukoianova, E., Papageorgiou, E., Che, N., Goel, A., Li, M., ... & Zhou, Y. S. (2022). *Towards central bank digital currencies in Asia and the Pacific*. Available at: https://www.imf.org/en/Publications/fintech-notes/Issues/2022/09/27/Towards-Central-Bank-Digital-Currencies-in-Asia-and-the-Pacific-Results-of-a-Regional-Survey-523914

Jaiyesimi, R. (2016). The challenge of implementing the sustainable development goals in Africa: The way forward. *African Journal of Reproductive Health, 20*(3), 13–18.

Jiang, J. (2023). Privacy implications of central bank digital currencies. *Seton Hall Law Review, 54*, 69.

Joiner, J., Okeleke, K., Borole, S., & Ballon, H. F. A. (2024). *The mobile economy 2024*. Available at: https://data.gsmaintelligence.com/research/research/research-2024/the-mobile-economy-2024

Kosse, A., & Mattei, I. (2023). *Making headway-results of the 2022 BIS survey on central bank digital currencies and crypto*. BIS Papers. Available at: https://www.bis.org/publ/bppdf/bispap136.pdf

Lannquist, A., & Tan, B. (2023). *Central bank digital currency's role in promoting financial inclusion* (pp. 1–33). International Monetary Fund.

Lovegrove, S. (2023). *BIS paper on central bank digital currencies: Ongoing policy perspectives*. Available at: https://www.regulationtomorrow.com/eu/bis-paper-on-central-bank-digital-currencies-ongoing-policy-perspectives/

Millard, J., & Fucci, V. (2023). The role of social innovation in tackling global poverty and vulnerability. *Frontiers in Sociology, 8*, 966918.

Mu, C. (2023). Theories and practice of exploring China's e-CNY. In *Data, digitalization, decentialized finance and central bank digital currencies: The future of banking and money* (pp. 179–190).

Nikolic, A. M. (2023). Global progress report on sustainable development goal 16 indicators: A wake-up call for action on peace, justice and inclusion (UNDOC, OHCHR, UNDP, 2023). *Drepturile Omului, 86*.

OECD. (2023). *OECD/INFE 2023 international survey of adult financial literacy*. Available at: https://www.oecd.org/en/publications/oecd-infe-2023-international-survey-of-adult-financial-literacy_56003a32-en.html

Ozili, P. K. (2023). Central bank digital currency, poverty reduction and the United Nations sustainable development goals. In *Embracing business sustainability through innovation and creativity in the service sector* (pp. 175–183). IGI Global.

Ree, J. (2023). *Nigeria's eNaira, one year after*. International Monetary Fund. Available at: https://www.imf.org/en/Publications/WP/Issues/2023/05/16/Nigerias-eNaira-One-Year-After-533487

Resilient Digital Africa. (2024). *Mobilizing $86 billion to connect the entire Africa*. Available at: https://resilient.digital-africa.co/en/blog/2024/05/02/mobilizing-86-billion-to-connect-the-entire-africa/

Ruch, F. U. (2020). Prospects, risks, and vulnerabilities in emerging and developing economies: Lessons from the past decade. *World Bank Policy Research Working Paper, 9181*.

Sachs, J. D. (2015). Achieving the sustainable development goals. *Journal of International Business Ethics, 8*(2), 53–62.

Sachs, J. D., Lafortune, G., & Fuller, G. (2024). The SDGs and the UN summit of the future. *Sustainable Development Report*.

Schumacher, L. V. (2024). Central Bank digital currencies (CBDCs): Exploring characteristics, risks and benefits. In *Decoding digital assets: Distinguishing the dream from the dystopia in stablecoins, tokenized deposits, and central bank digital currencies* (pp. 81–157). Cham.

Secretariat, C. P. M. I. (2022). *Options for access to and interoperability of CBDCs for cross-border payments*. Available at: https://policycommons.net/artifacts/4592296/options-for-access-to-and-interoperability-of-cbdcs-for-cross-border-payments/5416691/

Sethaput, V., & Innet, S. (2023). Blockchain application for central bank digital currencies (CBDC). *Cluster Computing, 26*(4), 2183–2197.

Suratini, S., El Wafa, F., Ariyanti, R., Hastomo, D. A., & Arif, D. (2024). The impact of digital currencies, trade policies, and climate change mitigation on global economic stability. *The Journal of Academic Science, 1*(3), 152–158.

Tan, B. (2023). *Central Bank digital currency adoption: A two-sided model*. International Monetary Fund.

Themistocleous, M., da Cunha, P. R., Tabakis, E., & Papadaki, M. (2023). Towards cross-border CBDC interoperability: Insights from a multivocal literature review. *Journal of Enterprise Information Management, 36*(5), 1296–1318.

Thompsett, L. (2024). *ECB releases digital Euro preparation report*. Available at: https://fintechmagazine.com/articles/ecb-releases-digital-euro-preparation-report

Transparency International. (2019). Resilient institutions. Available at: https://www.transparency.org/en/news/resilient-institutions.

UNESCO. (2024). *Global education monitoring report 2024*. UNESCO. Available at: https://www.unesco.org/gem-report/en/2024genderreport

United Nations Department of Economic and Social Affairs. (2016). *The sustainable development goals report 2016*. United Nations. Available at: https://mdgs.un.org/sdgs/report/2016/The%20Sustainable%20Development%20Goals%20Report%202016.pdf

United Nations. (2021). *Information economy report 2011*. Available at: https://unctad.org/system/files/official-document/tir2020_en.pdf

United Nations. (2024). The Sustainable Development Goals Report 2024. Available at: https://unstats.un.org/sdgs/report/2024/The-Sustainable-Development-Goals-Report-2024.pdf

United Nations Conference on Trade and Development. (2023). SDG Investment Trends Monitor. Available at: https://unctad.org/publication/sdg-investment-trends-monitor-issue-4

United Nations Population Division. (2023). World Population Prospects 2022: Summary of Results. Available at: https://www.un.org/development/desa/pd/content/World-Population-Prospects-2022

UN Women. (2023). Driving Digital Financial Transformation in Support of SDG 5: Recent Gains and Remaining Challenges. Available at: https://www.unwomen.org/sites/default/files/2022-12/EP.10_Yasmin%20Bin-Humam.pdf

Weiland, S., Hickmann, T., Lederer, M., Marquardt, J., & Schwindenhammer, S. (2021). The 2030 agenda for sustainable development: Transformative change through the sustainable development goals? *Politics and Governance, 9*(1), 90–95.

Wenker, K. (2022). Retail central bank digital currencies (CBDC), disintermediation and financial privacy: The case of the Bahamian sand dollar. *FinTech, 1*(4), 345–361.

Westermeier, C. (2018). The Bank of International Settlements as a think tank for financial policy-making. *Policy and Society, 37*(2), 170–187.

World Economic Forum. (2020). Central Bank Digital Currency Policy-Maker Toolkit. Available at: https://www.weforum.org/publications/central-bank-digital-currency-policy-maker-toolkit/

World Economic Forum. (2023). *Central bank digital currency policy-maker toolkit*. Available at: https://www3.weforum.org/docs/WEF_CBDC_Policymaker_Toolkit.pd

World Food Programme. (2024). Building Blocks: Blockchain network for humanitarian assistance. Available at: https://innovation.wfp.org/project/building-blocks.

Yang, J., & Zhou, G. (2022). A study on the influence mechanism of CBDC on monetary policy: An analysis based on e-CNY. *PLoS One, 17*(7), e0268471.

10 Ethical Banking and Growth of Small and Medium Enterprises in Emerging Economies

Khadijah Iddrisu, Eric Kofi Boadi, Stephanie Giamporcaro, and James Ntiamoah Doku

10.1 Introduction

The United Nations has introduced policies known as the Sustainable Development Goals (SDGs), which are targeted to be achieved by 2030. Numerous countries have also adopted strategies to help actualise these SDGs within the set timeframe. Small and medium enterprises (SMEs) have been recognised as crucial instruments for realising these goals in emerging economies. This is because SMEs tend to be the backbone of some of the emerging economies. For example, SMEs are significant drivers of employment and contribute about 70% to Ghana's gross domestic product (GDP) (Amoah et al., 2022; Abor & Quartey, 2010). In Côte d'Ivoire, about 98% of local businesses are comprised of SMEs, with about 18% of modern employment opportunities and about 18% of the GDP attributed to them (Hongbo et al., 2018). Even developed economies like the United Kingdom have recognised the crucial role of SMEs, which contribute 27 million jobs and £4.5 trillion in annual turnover; the UK declared 2024 as the Year of SMEs.[1] However, despite the potential of SMEs, they are confronted with many challenges that seek to affect their growth. Among these challenges, access to financial services, particularly credit facilities, has been a significant burden for them (Iddrisu, 2024; Iddrisu et al., 2023; Ayyagari et al., 2017).

Primarily, SMEs often face difficulties accessing financial services because they are perceived as high-risk clients, due to their lack of collateral, formal documentation and properly maintained accounts (Doku et al., 2023; Iddrisu et al., 2022; Abor & Quartey, 2010). However, efforts to advance environmental sustainability and equality have resulted in the development of new banking practice options which emerged in the twentieth century which seek to provide funding to the marginalised including SMEs (Wachira et al., 2023). One of these options includes ethical banking. Ethical banking, also known as

K. Iddrisu (✉)
School of Business, Simon Diedong Dombo University of Business and Integrated Development Studies, Wa, Ghana
e-mail: khadijahiddris@iug.edu.gh

E. K. Boadi
Accountancy Department, Faculty of Business and Management Studies, Koforidua Technical University, Koforidua, Ghana
e-mail: boadikofieric@ktu.edu.gh

S. Giamporcaro
Kedge Business School, Talence, France
e-mail: s.giamporcaro@uct.ac.za

J. N. Doku
Department of Banking and Finance, University of Professional Studies, Accra, Ghana
e-mail: jdoku@upsamail.edu.gh

[1] https://www.icaew.com/insights/viewpoints-on-the-news/2024/mar-2024/government-aims-for-year-of-the-sme-in-2024. Accessed July 24, 2024.

I. N. Yakubu (ed.), *Strategic Approaches to Banking Business and Sustainable Development Goals*, Sustainable Development Goals Series, https://doi.org/10.1007/978-3-031-80744-2_10

responsible banking, refers to banking practices that prioritise social, environmental and economic responsibility alongside financial performance (Wachira et al., 2023; Climent, 2018). Banks that adopt ethical banking practices aim to align their operations with values that support sustainable development, social equity and environmental stewardship. Ethical banking also aids SMEs by ensuring fair lending terms, emphasising transparency and investing in sustainable projects that positively impact communities and the environment (Wachira et al., 2023; Climent, 2018). The services offered by ethical banking highlight their potential to foster SME growth. Consequently, this chapter delves into how ethical banking contributes to the expansion of SMEs in emerging markets by narrowing the financing gap associated with SMEs' growth.

The rest of this chapter is structured as follows: Sect. 10.2 discusses the concept of ethical banking, whereas the principles are discussed in Sect. 10.3. Section 10.4 provides the ecosystem of SMEs in emerging markets. Section 10.5 discusses the role ethical banking plays in promoting the growth of SMEs. The role of central banks in encouraging ethical banking practices that enhance the growth of SMEs is discussed in Sect. 10.6. We end this chapter by providing the conclusion.

10.2 Ethical Banking

10.2.1 Concepts of Ethical Banking

Ethical banking does not have widely accepted definitions but has some common characteristics such as equality in the delivery of financial services. For example, Harvey (1995) explained the concept of ethical banking as the responsible sourcing and distribution of funds. Sasia and De la Cruz (2008) also perceived ethical banking as a financial practice prioritising the provision of services to marginalised groups, social economy enterprises, non-governmental organisations (NGOs) and companies committed to responsible behaviour towards people, social and natural environments. According to Barigozzi and Tedeschi (2019), ethical banking is the financing of socially responsible activities. San-Jose et al. (2011) also identified ethical banking as a situation where banks have integrity, responsibility and affinity. MicroBank (2023) also defines ethical banking as an institution offering financial products that create social value and are ethically responsible. From these definitions, it can be identified that ethical banking is linked to the concept of socially responsible investment (SRI) and corporate social responsibility (CSR) (Valls Martínez et al., 2020).

During the Renaissance, Italian banks, influenced by Catholic morals, prohibited usury and favoured marginalised individuals. Similarly, Montes de Piedad, established by religious orders, laid the foundation for the first savings banks in the early nineteenth century, emphasising a clear social mission (Valls Martínez et al., 2020). Thereafter, the establishment of mutual savings banks in the United States and Europe marked an early attempt at ethical banking. These institutions were created to encourage saving among the working class, offering a secure place for deposits and making loans to local communities (Bodenhorn, 2016). Today, mutual funds, and savings and loans are on the rise to serve the poor, although the focus of these institutions has shifted. The concept of ethical banking expanded with the advent of credit unions in mid-nineteenth-century Germany. Herman Schulze-Delitzsch's cooperatively owned bakery inspired the creation of the first People's Bank of Germany in 1850. Friedrich Raiffeisen established the Heddesdorf Credit Union in 1864 to help farmers purchase livestock and equipment, emphasising self-governance to combat poverty.[2] Credit associations spread to Quebec in the early twentieth century, where Alphonse Desjardins founded North America's first credit union to counter loan sharks and provide affordable financing. The first U.S. credit union was established in New Hampshire in 1909. By the 1920s, not-for-profit cooperatives grew, offering small loans for appliances.[2]

[2] https://news.americafirst.com/credit-union-history-steeped-cooperation/. Accessed July 26, 2024.

Ethical banking was also significantly influenced by the Quakers, or the Religious Society of Friends, who had risen to prominence in the eighteenth century. They underscored the importance of steering clear of investments in projects, such as the slave trade and conflicts, that were contrary to their ethical principles (MicroBank, 2023).

The consolidation of ethical investing into socially responsible investing in the 1960s and 1970s set the foundation for contemporary ethical banking. During this period, investors began integrating social, environmental and ethical criteria into their investment decisions, alongside financial returns (Townsend, 2020). Also, during this season, the Pax World was founded in 1971 by two United Methodist ministers who opposed the Vietnam War, and this was the first socially responsible mutual fund in the US Managed by Tony Brown, it focused on ethical investments for church assets. Countries like the Netherlands and the United States acknowledged the significance of equality in financial services and environmental sustainability, leading to the creation of ethical banks such as ShoreBank and Triodos Bank. ShoreBank was founded in Chicago in 1973 by former Hyde Park Bank colleagues and was among the first US community development banks, focusing on underserved areas (Post & Wilson, 2011). In the late 1960s, Ronald Grzywinski and his team initiated an urban development division at Hyde Park Bank, addressing redlining and urban decay through minority-owned business loans. The 1970 Bank Holding Company Act enabled investments in community development, facilitating ShoreBank's establishment with US$800,000 in equity and a US$2.25 million loan, merging banking with economic development (Post & Wilson, 2011). The Triodos Bank in the Netherlands was established in 1980 to finance eco-friendly projects that benefit people. It was founded with €540,000 in start-up share capital and received a full banking license from the Dutch Central Bank.[3] Triodos Bank was the first bank to issue the Green Fund on the Amsterdam Stock Exchange in 1980 to minimise pollution and finance environmentally friendly projects. Today, Triodos Bank has expanded its scope by establishing branches in the UK, Belgium, Spain and Germany. The activities of these banks led to the establishment of similar institutions while existing commercial banks also began adopting ethical banking practices. For instance, the establishment of additional ethical banks, such as Cooperative Bank in England (1992), Banca Popolare Etica in Italy (1995) and JAK Medlemsbank in Sweden (1997) (Post & Wilson, 2011; Valls Martínez et al., 2020).

The rise of ethical banks in a few developed economies spurred the global expansion of ethical banking, which in turn necessitated the development of standardised principles. Today, nearly every country has incorporated some form of ethical banking into its practices through guidelines, policies and standards. Global standards, in particular, are discussed in the following Sect. 10.3.

10.2.2 Ethical Banking Versus Traditional Banking

Ethical banking places a strong emphasis on balancing people, the environment and profitability in its operations. This approach integrates social responsibility and environmental sustainability into core banking activities. Ethical banks aim to support initiatives that promote positive social impact, environmental stewardship and sustainable development, ensuring that their financial practices contribute to the well-being of communities and the planet. In contrast, traditional banking primarily focuses on maximising shareholder wealth as its central objective. This often involves prioritising profit generation and financial returns, sometimes at the expense of social or environmental considerations. Traditional banks may concentrate on optimising their financial performance and shareholder value, which can lead to less emphasis on ethical practices or broader societal impacts. The difference between ethical and traditional banking is discussed in Table 10.1.

[3] https://www.triodos.co.uk/about-us. Accessed July 26, 2024.

Table 10.1 Difference between ethical and traditional banking

	Ethical banking	Traditional banking
Main objective	Focus on people, environment and profitability	Focus on profitability and maximising shareholder's wealth
Loan policy	Equality in allowing access to credit	Skim-creaming (i.e. allow credit to low-risk entities)
Geographical location	Encourages few physical branches and more smart branches (fintech)	Focus on more physical branches
Investment	Green and social projects	Profitable and low-risk projects

Source: Valls Martínez et al. (2020)

10.2.3 Ethical Banking Practices

As previously discussed, ethical banking aims to protect people and the environment through various practices, some of which are outlined below:

1. **Transparency**: Ensuring clear and honest communication with customers about fees, terms and the bank's investments and activities (San-Jose et al., 2011).
2. **Community investment**: Investing in local communities through grants, loans and development programs that support social and economic growth (Valls Martínez et al., 2020).
3. **Responsible investment**: Avoid investments in industries that harm society or the environment, such as fossil fuels, tobacco or weapons manufacturing.
4. **Fair labour practices**: Ensuring fair wages and safe working conditions for all employees and contractors.
5. **Customer protection**: Upholding high standards of data privacy and protection, and providing fair dispute resolution processes.
6. **Green banking**: Implementing internal policies that reduce the bank's environmental footprint, such as energy-efficient buildings and paperless banking solutions.
7. **Sustainable lending**: Providing loans to projects and businesses that promote environmental sustainability, such as renewable energy ventures and eco-friendly infrastructure.
8. **Ethical product offerings**: Creating and promoting financial products that meet the needs of underserved or disadvantaged communities.

Therefore, any bank that incorporates these practices into its operations is striving to engage in ethical banking.

10.3 Policies, Principles and Standardisation of Ethical Banking

There are numerous principles, both direct and indirect, that guide ethical banking practices. While national principles exist, this chapter will specifically concentrate on global policies. These policies, principles and standardisations include but are not limited to the United Nations Environment Programme Finance Initiative (UNEP FI) principles, Inclusive Green Finance (IGF), International Organisation for Standardisation (ISO)-14,000, Equator Principles (EPs), Global Alliance for Banking on Values (GABV), Sustainability Accounting Standards Board (SASB), Sustainable Banking and Finance Network (SBFN) and Network for Greening the Financial System (NGFS). It is important to note that SASB and NGFS are indirect initiatives that guide ethical banking.

10.3.1 The Policies of the United Nations Environment Programme Finance Initiative

The UNEP FI was established in 1992, following the visionary leaders' discussions at the United Nations Conference on Environment and Development (UNCED), also known as the Rio Earth Summit, on the necessity of reshaping private finance for sustainable development.[4] It collaborates with key stakeholders in the financial sector—such as banks, insurance companies and financial markets—as well as various environmental institutions, and it operates voluntarily. This principle is recognised as an early effort to safeguard both people and the environment in the provision of financial services.[4] It also helps countries to develop sustainable financial policies and principles. Currently, it has 542 members and encompasses four main components: the Principles for Responsible Banking (PRB), the Principles for Sustainable Insurance (PSI), the Principles for Responsible Investment (PRI) and the Sustainable Blue Economy Finance Principles (SBEFP).[4] These components illustrate the broad scope of this principle within the financial sector. The PRB is a set of six principles that guide the conduct of banks to be more ethical. These six principles include (i) alignment, (ii) impact and targeting setting, (iii) clients and customers, (iv) stakeholders, (v) governance and culture and (vi) transparency and accountability. Currently, it has 330 signatory banks and has established an educational centre, that is PRB, Academy that trains banks and bankers on banking that protects people and the environment.[5]

PSI was launched at the 2012 UN Conference on Sustainable Development and includes principles that were designed to provide a global framework for managing environmental, social and governance (ESG) risks and opportunities in the insurance industry.[6] The PSI initiatives include the Sustainable Insurance Facility of the V20, resources for net-zero insurance, research on nature-positive insurance, the development of ESG guides for insurers and support for implementing the recommendations of the Task Force on Climate-related Financial Disclosures (TCFD). This framework represents the largest collaboration between the UN and the insurance industry. PSI has 165 signatories including banks, investment companies and insurance firms.[6]

The PRI initiative and its set of six principles[7] were launched in 2006 by the UNEP FI and the UN Global Compact, which offers a voluntary framework for investors to incorporate ESG issues into their decision-making and ownership practices.[8] These investors such as the French BNP Paribas Asset Management or the South African Sanlam Investment are usually part of bigger banking or insurance groups (BNP Paribas Asset Management and Sanlam Group for our two examples). Devised in partnership with the investment community, the PRI believes that ESG issues impact investment portfolio performance and should be considered to fulfil fiduciary duties. By implementing the principles, signatories contribute to a more sustainable global financial system. Two years after the launch of PRI, they quickly gained support from over 400 institutions managing over US$ 15 trillion in assets, aligning investor goals with the UN's sustainable development objectives through collaboration with UNEP FI and the UN Global Compact. In July 2024, 5317 investment institutions including asset owners, asset managers and service providers, have become signatories, aligning their objectives with broader societal goals.[9]

SBEFP are set of 14 principles[10] that is the first global framework for banks, insurers and

[4] https://www.unepfi.org/about/. Accessed July 26, 2024.

[5] https://prbacademy.com/. Accessed July 26, 2024.

[6] https://www.unepfi.org/insurance/insurance/. Accessed July 26, 2024.

[7] These principles include (1) we will incorporate ESG issues into investment analysis and decision-making process, (2) we will be active owners and incorporate ESG issues into our ownership policies and practices, (3) we will seek appropriate disclosure on ESG issues by the entities in which we invest, (4) we will promote acceptance and implementation of the principles within the investment industry, (5)we will work together to enhance our effectiveness in implementing the principles, and (6) we will each report on our activities and progress towards implementing the principles.

[8] https://unglobalcompact.org/take-action/action/responsible-investment#:~:text=Launched%20in%202006%20by%20UNEP,those%20of%20society%20at%20large. Accessed July 26, 2024.

[9] https://www.unpri.org/signatories/signatory-resources/signatory-directory. Accessed July 30, 2024.

[10] SBEFP are (1) protective, (2) compliant, (3) risk-aware, (4) systemic, (5) inclusive, (6) cooperative, (7) transparent, (8) purpose, (9) impactful, (10) precautionary, (11) diversified, (12) solution-driven, (13) partnering and (14) science-led.

investors to finance a sustainable ocean economy and was launched in 2018. They support SDG 14 (Life Below Water) by setting ocean-specific standards and helping the financial industry integrate sustainability in ocean-based sectors. Developed by the European Commission, World Wildlife Fund (WWF), World Resources Institute (WRI) and European Investment Bank (EIB), these principles are hosted by UNEP FI under the Sustainable Blue Economy Finance Initiative.

10.3.2 Alliance for Financial Inclusion Policies

Policies such as Inclusive Green Finance (IGF) policy, Consumer Empowerment and Market Conduct (CEMC), Forcibly Displaced Persons (FDPs), National Financial Inclusion Strategy (NFIS), micro, small and medium enterprise (MSME), Digital Financial Services (DFS), Financial Inclusion Data (FID) and Gender Inclusive Finance (GIF) are all policies introduced by the Alliance for Financial Inclusion (AFI)[11] in collaboration with financial policymakers and regulators to address financial exclusion and climate change concerns. These policies mitigate climate change impacts through financial inclusion policies and regulations. They promote green products in savings, credit, insurance and digital channels, which supports resilience and aligns with the UN SDGs, particularly SDGs 1 (no poverty), 7 (affordable clean energy) and 13 (climate action). The Sharm El Sheikh Accord which was endorsed at the 2017 AFI Global Policy Forum, commits AFI members to collaboratively develop financial inclusion policies with positive environmental outcomes, particularly for vulnerable communities. These policies especially IGF are part of the International Climate Initiative (IKI) and supported by the German Federal Ministry for Economic Affairs and Climate Action, which advances this commitment. AFI comprises 89 central banks and financial regulators. Since 2019, AFI has completed 89 projects, with the majority in Sub-Saharan Africa (46%), followed by the Middle East and North Africa (14%), Asia (13%), Latin America and the Caribbean (12%), the Pacific (10%) and Eastern Europe and Central Asia (5%).[12] The majority of these projects focus on DFS (19%), IGF (18.1%), NFIS (16.4%), CEMC (15.5%), GIF (14.7%), FID (7.8%), FDPs (6%) and MSME (2.6%).[12]

10.3.3 International Organisation for Standardisation (ISO)-14000

The International Organisation for Standardisation (ISO) 14000 series comprises a comprehensive set of international standards designed to facilitate effective environmental management. These standards provide organisations with structured frameworks and practical tools to systematically identify, assess, manage and mitigate their environmental impacts, aiming to foster sustainable practices and enhance environmental performance. The most prominent standard within this series is ISO 14001, which outlines specific requirements for establishing, implementing, maintaining and continually improving an environmental management system (EMS). ISO 14001 helps organisations manage their environmental responsibilities, ensuring that they comply with legal and regulatory requirements while striving to reduce their ecological footprint. In addition to ISO 14001, the ISO 14000 family includes a range of supplementary standards and guidelines that cover various aspects of environmental management. These include:

[11] AFI is a registered organisation in Malaysia under the International Organisations (Privileges and Immunities) Act 1992. AFI promotes financial inclusion through evidence-based policies, peer learning, and collaboration among 81 central banks and regulators, focusing on green finance and global knowledge exchange.

[12] https://www.afi-global.org/activities/in-country-implementation/. Accessed July 26, 2024.

- ISO 14004: Provides general guidelines on the principles, systems and supporting techniques of environmental management.
- ISO 14010-14014: Offer guidelines on environmental auditing, including principles and procedures for auditing and evaluation.
- ISO 14015: Focuses on the environmental assessment of sites and organisations.
- ISO 14020-14025: Cover environmental labels and declarations, including guidelines for communicating environmental information to stakeholders.
- ISO 14040-14044: Address life cycle assessment, providing standards for evaluating the environmental impacts associated with all stages of a product's life cycle, from raw material extraction through to disposal.

10.3.4 Equator Principles

The EPs were established in June 2003 by a consortium of leading financial institutions, led by Citigroup, ABN AMRO, Barclays and WestLB, in collaboration with the International Finance Corporation (IFC) of the World Bank Group. The principles were developed to create a common framework and set of standards for assessing and managing environmental and social risks in project finance. The Equator Principles are a risk management framework adopted by financial institutions for determining, assessing and managing environmental and social risk in projects. They are primarily intended to provide a minimum standard for due diligence and monitoring to support responsible risk decision-making. The Equator Principles apply globally to all new project finance transactions with total project capital costs of US$10 million or more. They cover various sectors, including infrastructure, energy, mining and agriculture. The principles provide a framework for managing environmental and social risks through all stages of project development, from assessment to monitoring. The Equator Principles Association, consisting of member financial institutions, governs the EPs. Its governance structure includes a Steering Committee, which oversees strategic direction and implementation; Working Groups, which focus on specific issues, updates and guidance; and a Secretariat, which supports coordination and communication among members.

The EPs consist of 10 core principles:

1. Review and categorisation: Classifying projects by their environmental and social impacts.
2. Environmental and social assessment: Identifying potential risks and impacts.
3. Applicable standards: Using relevant international standards, like those of the IFC.
4. Management system and action plan: Developing plans to manage risks.
5. Stakeholder engagement: Engaging affected communities and stakeholders.
6. Grievance mechanism: Addressing concerns and grievances.
7. Independent review: Seeking expert review for high-risk projects.
8. Covenants: Including covenants in financial agreements.
9. Monitoring and reporting: Regularly monitoring and reporting performance.
10. Reporting and transparency: Committing to transparency in implementation.

Since their establishment, the EPs have been adopted by 130 financial institutions globally (i.e. as of July 2024), covering most international project finance debt in developed and emerging markets.[13] They have integrated environmental and social considerations into financial decisions, enhancing accountability and transparency. The principles have been revised to stay relevant, with the latest version, Equator Principles 4 (EP4), released in 2020, introducing enhanced requirements for human rights, climate change and

[13] https://equator-principles.com/signatories-epfis-reporting/#:~:text=130%20financial%20institutions%20globally%20are%20Signatories%20to%20the%20Equator%20Principles. Accessed July 30, 2024.

indigenous peoples, highlighting their growing importance in global sustainability.

10.3.5 Principles of the Global Alliance for Banking on Values

GABV was established in 2009 by a group of pioneering financial institutions dedicated to promoting sustainable economic, social and environmental development. Founding members include Triodos Bank (Netherlands, UK, Belgium, Spain and Germany), ShoreBank (USA), BRAC Bank (Bangladesh), Mibanco Peru (Peru), Alternative Bank Switzerland (Switzerland), Merkur Bank (Denmark), GLS Bank (Germany), Banca Etica (Italy), New Resource Bank (USA) and XacBank (Mongolia). These institutions shared a common vision of leveraging finance to contribute positively to communities and the environment. The membership has now grown from 10 banks to 73 banks and financial institutions operating in more than 45 countries across Africa, Latin America, North America and Europe.[14] The GABV promotes a sustainable banking model prioritising social and environmental impact alongside financial returns. It advocates for a banking system serving people and the planet, encourages practices supporting sustainable development and addressing global challenges and fosters collaboration and knowledge sharing among member banks to enhance their impact.

The GABV promotes six principles of values-based banking:

1. Triple bottom line approach: Integrating social, environmental and financial goals.
2. Real economy focus: Financing projects that have a tangible positive impact on the real economy.
3. Long-term resiliency: Prioritising long-term sustainability over short-term profits.
4. Transparency and inclusion: Ensuring transparent governance and inclusive decision-making processes.
5. Social and environmental impact: Measuring and reporting on their activities' social and environmental impact.
6. Mission-driven culture: Embedding the bank's mission in its culture and operations.

Therefore, banks that wish to adopt ethical banking can join this alliance and integrate their principles into their decision-making.

10.3.6 Sustainable Banking and Finance Network Policies

SBFN, formerly known as Sustainable Banking Network (SBN), was established in 2012. It was initiated by the IFC, a member of the World Bank Group, to serve as a platform for financial regulators, banking associations and financial institutions from emerging markets to exchange knowledge and collaborate on sustainable finance practices (SBFN, 2024). The primary purpose of the SBFN is to support the development and implementation of sustainable banking practices by integrating ESG considerations into financial sector policies. It aims to address global challenges like climate change and social inequality, promote sustainable finance, build the capacity of financial regulators in emerging markets and facilitate global knowledge sharing and best practices among its members. SBFN is governed collaboratively by its member institutions and the IFC. The governance structure includes a Steering Committee providing strategic direction, working groups focusing on specific sustainable finance themes like green bonds and climate risk management and the IFC serving as the Secretariat. The Secretariat coordinates activities, facilitates member communication and offers technical support and expertise.

SBFN supports its mission through various activities, including:

- Guideline development: Creating sustainable finance. guidelines tailored for emerging markets.

[14] https://www.gabv.org/about-us/. Accessed July 27, 2024.

- Capacity building: Offering training programs, workshops and seminars for financial regulators and institutions.
- Research and publications: Conducting research and publishing reports on sustainable finance trends and challenges.
- Peer learning: Facilitating peer learning and knowledge exchange via conferences and online platforms.
- Technical assistance: Providing support for developing and implementing sustainable finance policies.

SBFN's 91 members comprise 55 regulators and 36 industry associations. Of these, 80% are banks, 18% are capital market entities, 13% are insurance firms, 10% are other financial institutions, 5% are environmental agencies and 1% are pension fund institutions (SBFN, 2024). Banks seeking to adopt ethical banking practices can join the SBFN to align their operations with established ethical principles. By becoming a member, these banks can integrate sustainable and responsible practices into their decision-making processes, benefit from best practices shared within the network and contribute to advancing global standards for ethical and sustainable finance. Joining the alliance provides access to a collaborative platform for enhancing their commitment to environmental and social responsibility.

10.3.7 Sustainability Accounting Standards Board Policies

The Sustainability Accounting Standards Board (SASB) was established in 2011 by Jean Rogers, an engineer with a background in sustainability.[15] SASB was created as a response to the growing need for standardised reporting on environmental, social and governance (ESG) factors, which are increasingly recognised as critical to the financial performance of companies. The primary purpose of SASB is to develop and disseminate sustainability accounting standards that help public corporations disclose material ESG information to investors comparably and consistently. It was established due to investor demand for transparent and reliable ESG data, regulatory gaps in existing financial reporting standards regarding ESG impacts, and the need to enhance corporate accountability and encourage effective management of ESG factors. SASB operates under the Value Reporting Foundation (VRF), established by merging SASB and the International Integrated Reporting Council (IIRC) in 2021. In 2022, the SASB Foundation and IIRC Boards merged to form the Value Reporting Foundation Board, overseeing strategy, finances and operations. This board later consolidated into the International Financial Reporting Standards (IFRS) Foundation, aiding in the creation of the International Sustainability Standards Board (ISSB). Its governance includes a Board of Directors overseeing strategy, operations and finances, a Standards Board developing and maintaining standards and various advisory groups providing sector-specific input to ensure the standards' relevance and effectiveness.

SASB has developed standards for 77 industries, addressing unique sustainability issues relevant to each sector. These standards cover a wide range of ESG topics, including environmental (energy, water, waste management), social (labour practices, product safety, supply chain management) and governance (business ethics, risk management, regulatory compliance). SASB standards enhance corporate sustainability disclosures and inform investor decisions. SASB has significantly influenced sustainability reporting by providing a clear, consistent framework for disclosing material ESG information. Recognised for rigour and relevance, these standards integrate sustainability into mainstream financial analysis and decision-making. SASB standards can significantly support ethical banking by enhancing transparency, informing ethical investment decisions, aligning with sustainability principles, promoting stakeholder engagement, driving continuous improvement, attracting ethical investors and improving risk management. By adopting SASB standards, banks can strengthen their commitment to responsible and sustainable banking practices. This is because any bank sub-

[15] https://sasb.ifrs.org/about/governance-archive/. Accessed July 27, 2024.

scribing to SASB will report on ESG, encouraging ethical banking, as some of their activities will protect people and the environment.

10.3.8 Network for Greening the Financial System Policies

NGFS was established by eight central banks and financial supervisors, including the Banque de France, Bank of England, People's Bank of China, Monetary Authority of Singapore, De Nederlandsche Bank, Federal Reserve Board, Banca d'Italia and Banco de España in December 2017 at the One Planet Summit in Paris.[16] It was established to address the need for integrating climate and environmental risks into the financial system. Its objectives are to promote best practices for climate risk management, enhance financial resilience, facilitate knowledge sharing among central banks and supervisors and support the transition to a low-carbon, sustainable economy by encouraging alignment with climate goals. The NGFS is governed collaboratively with a Steering Committee providing strategic direction, working groups developing guidelines and tools on specific themes like risk management, and a Secretariat supporting operations and communication. The Secretariat, hosted by a member institution, manages administrative functions and facilitates coordination among members.

The NGFS engages in a range of activities to fulfil its mission:

1. Research and reports: Conducting research and producing reports on climate-related and environmental risks and their implications for financial stability.
2. Guidance and frameworks: Developing guidance and frameworks to help central banks and financial supervisors integrate climate risks into their policies and operations.
3. Capacity building: Offering workshops, seminars and training programs to build capacity among member institutions and stakeholders.
4. Knowledge sharing: Facilitating the exchange of knowledge and experiences among members through conferences, forums and collaborative projects.
5. Policy advocacy: Advocating for policy measures that support the integration of climate risks into financial regulation and supervision.

The NGFS has expanded to encompass a diverse membership of central banks, financial supervisors and international organisations from various regions, reflecting a wide range of financial and supervisory perspectives. Initially starting with just 8 members, the network has now grown to 141 members and includes 21 observers.[14] The network's inclusive approach enables it to address global challenges related to climate change and environmental sustainability effectively. Although primarily aimed at central banks, this initiative also fosters ethical banking by motivating central banks to incorporate best banking practices into their policy-making, ultimately influencing banks to adopt practices that protect people and the environment.

10.4 SMEs in Emerging Economies

SMEs have gained attention in recent years, where a lot of Development Financial Institutions (DFIs) such as the World Bank, African Development Bank (AfDB), Asian Development Bank (ADB) and others have shifted their focus to them. The definition of SMEs across the globe is debatable. For instance, Abor and Quartey (2010) define SMEs as micro-businesses with 1–9 employees, small at 10–99 and medium-large enterprises with 100–499 workers. The European Commission (2003) defines SMEs as any enterprise with fewer than 250 employees and having either an annual turnover not exceeding €50 million or a total annual balance sheet not exceeding €43 million. SMEs can also be cat-

[16] https://www.ngfs.net/en/about-us/governance/origin-and-purpose. Accessed July 27, 2024.

egorised into micro (i.e. <10 employees, ≤ €2 million annual turnover and ≤€2 million), small (i.e. <50 employees, ≤€10 million annual turnover and ≤€10 million) and medium (i.e. <250 employees, ≤€50 million annual turnover and ≤€43 million).

The Ghana Enterprises Agency (GEA), which was formally known as the National Board for Small Scale Industries (NBSSI), defined SMEs as small enterprises defined as any business with fewer than 10 employees, while firms with 10 or more employees are classified as medium or large enterprises (GCB, 2024). GEA further defines SMEs by both employee count and fixed assets. Specifically, the GEA considers a firm to be an SME if it employs fewer than 10 individuals and has fixed assets, including plant and machinery, valued at no more than ten million Ghana Cedis (GCB, 2024). However, this GEA definition has limitations and inconsistencies for universal application due to factors such as frequent depreciation of the local currency and challenges in asset valuation, as noted by the Ghana Enterprise Development Commission. A World Bank study by Zavatta (2008) defines SMEs as firms with up to 300 employees and annual sales not exceeding US$15 million. Therefore, it can be concluded that SMEs can be defined as any business organisation with fewer employees, revenue and assets at a certain threshold determined by institutions or in the country.

SMEs are integral to the global economy, especially in developing countries, where they play a vital role in driving economic growth and job creation. SMEs represent approximately 90% of all businesses worldwide and account for over 50% of global employment. They are key to fostering economic development and creating opportunities in both formal and informal sectors. In emerging markets, formal SMEs contribute up to 40% of GDP, with this contribution rising significantly when informal SMEs are considered. The significance of SMEs is underscored by the projected need for 600 million new jobs by 2030 to accommodate the expanding global workforce. This highlights the crucial role that SME development plays in economic planning and policy-making for governments. In these regions, SMEs are responsible for generating 70% of formal employment, illustrating their central role in labour markets. However, despite their importance, SMEs face considerable challenges, with access to finance being a major barrier. It ranks as the second most critical issue hindering SME growth in developing economies.

SMEs often encounter difficulties in securing bank loans due to limited credit history or collateral. As a result, they frequently rely on internal funding or financial support from friends and family to establish and maintain their operations. According to the IFC, approximately 65 million firms—constituting 40% of formal micro, small, and medium enterprises (MSMEs) in developing countries—face an annual financing gap of US$5.2 trillion (World Bank, 2019). This amount represents 1.4 times the current global lending levels for MSMEs. The distribution of this financing gap varies by region, with East Asia and the Pacific facing the largest share at 46%, followed by Latin America and the Caribbean at 23%, and Europe and Central Asia at 15%. The financing gap is particularly pronounced in Latin America and the Caribbean, as well as in the Middle East and North Africa, where unmet demand relative to potential demand reaches 87% and 88%, respectively (World Bank, 2019). Nearly half of formal SMEs in these regions lack access to formal credit, and the situation is even more challenging when considering micro and informal enterprises. This underscores the need for targeted financial solutions and policy interventions to support SME growth and enhance their contribution to economic development.

10.4.1 Challenges Faced by SMEs in Accessing Financial Services

SMEs in emerging economies encounter challenges when accessing financial services. This subsection discusses some of these challenges which include the following.

Limited Access to Credit

SMEs frequently encounter substantial difficulties in securing the collateral required to obtain loans from banks and other financial institutions (Abor & Quartey, 2010). A primary challenge is the often limited or non-existent credit history of these businesses, which complicates lenders' efforts to accurately assess their creditworthiness. Without a robust credit record, SMEs struggle to demonstrate their reliability and financial stability, making lenders wary of extending credit. Additionally, the cost of borrowing for SMEs tends to be relatively high because lenders view these businesses as higher risk compared to larger, more established companies. This perception of increased risk results in higher interest rates and more stringent lending terms, which can place a considerable financial burden on SMEs. The combination of inadequate collateral and elevated borrowing costs further exacerbates the financial strain on SMEs, impeding their ability to invest in growth, manage operations and navigate economic challenges effectively. Addressing these barriers is crucial for supporting the sustainability and development of SMEs, ensuring they have equitable access to the financial resources needed to thrive.

Risk Perception and Investment Reluctance

Financial institutions often perceive lending to SMEs in some emerging economies like Africa as inherently riskier due to several contributing factors. These include economic instability, which can lead to unpredictable business conditions; lower market demand that impacts the profitability and sustainability of local businesses; and limited collateral, which reduces the security for lenders. This heightened perception of risk can result in financial institutions being reluctant to extend credit or offer other financial services to these SMEs. This is shown in the study by Ayyagari et al. (2017), which identified that SMEs are perceived as higher-risk customers, leading financial institutions to either refrain from offering financial services to them or limit such services significantly. The problem is further compounded by the scarcity of financial institutions operating in rural or remote areas. This lack of presence not only restricts SMEs' access to essential financial services but also serves as a deterrent to potential investors or financial service providers who might otherwise consider entering the market. Consequently, the issue of limited access to finance becomes more pronounced, creating a challenging environment for SMEs in these regions to thrive and grow. The absence of adequate financial support hampers their ability to invest in new opportunities, expand their operations and contribute to local economic development.

Complex and Stringent Requirement

SMEs frequently encounter intricate and demanding documentation requirements when seeking financial assistance or complying with regulatory standards. These requirements can be both burdensome and costly, consuming valuable time and resources that SMEs might otherwise allocate to their core business operations. Additionally, the task of navigating regulatory requirements can be particularly challenging for SMEs, especially in dynamic and rapidly evolving environments where regulations may frequently change or become more stringent. The complexity of these processes often necessitates specialised knowledge or external assistance, further adding to the operational costs and administrative burden faced by these businesses.

Lack of Financial Literacy

SMEs frequently grapple with a limited understanding of the diverse financial products and services available, which hampers their ability to make informed decisions. This knowledge gap can lead to suboptimal choices, such as selecting unsuitable financial products, misinterpreting loan terms or overlooking beneficial financial instruments. The complexity of financial markets can be overwhelming for SME owners and managers, particularly those without a finance background, resulting in missed growth opportunities and ineffective risk management. Additionally, inadequate financial management skills further limit SMEs' ability to effectively utilise and manage resources. Essential financial management practices including budgeting, forecasting, cash flow management and strategic investment decisions are often underdeveloped. This can lead to problems such as cash flow issues, over-leveraging and poor capital allocation. These challenges underscore the need for improved financial education and support tailored to SMEs, helping them navigate financial obstacles, secure necessary funding and achieve sustainable growth.

10.5 Growth of Small and Medium Enterprises: The Role of Ethical Banking

Ethical banking practices enable support for SMEs in both developed and emerging economies. This chapter examines the impact of ethical banking on SMEs in emerging markets, highlighting solutions such as offering tailored financial products, fair financing, financial education, advisory services, support for sustainable practices and fostering long-term relationships and inclusive banking.

10.5.1 Tailored Financial Products and Fair Financing

Banks that embrace ethical banking practices often design financial products specifically tailored to meet the unique needs and challenges of SMEs. These customised solutions can encompass a variety of offerings, such as microloans, flexible repayment schedules and products that promote sustainable business practices. By providing these targeted financial services, ethical banks help SMEs thrive in competitive markets. Additionally, these banks prioritise transparency and fairness, offering loans with reasonable interest rates and clear terms, which make financing more accessible and affordable for SMEs. This approach not only supports SMEs in securing the capital they need but also instils confidence in the financial system. By making it easier for SMEs to access affordable financing, ethical banks enable them to invest in their growth and development more effectively, ultimately fostering a more inclusive and sustainable economic landscape. For example, Axis Bank in India has created a special banking group for SMEs called the Commercial Banking Group (CBG), which is in charge of SMEs, and has created special packages such as LCBN Power, Mpower, SME Plus and Zero-Collateral loans (see Box 10.1).

Box 10.1 Axis Bank in India Created Special Banking for SMEs called Commercial Banking Group

The Commercial Banking Group (CBG) a special banking created by the Axis Bank focuses on establishing and enhancing banking relationships with small and medium-sized enterprises (SMEs) nationwide. CBG addresses the unique and evolving financial needs of this vital economic sector through a broad array of tailored products and services, including working capital finance, term loans, trade finance, cash management, foreign exchange, salary accounts and trust services. CBG offers a range of tailored financial solutions for businesses:

- **LCBN power**: Enhance your firm's liquidity with quick and hassle-free bill discounting.
- **Mpower**: Property-backed overdrafts for meeting all business needs effortlessly.
- **Power OD**: Easy-to-use overdraft facilities to boost business liquidity.
- **SME plus**: Access a variety of customized products, including working capital finance, term loans, and trade products.
- **Zero collateral loans**: Obtain loans without locking in your assets.

Source: Axis Bank (2024).

10.5.2 Financial Education and Advisory Services

Ethical banking practices have paved the way for banks to offer comprehensive financial education and advisory services to SME clients. By providing guidance on financial management, investment strategies and risk mitigation, these banks empower SMEs to make well-informed decisions. This support enhances the financial health of SMEs, enabling them to navigate complex financial landscapes more effectively. With a solid foundation in financial literacy and strategic planning, SMEs can better manage their resources, reduce risks and identify opportunities for sustainable growth. This proactive approach not only helps SMEs thrive in the short term but also positions them for long-term success, contributing to overall economic stability and development. Through these initiatives, ethical banks foster a culture of responsible business practices, ensuring that SMEs are well-equipped to handle future challenges and seize new opportunities. For instance, Banco Popular de Honduras provides financial education and advisory services to SMEs through social media platforms such as WhatsApp, Facebook and YouTube on financial services and Fintech innovation tools (see Box 10.2).

Box 10.2 Banco Popular de Honduras Provides Financial Education to SMEs

Banco Popular de Honduras' mission is to support micro, small, and medium-sized businesses in Honduras, which are crucial for job creation, poverty reduction, and economic growth. The bank also addresses the working capital and asset needs of SMEs with a personalised and friendly approach. To aid these businesses, the bank produces and shares short educational videos on WhatsApp, Facebook, and YouTube. These videos offer practical tips on business management, technology use, financial education, and more, drawing on real experiences from small business owners. This content is tailored to the needs and interests of our clients.

Source: Banco Popular de Honduras (2024).

10.5.3 Support for Sustainable Practices

Banks that adopt ethical banking practices typically focus on supporting businesses that prioritise sustainability and social responsibility. By offering financing to SMEs committed to environmental sustainability and social impact, these banks encourage responsible business operations. This support not only helps SMEs enhance their long-term viability and reputation but also fosters a positive cycle of growth that benefits the wider community. Ethical banks play a crucial role in promoting sustainable economic development by providing the necessary resources and guidance to SMEs, enabling them to implement and maintain responsible practices. Consequently, these SMEs can become leaders in their industries, setting standards for others to follow and contributing to a more sustainable and equitable economy.

10.5.4 Long-Term Relationship

As a result of ethical banking practices, banks place a strong emphasis on building long-term relationships with their clients, fostering stable and reliable financial partnerships. This relationship-based approach goes beyond mere transactions, offering SMEs consistent support and flexibility as their businesses grow and evolve. By understanding the unique needs and aspirations of each SME, ethical banks can tailor their services and provide personalised guidance that adapts to changing circumstances. This ongoing support ensures that SMEs have a dependable financial ally, helping them navigate challenges, seize opportunities and plan for future expansion. Ultimately, the commitment to nurturing long-term relationships not only enhances the trust and loyalty between ethical banks and SMEs but also contributes to the sustained success and resilience of small businesses in a dynamic market environment. For instance, Axis Bank aims to establish a long-term relationship with customers through the creation of CBG (see Box 10.1).

10.5.5 Inclusive Banking Practices

Unlike traditional banking, which often prioritises less risky clients, ethical banking strives for inclusivity, extending services to underserved communities and sectors. This focus on financial inclusion allows ethical banks to help SMEs in rural or economically disadvantaged areas access the capital and support they need to grow and succeed. By reaching out to these often overlooked markets, ethical banks play a crucial role in bridging the financial gap and promoting equitable economic development. These banks offer tailored financial products, educational resources and advisory services, empowering SMEs to overcome barriers and achieve sustainable growth. Through their commitment to inclusivity, ethical banks contribute to the economic vitality of underserved regions, fostering job creation, innovation and improved quality of life for local communities. This approach not only supports individual SMEs but also stimulates broader economic progress, reinforcing the importance of ethical banking practices in achieving a more balanced and fair financial system. In Ghana, United Bank for Africa (UBA) Ghana does not only focus on companies but also includes SMEs and the marginalised in their financial service delivery (see Box 10.3).

Box 10.3 UBA Ghana Supports SMEs in Ghana

United Bank for Africa (UBA) empowers SMEs in Ghana by providing diverse lending options, such as term loans, overdraft facilities, and trade finance solutions. Leveraging its extensive network and financial expertise, UBA facilitates the necessary funding for SMEs to realize their growth ambitions. Additionally, the bank's partnership with the African Continental Free Trade Area (AfCFTA) includes a significant US$6 billion investment in African SMEs. UBA also collaborates with the United Nations Development Programme (UNDP) to support youth and women-led Micro, Small, and Medium-sized Enterprises (MSMEs) in Ghana. These initiatives are designed to help SMEs expand their operations, create employment opportunities, increase revenue, and promote environmental sustainability. UBA supports SMEs by providing funding, cross-border trade solutions, and comprehensive financial advisory services. Through UBA AfriTrade and UBA AfriCash, SMEs can receive payments and send remittances. UBA's partnership with the Ghana Export Promotion Authority (GEPA) under AfCFTA promotes trade liberalisation. These initiatives drive economic growth, job creation, innovation, and wealth generation, reflecting UBA's commitment to empowering SMEs and fostering economic development in Ghana and beyond.

Source: Dery (2024).

10.5.6 Community and Network Building

Ethical banks often cultivate a sense of community among their clients, creating networks of like-minded businesses that share common values and goals. These networks offer SMEs valuable opportunities for collaboration, allowing them to partner on projects, share resources and leverage each other's strengths. Knowledge sharing within these communities enables SMEs to learn from one another's experiences, gaining insights into best practices, innovative solutions and effective strategies for overcoming challenges. This mutual support system helps businesses navigate the complexities of their industries with greater confidence and resilience. By fostering these connections, ethical banks not only support individual SMEs in their growth and development but also contribute to a robust and interconnected business ecosystem. This communal approach enhances the overall stability and success of SMEs, reinforcing the importance of ethical banking in promoting sustainable and inclusive economic growth. For instance, BRAC Bank PLC, known for its ethical banking practices, has fostered a community among its clients by establishing networks of like-minded businesses that align with shared values and objectives (see Box 10.4).

Box 10.4 BRAC Bank PLC Dedicated to SMEs

BRAC Bank PLC was founded with the vision of providing banking solutions to 'unbanked' Small and Medium Enterprises (SMEs). Inspired by its parent organization, BRAC, the largest NGO globally, BRAC Bank introduced small ticket loans aimed at integrating grassroots entrepreneurs into formal banking. Today, nearly half of the bank's lending portfolio is dedicated to SMEs. Established in 2001, BRAC Bank distinguished itself by focusing on the neglected SME sector, which plays a crucial role in economic growth and employment. Unlike traditional banks, BRAC Bank stepped in to finance SMEs in Bangladesh when it was nearly impossible for these businesses to obtain bank financing. As the sole Bangladeshi member of the Global Alliance for Banking on Values (GABV), which includes 70 institutions across 45 countries, BRAC Bank shares the GABV's 3P Philosophy—People, Planet, and Prosperity. The bank is committed to the welfare of its clients and communities and partners with FMO to further support SMEs.

Source: BRAC Bank (2024).

10.6 The Role of Central Banks in Promoting Ethical Banking on Growth of SMEs

Central banks can significantly influence the adoption of ethical banking practices to foster SME growth. Key roles they can play: regulating and supervising banks, facilitating access to credit, providing education and training, supporting sustainable finance initiatives and conducting monitoring and evaluation.

10.6.1 Regulation and Supervision

Central banks play a crucial role in establishing ethical standards and best practices for the banking sector by setting the regulatory framework that governs financial institutions. By integrating ethical banking principles into these regulations, central banks can ensure adherence to standards of transparency, fairness and social responsibility. This fosters a conducive environment for ethical banking practices, which greatly benefit SMEs. Through promoting and enforcing these standards, central banks ensure that financial institutions operate in ways that support the ethical growth of SMEs, thereby fostering trust and stability within the financial system. For example, the Reserve Bank of India has implemented policies that encourage banks to adopt ethical banking and support SMEs (see Box 10.5).

Box 10.5 People's Bank of China Through Regulation Encourages Bank to be Ethical

In the first half of 2018, the People's Bank of China (PBC) implemented several policies and initiatives to strengthen financial stability and support economic development. Key actions included renewing RMB clearing agreements with various international banks, issuing notices to improve RMB cross-border business, and introducing new financial instruments like the

30-day Contingent Reserve Allowance and targeted reductions in reserve requirement ratios (RRR) for financial inclusion. The PBC also coordinated with other agencies to release guidelines aimed at enhancing financial services for the marine economy and SMEs, regulating asset management, and supporting green and agricultural finance. The PBC's efforts extended to improving financial support for poverty alleviation and enhancing the supervision of green financial bonds. Throughout the year, the PBC continued to foster international cooperation through bilateral currency swap agreements with various countries, including Australia, Albania, South Africa, Nigeria, Malaysia, and Indonesia. They also expanded collateral for the medium-term lending facility (MLF) to include SME and green bonds. Additionally, efforts were made to improve financial services for SMEs through policy measures and increased credit support. By the year's end, the PBC introduced the Targeted Medium-term Lending Facility (TMLF) to encourage financial institutions to allocate more capital to SMEs and private enterprises. They also worked on enhancing the management of internet gold business and gold accumulation, demonstrating a comprehensive approach to financial market development and economic support.

Source: PBC (2019).

10.6.2 Facilitating Access to Credit

Central banks can implement policies that make it easier for SMEs to access credit. This can include providing guarantees or establishing special funds that reduce the risk for banks' lending to SMEs. By lowering the barriers to obtaining credit, central banks enable SMEs to secure the financing they need for expansion and innovation. Additionally, central banks can collaborate with various agencies supporting SMEs to inform them about available fund by the Multilateral Development Banks (MDBs) and facilitate the mobilisation of these resources. Some of these funds are World Bank Group iSMEs' Finance which seeks to improve SMEs' access to finance and find innovative solutions to unlock sources of capital (see Box 10.6).

Box 10.6 World Bank Group iSME Finance

The World Bank Group focuses on improving SMEs' access to finance and unlocking capital sources through a holistic approach, combining advisory and lending services to enhance the economic contributions of SMEs, including those owned by women. The advisory and policy support for SME finance encompasses diagnostics, implementation support, global advocacy, and knowledge sharing of best practices. Key advisory services include financial sector assessments to identify regulatory and policy improvements, support for developing credit guarantee schemes, enhancing credit infrastructure, and introducing innovative financing methods such as e-lending platforms, alternative data for credit decisions, and supply chain financing. Additionally, the World Bank Group engages in policy and analytical work to support SME finance activities and advocates for SME finance at global forums like the G20 and the Financial Stability Board. Knowledge management tools and flagship publications are also provided to share successful models and policy frameworks. Several key initiatives highlight the World Bank Group's efforts. In Lebanon, the Innovative Small and Medium Enterprises (iSME) project has invested US$10.23 million across 22 investments, leveraging US$25.47 million in co-financing and expanding early-stage equity finance. The

(continued)

Box 10.6: (continued)

MSME Growth, Innovation, and Inclusive Finance Project in India provided a $500 million credit line to SIDBI, improving access to finance for underserved MSMEs and introducing innovative lending methods, which attracted US$1.9 billion in private sector financing. In Jordan, lines of credit projects have financed 8149 MSMEs, creating 7682 jobs, with significant benefits for women and youth. In Nigeria, the Development Finance Project supports the Development Bank of Nigeria (DBN), which has disbursed US$243.7 million to nearly 50,000 end-borrowers, with 70% of them being women. In Morocco, the MSME Development project has significantly increased MSME lending supported by guarantees, facilitating access to loans for first-time borrowers. The Women Entrepreneurship Development Project (WEDP) in Ethiopia has provided loans and business training to over 14,000 women entrepreneurs, significantly increasing average loan sizes and reducing collateral requirements. In Bangladesh, the Access to Finance for Women SMEs Project has supported the establishment of a credit guarantee scheme and the issuance of the country's first comprehensive SME Policy, aimed at expanding finance access to women SMEs. Additionally, in Ethiopia and Guinea, the World Bank Group supports the creation of enabling frameworks for leasing operations to increase SME finance access. Through these efforts, the World Bank Group collaborates with public and private sector stakeholders, as well as other multilateral and bilateral development organizations, to support SME finance development in emerging markets and developing countries.

Source: World Bank (2019).

Another significant financing initiative that central banks, particularly in Africa, can leverage is the AfDB's Africa SME Support Programme. This programme aims to offer technical assistance and capacity building to up to 25 smaller financial institutions focused on supporting SMEs (see Box 10.7). ADB has diverse programs dedicated to improving upon the growth of SMEs that central banks in the Asian country can take advantage of and enhance access to credit facilities for the SMEs. For instance, as shown in Box 10.8, ADB has partnered with Citi Bankto support SMEs in Asia.

Box 10.7 African SME Support Programme of the African Development Bank

The project seeks USD 3.98 million from FAPA resources to provide Technical Assistance (TA) and capacity building to up to 25 smaller SME-focused financial institutions (FIs) in Africa, particularly in Low-Income Countries and Fragile States. The AfDB's Africa SME Support Program (ASME) comprises a funding component and a TA component. The TA component aims to strengthen FIs in areas such as credit assessment, risk management, and management information systems (MIS) to enhance their support for SMEs. Concurrently, the funding component offers medium to long-term loans totalling US$125 million to approximately 25 SME-focused FIs. This programme started in 2013 and is still ongoing. The TA component, linked to EPSA (Enhanced Private Sector Assistance), represents a strategic collaboration between AfDB and the Government of Japan. It enhances the impact of the funding component, maximizing private sector development in Africa. The TA component is divided into five sub-components:

1. Project planning and preparation, including consultant procurement.

2. Needs assessment and drafting of Terms of Reference (ToRs).
3. Capacity building of FIs, including training and strengthening operational policies.
4. TA Program management, monitoring, and quality assurance.
5. Program communication and publicity activities.

The objective is to address specific weaknesses in the operational structures of smaller SME-focused FIs, including micro-finance institutions up-scaling to service SMEs. Beneficiaries will include 20 to 25 SME-focused FIs, encompassing banks, micro-finance institutions, and non-bank financial institutions.

Source: AfDB (2024).

Box 10.8 Asian Development Bank and Citi Bank Partnership to Support SMEs in Developing Asia

The Asian Development Bank (ADB) and Citi have entered into a master risk participation agreement aimed at enhancing supply chain financing for small and medium-sized enterprises (SMEs) in Asia and the Pacific. This collaboration seeks to support over US$100 million in additional annual trade, leveraging Citi's supply chain finance solutions through ADB's Trade and Supply Chain Finance Programme (TSCFP). ADB Vice-President for Market Solutions, Bhargav Dasgupta, emphasized the partnership's potential to amplify trade and supply chain finance's developmental impact by combining Citi's expansive network with ADB's presence in developing Asia. Supply chain financing is essential for boosting trade by providing vital capital to suppliers through their relationships with larger corporate partners. Since 2009, ADB and Citi have collaborated on trade finance initiatives, facilitating US$6.2 billion worth of trade. This partnership has enhanced SME access to finance and contributed to regional resilience. The new master risk participation agreement is set to support over US$100 million in additional trade annually in Asia and the Pacific. Since its inception in 2009, TSCFP has supported US$57 billion in trade, covering more than 45,000 transactions in emerging markets. This agreement between ADB and Citi aims to further empower SMEs in developing Asia by providing them with the necessary financial tools to participate in and benefit from regional and global trade, thereby promoting economic growth and resilience across the region.

Source: Trade Finance Global (2024).

10.6.3 Providing Education and Training

Central banks have a pivotal role in advancing ethical banking practices within the financial sector. They can offer a range of education and training programs to banks and financial institutions, focusing on the importance of ethical banking and practical strategies for implementing these practices effectively. By enhancing the capacity of the banking sector through such initiatives, central banks ensure that financial institutions are well-prepared to support the growth of SMEs in an ethical manner. Furthermore, central banks can form strategic partnerships with financial institutions that are specifically responsible for SME development. These partnerships can facilitate specialised training on ethical banking products and innovations derived from development banks. This collaborative approach helps integrate ethical considerations into the core operations of financial institutions, thereby fostering a more responsible and supportive environment for SME growth (see Boxes 10.6, 10.7 and 10.8).

Box 10.9 Central Banks Adopting Global Policies Have Advanced Ethical Banking

Some central banks who adopted some global policies Inclusive green finance have improved on their financial inclusion level. For example, India's move from indirect to direct payments, deposited into bank accounts with associated debit cards, reduced corruption and saved US$7 billion in 2.5 years. Bank of Ghana an active member of AFI initiated its first Mobile Money Interoperability System in May 2018, and the policy promoted financial inclusion. By 2019, mobile money transactions totalled GH₵308 million. Ghana is recognized as one of the leading mobile money markets in Africa. Ghana is recognized as one of the leading mobile money markets in Africa. In 2021, mobile money accounts in Ghana increased from 32.7 million to 40.9 million.

Source: Bank of Ghana (2021), Pazarbasioglu et al. (2020) and Ozyurt (2019).

10.6.4 Supporting Sustainable Finance Initiatives

Central banks play a crucial role in promoting sustainable finance by incentivising banks to back environmentally friendly and socially responsible projects. They can implement policies that offer financial incentives, such as reduced reserve requirements or favourable regulatory treatment, to banks that finance green projects or SMEs committed to sustainable practices. These incentives can be designed to encourage banks to integrate environmental and social considerations into their lending and investment decisions. By promoting such practices, central banks help create an ecosystem where ethical banking practices are not only encouraged but also rewarded, thereby supporting the growth of SMEs that are dedicated to sustainability. This approach not only aligns financial activities with broader environmental and social goals but also strengthens the overall resilience and responsibility of the financial sector. Central banks can adopt global ethical banking frameworks, such as those set by UNEP FI, NGFS and AFI. By aligning with these principles, central banks can incorporate them into their own policies, thereby influencing commercial banks to adhere to these standards and integrate ethical practices into their operations.

10.6.5 Monitoring and Evaluation

Central banks can play a crucial role in monitoring and evaluating the impact of ethical banking practices on SME growth. By systematically assessing the outcomes of different initiatives and policies, central banks can gain insights into how these practices influence the development of SMEs. This evaluation process enables central banks to identify successful strategies and areas needing improvement. By refining their approaches based on these assessments, central banks can ensure that ethical banking practices are not only effectively supporting SMEs but also contributing to their sustainable growth. This ongoing oversight helps in adapting policies to better meet the needs of SMEs while promoting responsible financial practices.

10.7 Conclusion

The chapter examines how SME growth can be fostered through better access to finance, addressing the financial constraints commonly faced by these enterprises. It explores the role of banking innovations, particularly ethical banking, in supporting SME development. The chapter delves into the principles of ethical banking and provides an overview of SMEs, highlighting their key financial challenges. It further investigates how ethical banking practices can drive SME growth and outlines ways in which central banks can bolster ethical banking to enhance SME expansion. The chapter identified that despite ethical banking not having a widely accepted definition, ethical banking can be seen as equality in financial service delivery. Ethical banking differs from traditional banking in that it prioritises people and the environment over mere profitability. Unlike traditional banking, which primarily aims to maximise shareholder wealth, ethical banking emphasises social responsibility and environmental sustainability in its financial practices. This form of banking started from some idea of religion to avoid extending finance that does not meet their beliefs and also promotes financing services towards the marginalised.

Ethical banking has gained prominence, necessitating principles and standardisation. The chapter identifies 8 key global principles shaping ethical banking: UNEP FI, AFI policies, ISO-14000, EPs, GABV, SASB, SBFN and NGFS. Despite SMEs being vital for job creation and GDP growth, they encounter significant financing challenges. These include limited credit access, risk aversion, stringent requirements and low financial literacy. Ethical banking offers solutions by providing tailored financial products, fair financing, financial education, advisory services, support for sustainable practices, and fostering long-term relationships and inclusive banking. However, some SMEs still struggle with financing. This chapter explores how central banks can enhance ethical banking to support SME growth. It highlights the roles central banks can play, such as regulation and supervision, facilitating credit access, offering education and training, backing sustainable finance initiatives and conducting monitoring and evaluation. Effective central bank involvement can improve SMEs' access to financial services, thereby boosting their growth.

References

Abor, J., & Quartey, P. (2010). Issues in SME development in Ghana and South Africa. *International Research Journal of Finance and Economics, 39*(39), 218–228.

AfDB. (2024). *Multinational - Africa SME Program - Inclusive Growth and Job Creation FAPA - Technical Assistance Request.* https://mapafrica.afdb.org/en/projects/46002-P-Z1-KB0-007

Amoah, J., Belas, J., Dziwornu, R., & Khan, K. A. (2022). Enhancing SME contribution to economic development: A perspective from an emerging economy. *Journal Of International Studies, 15*(2), 63–76. https://doi.org/10.14254/2071-8330.2022/15-2/5

Axis Bank. (2024). *SME (Commercial Banking Group).* https://www.axisbank.com/business-banking/sme

Ayyagari, M., Demirgüç-Kunt, A., & Maksimovic, V. (2017). *SME Finance* (8241; Policy Research Working Paper).

Banco Popular de Honduras. (2024). *Letr's learn together with Banco Popular.* https://bancopopular.hn/aprendamos-juntos/

Bank of Ghana. (2021). *Summary of economic and financial data.* https://www.bog.gov.gh/wp-content/uploads/2021/09/Summary-of-Economic-Financial-Data-September-2021-1.pdf

Barigozzi, F., & Tedeschi, P. (2019). On the credibility of ethical banking. *Journal of Economic Behavior & Organization, 166*, 381–402. https://doi.org/10.1016/j.jebo.2019.07.008

Bodenhorn, H. (2016, October). *The forgotten half of finance: Nineteenth-century working-class saving, public investment, and long-run development.*

BRAC Bank. (2024). *At a Glance: BRAC Bank is a performance-driven organization, where its values are at the core of every activity.* https://www.bracbank.com/en/about-us/

Climent, F. (2018). Ethical versus conventional banking: A case study. *Sustainability, 10*(7), 2152. https://doi.org/10.3390/su10072152

Dery, P. (2024, April 18). *UBA's role in SME financing; empowering businesses for growth.* https://thebftonline.com/2024/04/18/ubas-role-in-sme-financing-empowering-businesses-for-growth-2/#:~:text=UBA%20provides%20SMEs%20in%20Ghana,to%20fuel%20their%20growth%20ambitions

Doku, J. N., Iddrisu, K., Bortey, D. N. A. S., & Ladime, J. (2023). Impact of digital financial technology on financial inclusion in Sub- Saharan Africa: The moderation role of institutional quality. *African Finance Journal, 25*(2).

European Commission. (2003). *SME definition.* https://single-market-economy.ec.europa.eu/smes/sme-fundamentals/sme-definition_en

GCB. (2024). *Ghana SME Sector Report 2023.* https://www.gcbbank.com.gh/research-reports/sector-industry-reports/361-sme-sector-in-ghana-2023-v1/file

Harvey, B. (1995). Ethical banking: The case of the Co-operative bank. *Journal of Business Ethics, 14*(12), 1005–1013. https://doi.org/10.1007/BF00872116

Hongbo, L., Lucien, K. A., Raphael, Y. K., & Boris, A. A. (2018). Contribution of small medium enterprises (SMEs) to economic development: Comparative study of China and Cote d'Ivoire. *International Journal of Academic Research in Business and Social Sciences, 8*(11). https://doi.org/10.6007/IJARBSS/v8-i11/5393

Iddrisu, K. (2024). Foreign bank presence and income inequality in Africa: What role does economic freedom play? *Future Business Journal, 10.* https://doi.org/10.1186/s43093-024-00357-x

Iddrisu, K., Abor, J. Y., & Banyen, K. T. (2022). Fintech, foreign bank presence and inclusive finance in Africa: Using a quantile regression approach. *Cogent Economics & Finance, 10*(1). https://doi.org/10.1080/23322039.2022.2157120

Iddrisu, K., Abor, J. Y., & Banyen, K. T. (2023). Foreign bank presence and inclusive growth in Africa: The moderating role of financial development. *African Journal of Economic and Management Studies*. https://doi.org/10.1108/AJEMS-11-2022-0444

MicroBank. (2023, July 7). *What is ethical banking?*

Ozyurt, S. (2019, July 9). *Ghana is now the fastest-growing mobile money market in Africa*. https://qz.com/africa/1662059/ghana-is-africas-fastest-growing-mobile-money-market

Pazarbasioglu, C., Mora, A. G., Uttamchandani, M., Natarajan, H., Feyen, E., & Saal, M. (2020). *Digital financial services*. https://pubdocs.worldbank.org/en/230281588169110691/Digital-Financial-Services.pdf

PBC. (2019, March 7). *Highlights of monetary policies in 2018*.

Post, J. E., & Wilson, F. S. (2011). Too good to fail. *Stanford Social Innovation Review, 9*(4), 66–71. https://doi.org/10.48558/VXAZ-0494

San-Jose, L., Retolaza, J. L., & Gutierrez-Goiria, J. (2011). Are ethical banks different? A comparative analysis using the radical affinity index. *Journal of Business Ethics, 100*(1), 151–173. https://doi.org/10.1007/s10551-011-0774-4

Sasia, P., & De la Cruz, C. (2008). *La Banca Ética y Ciudadania*. Trotta.

SBFN. (2024, April 16). *Latest News—April 16, 2024: SBFN launched its 2024 Global Progress Brief and new Data Portal*. SBFN. https://www.sbfnetwork.org/

Townsend, B. (2020). From SRI to ESG: The origins of socially responsible and sustainable investing. *The Journal of Impact and ESG Investing, 1*(1), 1–16.

Trade Finance Global. (2024, May 15). *ADB, Citi sign agreement to boost SME supply chain finance in Asia-Pacific*. https://www.tradefinanceglobal.com/posts/adb-citi-sign-agreement-to-boost-sme-supply-chain-finance-in-asia-pacific/

Valls Martínez, M. d. C., Cruz Rambaud, S., & Parra Oller, I. M. (2020). Sustainable and conventional banking in Europe. *PLoS One, 15*(2), e0229420. https://doi.org/10.1371/journal.pone.0229420

Wachira, M. M., Iddrisu, K., Abban, G., & Abor, J. Y. (2023). Sustainable financing in developing economies. In J. Y. Abor (Ed.), *Sustainable and responsible investment in developing markets* (pp. 159–171). Edward Elgar Publishing. https://doi.org/10.4337/9781803927060.00019

World Bank. (2019, October 16). *Small and Medium Enterprises (SMEs) Finance: Improving SMEs' access to finance and finding innovative solutions to unlock sources of capital*. https://www.worldbank.org/en/topic/smefinance

Zavatta, R. (2008). *Financing technology entrepreneurs and SMEs in developing countries*. http://www.infodev.org/publications

11 Gender Equality and Women's Empowerment in Banking

Khadijah Iddrisu, Isaac Ofoeda, Osman Issah, and Marijana Iddrisu

11.1 Introduction

Inclusive growth strategies stress providing equal opportunities, particularly for the poor and women, who are more vulnerable to marginalisation (Iddrisu et al., 2022; Iddrisu et al., 2023a, b, c; Ofori & Asongu, 2021). Therefore, policymakers and development institutions encourage the involvement of women in most sectors, including the banking sector. For example, the United Nations' Sustainable Development Goals (SDGs), particularly Goal 5 (Gender Equality), Goal 8 (Decent Work and Economic Growth), and Goal 10 (Equality), underscore the need for inclusive financial systems that empower women and promote their economic participation. Similarly, the World Bank, International Monetary Fund (IMF), and other regional development banks such as the African Development Bank (AfDB), European Investment Bank (EIB), and Asian Development Bank (ADB) advocate for gender-responsive financial policies and investments since it can help to reduce poverty and attain sustainable development (AfDB, 2024a; Appaya & Abbas, 2024; Comunale et al., 2023).

Gender equality in banking fundamentally entails ensuring that women enjoy equal access to financial products and services compared to men (Appaya & Abbas, 2024). Throughout history, women have faced obstacles such as restricted access to credit, savings, and insurance, driven by discriminatory practices, cultural norms, and legal restrictions (Comunale et al., 2023; Morsy, 2020; Morsy & Youssef, 2017). These barriers exacerbate economic inequalities and restrict women's full participation in the economy. Globally, women's access to finance lags behind men's despite significant progress. There was a substantial increase in bank account ownership overall from 2011 to 2017, yet a persistent 9% gender gap remains (Morsy, 2020). At the regional level, for instance, in sub-Saharan Africa (SSA), 37% of women have bank accounts compared to 48% of men, with a widening gap (Morsy, 2020). Also, North Africa faces more significant challenges, with two-thirds of adults unbanked and an 18% gender gap, the world's largest (Morsy, 2020).

These challenges have prompted many entities to create policies and banking products aimed

K. Iddrisu (✉)
School of Business, Simon Diedong Dombo University of Business and Integrated Development Studies, Wa, Ghana
e-mail: khadijahiddris@iug.edu.gh

I. Ofoeda
Department of Accounting, University of Professional Studies, Accra, Ghana
e-mail: isaac.ofoeda@upsamail.edu.gh

O. Issah
Department of Accounting, School of Business, University for Development Studies, Tamale, Ghana
e-mail: oissah@uds.edu.gh

M. Iddrisu
School of Business, University for Development Studies, Tamale, Ghana

I. N. Yakubu (ed.), *Strategic Approaches to Banking Business and Sustainable Development Goals*, Sustainable Development Goals Series, https://doi.org/10.1007/978-3-031-80744-2_11

at promoting gender equality and empowering women within the banking sector. For instance, there is the establishment of Women's World Banking (WWB) to ensure women's financial inclusion. Additionally, multilateral development banks (MDBs), including the African Development Bank (AfDB), have introduced policies such as the Affirmative Finance Action for Women in Africa (AFAWA) to advance gender equality within the banking sector (AfDB, 2024a, b). Some central banks, including the Reserve Bank of India (RBI), have implemented policies such as the RBI's Gender-Disaggregated Data Collection Policy, which aim to enhance gender equality within the banking sector (Das Barwa, 2015). Also, there is the integration of gender equality in banking in most banking institutions. Banking institutions such as the Ecobank Group, Prudential Bank Limited, and EIB have incorporated gender equality into their decision-making processes[1,2,3] This chapter explores whether these developments have genuinely enhanced gender equality in banking or if they are merely superficial measures aimed at garnering global attention.

The rest of this chapter is presented as follows: Section 11.2 discusses the overview of gender equality and women's empowerment in banking. Section 11.3 identifies the main barriers to women's participation in banking. Section 11.4 identifies the importance, challenges, and opportunities in promoting gender equality and women's empowerment in banking. Section 11.5 identifies the role central banks play in enhancing gender equality and women's empowerment in banking. Finally, this chapter concludes with a summary and practical recommendations for policymakers.

11.2 Overview of Gender Equality and Women's Empowerment in Banking

This section defines gender equality and women's empowerment in banking (GEWEB) and explores the underlying rationale for GEWEB. It also traces the development of key policies and institutions that have supported and contributed to the advancement of GEWEB.

The GEWEB lacks a universally accepted definition but embodies the principle of providing equal banking opportunities for women. Essentially, GEWEB encompasses initiatives and strategies designed to ensure women have equal access to financial services, encourage their involvement in financial decision-making, and support their career progression within the banking sector (Appaya & Abbas, 2024; Morsy, 2020; Morsy & Youssef, 2017). It focuses on the inclusion of women as both customers and professionals in the financial industry. GEWEB is crucial for driving economic development, minimising gender disparities, and enhancing social and economic inclusion.

The concept of GEWEB emerged from the acknowledgement that women frequently encounter vulnerabilities and obstacles that limit their ability to make crucial decisions for their growth and development. Just as women faced restrictions in community decision-making and marital choices, similar limitations existed in the banking sector. For instance, in the nineteenth century, coverture laws in the United States and the United Kingdom subsumed married women's legal rights and obligations under those of their husbands, effectively stripping them of independent financial agency (Auchmuty, 2016). Women often required male co-signers for loans, and banks were reluctant to provide financial services to women, viewing them as financially dependent on men (Auchmuty, 2016). In Africa, a combination of cultural, legal, and socio-economic factors has historically marginalised women from banking and financial services. Traditional gender roles often designated men as the main financial decision-makers, while women were expected to focus on household management and childcare,

[1] https://ecobank.com/group/news-and-media/news?news=20240322123810022H2XQEB62Z6. Accessed 26 June 2024.

[2] https://www.prudentialbank.com.gh/news/306-prudential-bank-ltd-holds-maiden-women-empowerment-seminar.html. Accessed 26 June 2024.

[3] https://www.eib.org/en/projects/topics/social-sustainability/gender/index. Accessed 26 June 2024.

restricting their engagement in financial activities. For instance, in Nigeria, women were typically required to defer to their husbands or male relatives for financial decisions, limiting their ability to independently open bank accounts or access credit (Akanle, 2011). Customary laws frequently favoured men in matters of property and inheritance rights, depriving women of the collateral needed for loans or business ventures. In regions of Kenya and Uganda, customary land tenure systems predominantly assigned land ownership and inheritance rights to men, creating significant barriers for women seeking to use land as collateral for loans (Jack & Suri, 2011).

Subsequently, governments and community leaders recognised the need to ensure gender equality, acknowledging that widowed women often faced significant challenges in surviving without equitable access to financial resources. Therefore, various laws and decrees made women to be recognised. For instance, in the United States, the Equal Credit Opportunity Act (ECOA) of 1974 was a landmark in prohibiting credit discrimination based on gender (see Box 11.1), but before this, women frequently faced rejection for loans or credit cards unless they had a male co-signer in the twentieth century.

Box 11.1: Equal Credit Opportunity Act (1974) in the United States

The Equal Credit Opportunity Act (ECOA) was initially passed as part of the Consumer Credit Protection Act and signed into law on 28 October 1974. It prohibits any creditor or card issuer from discriminating against individuals based on their sex or marital status when approving or denying credit. It establishes that any entity engaging in such discrimination is civilly liable to the affected individual for monetary and punitive damages. Federal district courts, alongside state courts, have jurisdiction to hear cases under this act, regardless of the amount in dispute. ECOA facilitates electronic disclosures and enhances regulatory oversight, with significant contributions from the 1976, 1985, 1991, 1994, 1996, 2003, 2009, and 2010 amendments. The Department of Justice (DOJ) can sue under the Equal Credit Opportunity Act (ECOA) for patterns of discrimination, particularly in home loans, alongside the Fair Housing Act. Victims of unfair credit practices in residential property can file complaints with the Department of Housing and Urban Development (HUD) or pursue lawsuits. Federal agencies regulate lenders for ECOA compliance and refer suspected discrimination cases to the DOJ, which guides the handling of these referrals. Annually, the DOJ reports its ECOA activities to Congress. The Consumer Financial Protection Bureau (CFPB) oversees Regulation B for fair lending. The CFPB, Office of the Comptroller of the Currency (OCC), Federal Reserve Board (FRB), Federal Deposit Insurance Corporation (FDIC), National Credit Union Administration (NCUA), and Federal Trade Commission (FTC) each monitor different lender types for ECOA compliance.

Source: U.S. Department of Justice (2024)

Women's advocacy groups began to push for greater access to financial services and equitable treatment within the banking sector. The formation of women-centric financial institutions, like

Box 11.2: Women's World Banking

Women's World Banking (WWB) emerged from the 1975 World Conference on Women in Mexico City, where visionaries Michaela Walsh, Ela Bhatt, Esther Ocloo, and Mary Okelo identified financial access as a critical barrier for women to fully participate in the global economy. Michaela Walsh became the organisation's first president in 1979.

(continued)

Box 11.2: (continued)

Despite decades of progress, nearly 1 billion women worldwide remain underserved by the financial sector. For over 40 years, WWB has promoted women's financial inclusion and economic empowerment. WWB collaborates with key partners such as Elevandi, which fosters international collaborations on technological innovation; Giving Women, a network supporting social change projects for vulnerable women; and the Women Entrepreneurs Finance Code, a coalition aiming to increase funding for women-led businesses. WWB operates in Africa, Latin America, South Asia, and Southeast Asia. Women's World Banking collaborates with financial service providers to develop solutions that cater to women's specific needs, capabilities, and aspirations throughout the entire customer journey. This includes activities from marketing and awareness to onboarding, product experience, and encouraging repeat usage to enhance user engagement. The aim isn't just access to financial services but ensuring usage that leads to substantial empowerment for women.

Source: WWB (2024)

WWB, aimed to promote financial inclusion for women globally (see Box 11.2).

In the late 1990s, the United Nations acknowledged the importance of advancing gender equality through various policies, including the UN Conference on Environment and Development (UNCED) in Rio de Janeiro in 1992, the UN Conference on Human Rights in Vienna in 1993, the International Conference on Population and Development in Cairo in 1994, and the World Summit for Social Development in 1995. However, it was the creation of the UN Beijing Declaration and Platform for Action in 1995 that marked a pivotal moment, introducing comprehensive policies that directly addressed gender equality without limitations (see Box 11.3).

Box 11.3: Beijing Declaration and Platform for Action

The Beijing Declaration and Platform for Action, established during the Fourth World Conference on Women in September 1995. It identified 12 critical areas requiring urgent action to achieve greater gender equality and opportunities. These areas include women and poverty, education and training of women, women and health, violence against women, women and armed conflict, women and the economy, women in power and decision-making, institutional mechanism for the advancement of women, human rights of women, women and the media, women and the environment, and the girl child. This agenda aims to accelerate the implementation of strategies for women's advancement and eliminate barriers to their full participation in economic, social, cultural, and political life. It advocates for shared power and responsibility between women and men in all spheres, affirming equality as a human rights principle essential for social justice, development, and peace. The platform underscores the indivisibility of women's rights as human rights, aligning with the Vienna Declaration and Programme of Action. It highlights partnership and solidarity and emphasises collective action to address common concerns and diverse challenges women face globally. Success hinges on robust commitment from governments, international organisations, and institutions at all levels, backed by sufficient resources and mechanisms for accountability. This comprehensive framework underscores the necessity of sustainable economic growth and development to support social justice and gender equality worldwide.

Source: Philippine Commission on Women (2024)

Box 11.4: United Nations Women

UN Women, established in July 2010 by the United Nations General Assembly, addresses longstanding challenges in promoting global gender equality. Before its creation, the UN faced funding inadequacies and lacked a centralised authority to steer gender equality efforts. The formation of UN Women was a pivotal part of the UN reform agenda, consolidating resources and mandates from four previous entities: the Division for the Advancement of Women (DAW), the International Research and Training Institute for the Advancement of Women (INSTRAW), the Office of the Special Adviser on Gender Issues and Advancement of Women (OSAGI), and the United Nations Development Fund for Women (UNIFEM). UN Women functions as the leading UN entity dedicated to advancing gender equality and empowering women worldwide. It champions the rights and needs of women and girls, striving to accelerate progress towards their empowerment globally. The entity collaborates with governments and civil society to develop and implement laws, policies, programmes, and services that uphold global standards for gender equality.

UN Women focuses on four strategic priorities:

1. Ensuring women's equal participation and leadership in governance
2. Promoting income security, decent work, and economic autonomy for women
3. Ending all forms of violence against women and girls
4. Enhancing women's roles in peace-building, resilience-building, and humanitarian action

Additionally, UN Women coordinates and advocates for gender equality within the broader UN system, particularly in relation to the Sustainable Development Goals (SDGs). It emphasises the fundamental role of gender equality in achieving sustainable development and monitors progress across the UN system. Gender equality remains crucial as it is not only a human rights imperative but also pivotal for socio-economic development. Despite progress through agreements like the Beijing Declaration and Platform for Action and the Convention on the Elimination of All Forms of Discrimination against Women (CEDAW), entrenched gender inequalities persist globally. UN Women continues to support inter-governmental bodies, such as the Commission on the Status of Women, in formulating policies and norms while providing technical and financial assistance to member states and fostering partnerships with civil society to advance gender equality worldwide.

Source: UN Women (2024)

Subsequently, additional policies like Habitat II in Istanbul in 1996 continued to endorse gender equality. Despite these early initiatives, there was a notable absence of a dedicated body to ensure consistent oversight and commitment to implementing these policies. Therefore, in 2010, the UN established UN Women with the specific mandate of enhancing the effective implementation of UN policies regarding gender equality (see Box 11.4).

The creation of UN Women encouraged a lot of countries to also create an agency and/or ministries that cater for women's empowerment and gender equality. Some of these agencies include the Ministry of Women and Child Development (India), the Ministry of Gender, Children, Disability and Social Welfare (Malawi), the Department of Women, Youth and Persons with Disabilities (South Africa), the Office for Gender Equality and Family (Norway), the Department for Gender Equality (Sweden), the Ministry of Women (Brazil), the Ministry of Women, Family and Human Rights (Brazil), the Ministry of

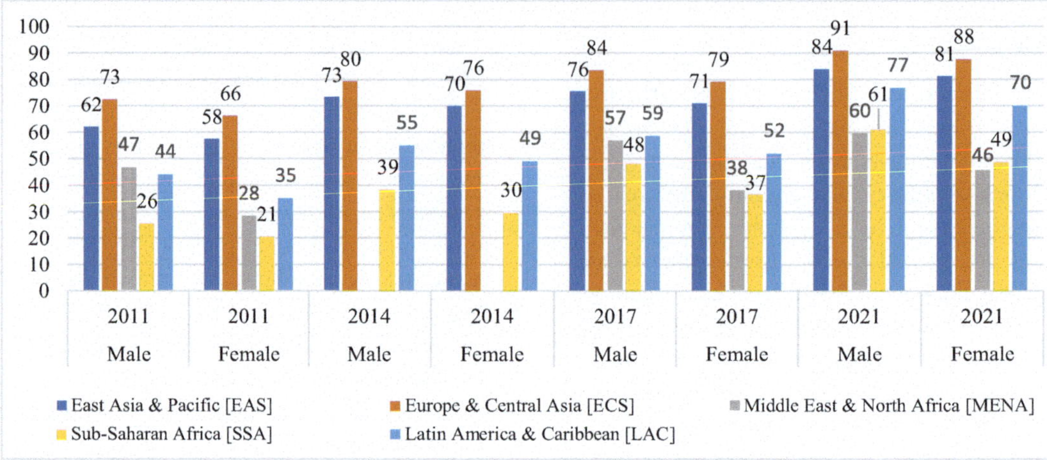

Source: Authors' Construct from Global Findex Data, World Bank 2024

Fig. 11.1 Regional gender account holding (% age 15+), 2011–2021. (*Source*: Authors' Construct from Global Findex Data, World Bank (2024))

Women Affairs and Social Development (Nigeria), the Ministry of Gender, Labour and Social Development (Uganda), the Ministry of Gender Equality and Family (Republic of Korea), the Ministry of Gender, Children and Social Protection (Ghana), and the Philippine Commission of Women (Philippines). Many of these agencies have influenced policymakers to integrate women into policy-making processes. For example, the UN Agenda 2030 established SDG 5 and SDG 10, which directly advocate for women's protection and equal access to development, alongside other SDGs indirectly supporting women. The African Development Agenda 2063, known as 'The Africa We Want', includes Aspiration 6.1, specifically focusing on empowering women in the region.[4] Additionally, other regional policies such as the Association of Southeast Asian Nations (ASEAN) Declaration on the Advancement of Women, the Pacific Islands Forum Gender Equality Declaration, the European Union Gender Equality Strategy, and the African Union's gender architecture such as Article 4 (L) of the Constitutive Act of the African Union, the Protocol to the African Charter on Human and People's Rights on the Rights of Women in Africa, the Solemn Declaration on Gender Equality in Africa, the Fund for African Women, and the African Women's Decade also prioritise women. Today, numerous advocacy efforts by various pressure groups and development institutions aim to advance gender equality across workplaces, homes, social settings, and beyond.

Despite numerous policies aimed at addressing gender inequality, the issue persists globally. For instance, 10.3% of women worldwide live in extreme poverty, experiencing greater economic deprivation than men (UN Women, 2024a). Access to formal banking services remains crucial for women's financial inclusion, as over 700 million women lack such access globally. In Africa, despite a growing number of women-owned businesses—ten million in developing countries alone—women entrepreneurs still face significant challenges, with 70% reporting inadequate access to growth capital compared to their male counterparts (UN Women, 2024a).

Available data from the Global Financial Inclusion (GFI) of the World Bank also revealed that there remains a disparity in account ownership between men and women (see Fig. 11.1). Therefore, the next section discusses the chal-

[4] https://au.int/en/agenda2063/aspirations#:~:text=Aspiration%206%3A%20An%20Africa%2C%20whose,decision%20making%20in%20all%20aspects. Accessed 26 June 2024.

lenges women face in participating in or accessing finance.

11.3 Barriers to Women's Participation in Banking

As previously mentioned, GEWEB has not been progressing as anticipated in light of various policies and advocacy efforts. However, some challenges have not yet been adequately addressed. This section explores these challenges.

Lack of Financial Independence
At prime working age, only 61% of women participate in the labour force compared to 90% of men (ILO, 2024). With a lower participation rate in the labour force, many women have limited or no personal income (ILO, 2024; UN Women, 2024b). This makes it difficult for them to save money, as they do not have sufficient funds beyond their basic needs. A woman with irregular or no personal income might prioritise immediate expenses over savings, making formal banking seem less relevant to her day-to-day financial management (UN Women, 2024b). Also, women who are not actively participating in the labour market may have fewer opportunities to develop financial literacy or experience with managing money. This can create a lack of confidence in dealing with financial institutions and a preference for informal financial practices. Additionally, in many societies, traditional gender roles and cultural expectations contribute to women's financial dependence (Nartey et al., 2023). These norms can discourage women from seeking financial independence or managing their own finances, perpetuating reliance on male family members. In some cases, women rely on their husbands if married, or their fathers if unmarried, for financial support. A married woman who depends on her husband's income may not feel empowered to open her own bank account or save independently, as household funds are managed collectively or by her spouse. An unmarried woman living with her parents may receive an allowance or financial support from her father, reducing her incentive to engage with banking systems on her own.

Inadequate Knowledge of Various Empowerment Programmes
Recently, there has been an increase in women's empowerment programmes in banking organised by development partners like UN Women, MDBs, and other financial institutions. For instance, the UN Women has numerous programmes focused on empowering women to engage in economic activities, including banking (see Box 11.4). At the regional level, the AfDB has a programme like AFAWA, which aims to fund, offer technical support, and create a conducive business environment for women in Africa (see Box 11.5). Several local banking institutions, including Ecobank Group and Prudential Bank Limited, have imple-

Box 11.5: Affirmative Finance Action for Women in Africa

Affirmative Finance Action for Women in Africa (AFAWA) is a transformative initiative by the African Development Bank (AfDB) aimed at enhancing women's participation in Africa's economy through strategic financial and supportive measures. Central to AFAWA's approach are innovative financial instruments and comprehensive support mechanisms designed to address the systemic barriers women face in accessing finance and achieving business success. One of AFAWA's key financial instruments is a $300 million risk-sharing facility, which aims to mobilise US$3 billion in credit for women-owned businesses across Africa. This initiative leverages partnerships with commercial banks and microfinance institutions to foster structural changes that benefit women entrepreneurs.

(continued)

Box 11.5: (continued)

Additionally, AFAWA has introduced a rating system that evaluates financial institutions based on their lending practices to women and their socio-economic impact. Institutions performing well under this system receive preferential financing terms from the African Development Bank, incentivising increased lending to women.

AFAWA also provides technical assistance to financial institutions, ensuring they effectively implement products tailored for women and enhance the capacity of women entrepreneurs through training in business development and financial planning. Moreover, AFAWA engages with African governments and stakeholders to advocate for legal, policy, and regulatory reforms that dismantle barriers hindering women's economic participation. In practice, AFAWA has made substantial investments, including dedicating over US$50 million in credit lines to women through partnerships with institutions like Fidelity Bank and Kenya Commercial Bank. It has also invested US$12.5 million in Alitheia IDF Managers, a pioneering private equity fund managed by women that targets high-growth small and medium enterprises (SMEs) owned or led by women in Southern and Western Africa. This fund aims to raise US$100 million across ten countries, catalysing significant economic empowerment for women entrepreneurs. Through initiatives like technical assistance and policy dialogue with central banks and governments, AFAWA is actively fostering an enabling environment for women in business. Additionally, the development of the 50 Million African Women Speak digital platform is set to connect female entrepreneurs continent-wide, further enhancing their access to markets and resources.

Source: AfDB (2024a, b)

mented policies to promote women's empowerment within the banking sector[1,2,3]. However, many women remain unaware of these programmes, often due to corruption, limited education, and inequality. In Africa, for instance, such programmes are frequently accessed by individuals connected to those in power, leaving many women without access. Additionally, many women lack education, making it difficult for them to understand how these programmes operate. Moreover, some women face corruption, needing to pay bribes to access these opportunities.

Poverty

Globally, women experience higher poverty rates compared to men, with a gap of 10.3% (UN Women, 2024b). This economic disparity means that women living in poverty often cannot afford the minimum deposits or associated costs required by many financial institutions to open and maintain accounts. Additionally, transaction fees, account maintenance charges, and automated teller machine (ATM) usage fees can be prohibitive, further deterring women with limited financial resources from accessing formal banking services. Many women in developing countries live hand-to-mouth, where their income is barely sufficient to cover daily expenses, leaving little or nothing for savings. For example, a World Bank (2024) survey on SSA revealed that the predominant barrier to women's participation in banking is a lack of money.

Access Barrier

To comply with Know Your Customer (KYC) regulations, financial institutions generally require identification documents such as national IDs, passports, or proof of address. Also, to obtain a loan, documentation like credit histories, proof of income, or collateral is needed. However, most women do not have all or some of these documentations. For instance, a World Bank (2024) survey showed that women, especially those in SSA, do not have the necessary documents to own a formal account or obtain a loan. Women lacking these records face challenges in accessing credit, thereby limiting their financial opportunities. As a result, they often turn to informal financial systems, which do not offer the protections and benefits of formal banking, such as regulated interest rates or secure savings options. Banks may lack gender-sensitive practices or culturally appropriate services that cater to the needs of women, further alienating female customers. A bank with predominantly male staff may not provide privacy or respect cultural norms, dissuading women from seeking financial assistance.

Women in rural or remote areas often encounter obstacles in accessing physical bank branches due to considerable distances and insufficient transportation infrastructure. Given their predominant roles in household and caregiving responsibilities, they often struggle to find the time needed to commute to distant banks, especially given the constraints of limited opening hours. Moreover, banks located in rural or underserved regions typically offer fewer services compared to their urban counterparts, thereby restricting women's access to a wide range of financial products. For instance, rural banks may lack investment services or digital banking options, further limiting women's opportunities to leverage these essential financial tools. For example, in certain economies in SSA, women were more prone than men to identify inadequate documentation as a significant barrier (World Bank, 2024). Distance to physical bank branches and the cost associated with financial services were also frequently cited as reasons for difficulty.

Lack of Confidence in the Financial Sector

Some women, especially those in the SSA where there is a high gap, do not have confidence in the bank (Abebe et al., 2017; World Bank, 2024). Women who distrust financial institutions are less likely to open bank accounts, use credit services, or engage with financial products. This stems from concerns about the security of their funds, privacy, or fears of being exploited. Women might avoid savings accounts due to past experiences of mismanagement or hidden fees by banks (Iddrisu et al., 2022). In Ghana, the collapse and mergers of banks have undermined women's confidence and trust in the banking sector, leading them to prefer keeping their money at home (Yomboi et al., 2021).

11.4 Gender Equality and Women's Empowerment in Banking: Importance and Opportunities

Promoting gender equality and empowering women within the banking sector are essential for promoting comprehensive economic growth and social development. Ensuring that women have equitable access to financial services and opportunities is not just a fundamental human rights issue but also a critical strategic necessity for fostering sustainable development.

11.4.1 Importance

11.4.1.1 Financial Inclusion

As of 2021, women comprise 55% of the global population without access to formal banking services, meaning they lack access to essential financial products such as banking and insurance. Globally, women encounter persistent disparities in opportunities compared to men, particularly evident in micro, small, or medium enterprise (MSME) finance, where there exists a significant US$1.7 trillion gap. Despite women owning 22% of micro-enterprises and 32% of small and medium enterprises (SMEs), 740 million women worldwide still lack access to banking services (UN Women, 2024a). Additionally, 2.4 billion women do not enjoy equal economic rights relative to men. Enabling women's access to banking services empowers them to participate in savings initiatives, invest in economic opportunities, and establish financial security. For example, the efforts by XA Bank to develop a credit facility for women have led to improvements in the business prospects of some women in China (see Box 11.6). This access not only enhances their ability to manage and accumulate wealth but also fosters greater engagement in economic activities, thereby contributing to global economic stability and Sustainable Development Goals.

Box 11.6: XA Bank Develops Credit Facility for Women in China

XA Bank, a newly licensed private bank, identified both a significant challenge and an opportunity to advance gender equality in Anhui Province. Under the leadership of CEO Mr. Sun, who had previously worked with Ma'anshan Rural Commercial Bank and witnessed the success of the World Bank-sponsored inter-process communication (IPC) technology, XA Bank sought to empower women entrepreneurs using modern technology. Mr. Sun's experience with IPC, which effectively facilitated credit provision through offline loan officers, inspired XA Bank to develop a more tech-driven approach to inclusive financing. Recognising the potential for technological innovation, XA Bank, supported by the i3 program, conducted thorough market research and developed a natural language processing (NLP)-based digital application tailored for women entrepreneurs. This new product aims to address credit demands ranging from 200,000 RMB to 1,000,000 RMB, offering interest rates between 7 and 8%. The digital application streamlines the lending process by allowing users to interact with a virtual 'investigator', answering questions and uploading necessary information through the app. The backend analytics then perform automatic comparisons and assessments. The process is designed to be efficient and user-friendly, requiring only one field visit from the bank for contract signing. Loans are disbursed within two days, significantly reducing the traditional barriers and delays associated with securing financing. This approach not only improves access to credit for women entrepreneurs but also aligns with XA Bank's commitment to fostering gender equality and supporting sustainable economic development in the region.

Source: Yi and Singh (2022)

11.4.1.2 Economic Empowerment

Access to credit and financial resources plays a pivotal role in empowering women economically (Iddrisu et al., 2022). When women have access to credit, they can utilise these financial resources to start new businesses or expand existing enterprises. This capability not only enhances their entrepreneurial opportunities but also allows them to participate more actively in economic activities (Orser et al., 2006). By starting businesses or scaling up their enterprises, women can generate income, create employment opportunities for others, and contribute to economic growth within their communities and beyond. This process is instrumental in poverty reduction as it helps lift households out of poverty, improves living standards, and promotes socio-economic development (Simba et al., 2023). Furthermore, empowered women who are economically active can reinvest their earnings back into their families, particularly in areas such as education and healthcare. This holistic approach contributes to building more resilient and sustainable communities over the long term. Therefore, facilitating access to credit and financial resources for women not only empowers them individually but also has far-reaching positive impacts on broader economic development and poverty alleviation efforts.

11.4.1.3 Social Impact

When women have control over their finances, they are more likely to invest in their own education and that of their children (Forsgren et al., 2019). They can afford school fees, books, uniforms, and other educational expenses. Educated women tend to make informed decisions about their families' health, nutrition, and overall well-being. They also become role models for their children, emphasising the importance of education and fostering a culture of learning within the household (Guiso & Zaccaria, 2023). Financially independent women can access quality healthcare services for themselves and their families. They can afford regular check-ups, vaccinations, and treatments when needed, reducing the risk of illness and improving overall health outcomes

(Struckell et al., 2022). Moreover, women who control household finances are more likely to prioritise preventive healthcare measures and nutrition, leading to healthier lifestyles for their families. Improved education and healthcare outcomes contribute significantly to breaking the cycle of intergenerational poverty. Children of educated mothers tend to have better educational attainment and employment prospects, breaking free from the constraints of poverty in future generations. Similarly, better health outcomes lead to increased productivity and economic opportunities for families, contributing to sustained economic growth and development.

11.4.2 Opportunities

11.4.2.1 Technology and Innovation

Digital banking solutions have the potential to surmount traditional physical barriers and extend financial services to women living in remote and underserved areas (Appaya & Abbas, 2024). Technologies such as mobile banking, digital wallets, and fintech innovations play a pivotal role in enhancing the accessibility and affordability of financial services for women. For example, mobile banking enables women to conduct financial transactions using their smartphones, bypassing the need for physical bank branches. This convenience is particularly beneficial in rural areas where banks are scarce or distant, allowing women to manage their savings, make payments, and access credit without travelling long distances. Digital wallets provide a secure and convenient platform for women to store money digitally and make transactions electronically (Caron, 2022). They can facilitate peer-to-peer payments, bill payments, and online purchases, offering flexibility and ease of use, especially for women who may not have access to traditional banking infrastructure. Digital wallets provide a secure and convenient platform for women to store money digitally and make transactions electronically. They can facilitate peer-to-peer payments, bill payments, and online purchases, offering flexibility and ease of use, especially for women who may not have access to traditional banking infrastructure (Iddrisu et al., 2022). Financial technology innovations such as peer-to-peer lending platforms, crowdfunding, and blockchain-based financial services create new opportunities for women to access credit and investment opportunities. These platforms often have lower barriers to entry and streamlined processes, making financial services more inclusive and accessible to women entrepreneurs and small business owners.

11.4.2.2 Capacity Building

Investing in women's financial education and entrepreneurship training is vital as it provides women with the essential knowledge, skills, and confidence to effectively manage their finances and access financial services. While various initiatives aim to invest in women, these efforts are still insufficient. For instance, according to UN Women (2024b), programmes dedicated to gender equality account for only 4% of official development assistance. To achieve gender equality and women's empowerment, an additional USD 360 billion per year is needed in developing countries. This amount is a fraction of the US$2.2 trillion spent globally on military expenditures in 2022. Therefore, significant investments are required, such as in care services like daycare and elderly care, which could create nearly 300 million jobs by 2035. Additionally, closing gender employment gaps could increase gross domestic product per capita by 20% across all regions (UN Women, 2024b).

11.4.2.3 Additional Policy Implementation

Governments can enact policies that promote gender equality and address barriers hindering women's access to financial services (Guiso & Zaccaria, 2023; Struckell et al., 2022). This includes legal reforms to ensure women's property rights, inheritance rights, and access to credit are protected and enforced (Simba et al., 2023). Policy frameworks that mandate gender-disaggregated data collection and reporting also help in monitoring progress towards gender equality in finance. For example, the RBI introduced a policy that is gender-disaggregated data collection (see Das Barwa, 2015). Financial institutions can implement regulations and guidelines that are sensitive to gender disparities. This may involve designing financial products and services that cater specifically to women's needs and preferences, such as flexible loan repayment schedules, savings products tailored for irregular income streams, or microfinance initiatives targeting women entrepreneurs. Non-Governmental Organisations (NGOs) and development organisations often lead targeted programmes aimed at enhancing women's financial literacy, entrepreneurship skills, and access to financial services. These programmes provide training and capacity-building initiatives that empower women to effectively manage their finances, make informed financial decisions, and participate more actively in economic activities. Advocacy efforts by civil society organisations and NGOs raise awareness about the importance of women's financial inclusion among policymakers, financial institutions, and communities. They advocate for gender-responsive practices and policies within the financial sector, promoting greater accountability and commitment to gender equality goals.

11.5 The Role of Central Banks in Empowering Women in Banking

Monitoring

Central banks should actively monitor financial institutions, especially microfinances, to ensure they stay aligned with their foundational objectives of empowering women and petty traders (Abor et al., 2024). Over time, some microfinance institutions (MFIs) may shift their focus from their mission, potentially prioritising profit over their social goals. Regular audits and compliance checks can verify whether financial institutions are adhering to regulations that promote financial inclusion and support for women and small traders (McKee et al., 2011). Central banks can analyse data on lending patterns, interest rates, and customer demographics to ensure that MFIs are serving their intended beneficiaries. Gathering feedback from customers, particularly women and small traders, provides insights into whether financial institutions are meeting their needs effectively.

Penalties and Awards

Sanctions and penalties act as deterrents to prevent financial institutions from straying from their fundamental mission of supporting vulnerable groups. These measures help ensure that institutions remain dedicated to their purpose. Central banks can impose fines on financial institutions that do not adhere to regulations or engage in practices misaligned with their mission. Additionally, they can restrict these institutions from offering specific financial products or services until they achieve compliance. In extreme cases, central banks may revoke the licences of institutions that repeatedly violate regulations or engage in harmful practices.

However, central banks should develop gender equality award programmes with well-defined criteria, a transparent selection process, and substantial rewards. To enhance these programmes, central banks can partner with various stakeholders, including government agencies, NGOs, and industry associations, ensuring the initiatives have widespread support and significant impact. Continuous evaluation of awarded institutions is crucial to guarantee they maintain and progress their gender equality initiatives, thereby validating the awards and ensuring accountability.

Education on the Existing Programmes

Central banks can partner with commercial banks, microfinance institutions, and credit unions to create and distribute educational materials about women-specific banking products and international empowerment programmes. Financial institutions have direct access to customers and can effectively communicate information through their branches, customer service representatives, and online platforms. Joint awareness campaigns can be organised to highlight the benefits and availability of women-tailored financial products such as savings accounts with no minimum balance, loans with flexible repayment terms, and insurance products designed for women's unique needs. These campaigns can also inform women about empowerment programmes such as those offered by UN Women and regional initiatives aimed at enhancing women's economic participation. Central banks can facilitate workshops and training sessions in collaboration with financial institutions to educate women about managing finances, understanding credit, and making informed investment decisions. These sessions can be tailored to the specific needs of women, addressing common financial challenges they face.

Policy and Regulatory Frameworks

Central banks play a crucial role in fostering gender equality in financial services by crafting targeted policies that address the distinct financial needs of women and dismantle barriers to their inclusion. They develop regulations that specifically aim to improve women's access to financial services, such as requiring banks to offer products tailored to women's needs. By mandating the collection and reporting of gender-disaggregated data from financial institutions, central banks can pinpoint gaps in service provision. This detailed data enables central banks to adjust policies and implement targeted measures to enhance women's access to financial services, ensuring that financial products and services reach women effectively. For example, RBI mandates gender-disaggregated data collection from banks, which has helped in shaping policies that support women's financial inclusion (see Box 11.7).

Box 11.7: The Reserve Bank of India's (RBI) Gender-Disaggregated Data Collection Policy

The Reserve Bank of India's (RBI) Gender-Disaggregated Data Collection Policy, initiated in 2014, mandates that banks systematically collect and report detailed gender-specific data on various financial services and products. This includes information on account ownership, loan applications, credit disbursements, repayment rates, and the use of savings and investment products. The policy aims to identify, analyse, and address the disparities between men's and women's access to financial services, thereby fostering targeted interventions that enhance women's financial inclusion. Banks must regularly submit this data to the RBI, facilitating timely analysis and intervention and promoting transparency and accountability within the banking sector. By revealing gender gaps in financial access and usage, the data collected under this policy supports the development of gender-responsive financial products and services, contributing to a more inclusive financial system.

The implementation of this policy was influenced by the recommendations of the Nachiket Mor Committee on Comprehensive Financial Services for Small Businesses and Low-Income Households in 2013, which highlighted the need for gender-specific financial data to understand and address the unique financial behaviours and needs of women. Globally, such data collection aligns with recommendations from international bodies like the World Bank and the UN. The insights gained from gender-disaggregated data have enabled the RBI and other stakeholders to craft policies and initiatives that support women entrepreneurs, improve access to credit, and enhance financial literacy among women. This approach has increased the number of women accessing formal financial services, encouraged banks to develop more inclusive practices, and ensured continuous monitoring and accountability for gender equality in financial inclusion. Examples include the formulation and adjustment of financial inclusion plans and evaluating the impact of initiatives like the Pradhan Mantri Jan Dhan Yojana (PMJDY), tailored to better serve women's financial needs.

Source: Das Barwa (2015)

Additionally, central banks can advocate for or establish legal frameworks that protect women's economic rights, such as property and inheritance rights, and ensure equal access to credit. These legal protections are essential for empowering women to participate fully in the financial system. For example, securing women's property rights enables them to use assets as collateral for loans, while anti-discrimination laws in lending practices ensure fair access to credit. By addressing these structural barriers, central banks help integrate women into the financial system, promoting broader financial inclusion and economic empowerment for women.

11.6 Conclusion

Gender equality and women's empowerment in the banking sector are vital for achieving inclusive economic growth and sustainable development. Despite advances in policy and advocacy, women continue to face significant barriers in accessing financial services, which undermines their economic participation and potential. Addressing these barriers is essential to fully leverage women's roles as clients, employees, and leaders in financial institutions. This chapter identified that women globally face persistent challenges such as financial dependence, limited knowledge of empowerment programmes, poverty, access barriers, and lack of confidence in financial systems. These issues are more severe in regions like Sub-Saharan Africa and North Africa due to cultural, legal, and socio-economic factors. Ensuring financial inclusion is crucial for women's economic empowerment, as it allows them to engage in economic activities, invest in businesses, and enhance their socio-economic status. Financially empowered women contribute to economic growth, poverty reduction, and social development, investing more in education, healthcare, and the well-being of their families.

Opportunities for enhancing women's financial inclusion lie in technology and innovation, such as digital banking solutions and fintech. These tools can make financial services more accessible, especially in underserved areas. Capacity building through financial education and entrepreneurship training is also crucial to boost women's confidence and skills in managing finances. Central banks play a pivotal role in promoting gender equality in banking by enforcing regulations, educating on existing programmes, and developing gender-sensitive policies. Recommendations include strengthening policy implementation, expanding financial literacy programmes, leveraging technology, enhancing monitoring, and increasing awareness and access to financial services for women.

References

Abebe, J. O., Maina, L., Ondiek, J., & Ogolla, C. (2017). *Background paper: Driving gender-responsive financial inclusion models in Africa*. https://africa.unwomen.org/sites/default/files/Field%20Office%20Africa/Attachments/Publications/2018/03/Background%20Paper%20on%20Gender%20Responsive%20Financial%20Inclusion%20in%20Africacompressed.pdf

Abor, J. Y., Subramanian, L., Iddrisu, K., & Nsor-Ambala, R. (2024). Multilateral development banks: Contributions and challenges. In *Perspectives on development banks in Africa* (pp. 277–298). Springer Nature Switzerland. https://doi.org/10.1007/978-3-031-59511-0_12

AfDB. (2024a). *Gender*. https://www.afdb.org/en/topics-and-sectors/sectors/gender

AfDB. (2024b). *What does AFAWA do?* https://www.afdb.org/en/topics-and-sectors/initiatives-partnerships/afawa-affirmative-finance-action-for-women-in-africa/what-does-afawa-do

Akanle, O. (2011). The sociology of gender equality and development in democratizing Nigeria. *The Nigerian Journal of Sociology and Anthropology, 9*(1). https://doi.org/10.36108/NJSA/1102/90(0110)

Appaya, S., & Abbas, S. B. (2024, March 6). *Empowering women through digital financial inclusion*. World Bank Blog: Private Sector Development Blog. https://blogs.worldbank.org/en/psd/empowering-women-through-digital-financial-inclusion

Auchmuty, R. (2016). In T. Stretton & K. J. Kesselring (Eds.), *Review of the book Married Women and the Law: Coverture in England and the Common Law World*. University of Toronto Press.

Caron, L. (2022). Empty digital wallets: New technologies and old inequalities in digital financial services among women. *Oxford Open Economics, 1*. https://doi.org/10.1093/ooec/odac001

Comunale, M., de Bruxelles, P., & Unsal, D. F. (2023, May 9). *Women struggle for equal pay and progression at*

Central Banks: A new survey reveals the gender gap persists. IMF Blog. https://www.imf.org/en/Blogs/Articles/2023/05/09/women-struggle-for-equal-pay-and-progression-at-central-banks

Das Barwa, S. (2015). *Women's financial inclusion case study using sex-disaggregated data to promote women's financial inclusion in India*.

Forsgren, H., Haslam, A., Hunt, S., Wirkus, A., & Heim, N. (2019). *Girls' Access to Education in Ghana*. Ballard Brief. https://ballardbrief.byu.edu/issue-briefs/girls-access-to-education-in-ghana

Guiso, L., & Zaccaria, L. (2023). From patriarchy to partnership: Gender equality and household finance. *Journal of Financial Economics, 147*(3), 573–595. https://doi.org/10.1016/j.jfineco.2023.01.002

Iddrisu, K., Abor, J. Y., & Banyen, K. T. (2022). Fintech, foreign bank presence and inclusive finance in Africa: Using a quantile regression approach. *Cogent Economics & Finance, 10*(1). https://doi.org/10.1080/23322039.2022.2157120

Iddrisu, K., Abor, J. Y., & Banyen, K. T. (2023a). Foreign bank presence and inclusive growth in Africa: The moderating role of financial development. *African Journal of Economic and Management Studies, 15*, 349. https://doi.org/10.1108/AJEMS-11-2022-0444

Iddrisu, K., Doku, J. N., Abor, J. Y., & Dziwornu, R. (2023b). Financial inclusion and inclusive growth in Africa: What is the moderation role of financial stability? *Cogent Economics & Finance, 11*(2). https://doi.org/10.1080/23322039.2023.2267857

Iddrisu, K., Ofoeda, I., & Abor, J. Y. (2023c). Inward foreign direct investment and inclusiveness of growth: Will renewable energy consumption make a difference? *International Economics and Economic Policy, 20*(3), 367–388. https://doi.org/10.1007/s10368-023-00562-z

ILO. (2024, May 14). *Women with young children have much lower labour force participation rates*. https://ilostat.ilo.org/blog/women-with-young-children-have-much-lower-labour-force-participation-rates/

Jack, W., & Suri, T. (2011). *Mobile Money: The Economics of M-PESA*. https://doi.org/10.3386/w16721

McKee, K., Lahaye, E., & Koning, A. (2011). Responsible finance: Putting principles to work. *Focus Note, 73*, 1–16.

Morsy, H. (2020, March). *Access to Finance: Why Aren't Women Leaning in?* IMF Blog: Finance and Development Magazine. https://www.imf.org/en/Publications/fandd/issues/2020/03/africa-gender-gap-access-to-finance-morsy

Morsy, H., & Youssef, H. (2017). *Access to Finance—Mind the Gender Gap* (202; EBRD Working Paper).

Nartey, P., Sensoy Bahar, O., & Nabunya, P. (2023). A review of the cultural gender norms contributing to gender inequality in Ghana: An ecological systems perspective. *Journal of International Women's Studies, 25*(7).

Ofori, I. K., & Asongu, S. A. (2021). ICT diffusion, foreign direct investment and inclusive growth in Sub-Saharan Africa. *Telematics and Informatics, 65*, 101718. https://doi.org/10.1016/j.tele.2021.101718

Orser, B. J., Riding, A. L., & Manley, K. (2006). Women entrepreneurs and financial capital. *Entrepreneurship Theory and Practice, 30*(5), 643–665. https://doi.org/10.1111/j.1540-6520.2006.00140.x

Philippine Commission on Women. (2024, June 26). *Beijing Platform for Action*. https://pcw.gov.ph/beijing-platform-for-action/

Simba, A., Martins Ogundana, O., Braune, E., & Dana, L. (2023). Community financing in entrepreneurship: A focus on women entrepreneurs in the developing world. *Journal of Business Research, 163*, 113962. https://doi.org/10.1016/j.jbusres.2023.113962

Struckell, E. M., Patel, P. C., Ojha, D., & Oghazi, P. (2022). Financial literacy and self employment – The moderating effect of gender and race. *Journal of Business Research, 139*, 639–653. https://doi.org/10.1016/j.jbusres.2021.10.003

U.S. Department of Justice. (2024, January 25). *The Equal Credit Opportunity Act*. https://www.justice.gov/crt/equal-credit-opportunity-act-3

UN Women. (2024a). *About UN Women*. https://www.unwomen.org/en/about-us/about-un-women

UN Women. (2024b, March 8). *1 in every 10 women in the world lives in extreme poverty*. https://www.unwomen.org/en/news-stories/press-release/2024/03/1-in-every-10-women-in-the-world-lives-in-extreme-poverty#:~:text=New%20York%2C%208%20March%202024,build%20more%20prosperous%2C%20equitable%20societies

World Bank. (2024, April 17). *Financial Inclusion in Sub-Saharan Africa—Overview*. The Global Findex Database 2021. https://www.worldbank.org/en/publication/globalfindex/brief/financial-inclusion-in-sub-saharan-africa-overview

WWB. (2024). *Making finance work for women for 45 years*. Our Stories. https://www.womensworldbanking.org/about-us/45-years/

Yi, N., & Singh, A. (2022, December 22). *How a bank in China uses technology to empower women-led SMEs: Examining financial health in practice*. Impact Capital for Development. https://www.uncdf.org/article/8092/how-a-bank-in-china-uses-technology-to-empower-women-led-smes

Yomboi, J., Nangpiire, C., Kutochigaga, E. A., & Majeed, M. (2021). The Impact of the Collapsed Banks on Customers in Ghana. *Asian Journal of Economics, Business and Accounting, 21*, 15–25. https://doi.org/10.9734/ajeba/2021/v21i1730487

Part IV

Remittances, Corporate Social Responsibility, and Sustainable Development Goals

Driving Change: Examining the Role of Digital Remittances in Achieving SDG 10.c

12

Umar Mohammed

12.1 Introduction

The 2030 Agenda for Sustainable Development, adopted by United Nations member countries in 2015, serves as a blueprint for achieving peace and prosperity for people and the planet, now and in the future (UN, 2015). The integration of migration and remittances into the SDGs and their targets highlights the significant role they play in the development process of countries (Nurse, 2018). The UN (2019) estimates that 200 million migrants support 800 million people in their home countries, meaning that 1 in 7 people are directly impacted by remittances, which constitutes 15% of the income earned by migrants. While the SDGs agenda does not explicitly detail the broader contributions of remittances, many of its goals are directly and indirectly influenced by remittances. Indeed, a growing body of empirical evidence demonstrates the multifaceted developmental benefits of remittances that promote productive investment, spur savings, and drive consumption in recipient countries (Mohammed & Karagol, 2023; Přívara & Trnovský, 2021).

SDG Target 17.3 aims to gather additional financial resources for developing countries, with Indicator 17.3.2 focusing on increasing the volume of remittances as a share of GDP (UN, 2015). Also, SDG 10 aims to reduce income inequalities among countries, with Target 10.c explicitly aiming to reduce remittance transaction costs to below 3% and eliminate remittance corridors costing more than 5%. Remittances are crucial for reducing income inequality and achieving SDG 10. However, their full potential is limited by high transaction fees. Remittances have a countercyclical effect at the macroeconomic level, reducing growth volatility and helping countries respond to policy shocks. At the microeconomic level, remittances help low-income families improve children's health and educational outcomes, build savings, and increase spending on consumer durables and human capital (Malpass, 2022). Likewise, Portes (2009) reveals that remittances positively impact the bottom 70% of the population in low-income countries while negatively affecting the top 20%, potentially reducing overall inequality. Kratou and Khlass (2023) also found that remittances from low-skilled migrants are more effective in alleviating inequality. These studies affirm remittances are critical in reducing income inequality and achieving SDG 10. Kratou and Khlass (2023) recommend keeping transaction fees below 1.93%, even lower than the United Nations' target of 3%. Lowering remittance costs will also boost remittance inflows, thus supporting SDG Target 17.3 by mobilizing financial resources from diverse sources beyond traditional aid, loans, and grants.

U. Mohammed (✉)
Department of Economics, Ankara Yildirim Beyazit University, Ankara, Türkiye

I. N. Yakubu (ed.), *Strategic Approaches to Banking Business and Sustainable Development Goals*, Sustainable Development Goals Series, https://doi.org/10.1007/978-3-031-80744-2_12

The World Bank data suggest that remittances now constitute a primary external source of finance for SSA countries. In 2019, remittances overtook some traditional sources of finance, such as FDI and inched closer to ODA (World Bank, 2020). Despite the substantial flow of remittance into SSA, it is still minimal compared to the flows to other world regions.

Looking at remittances paid, as depicted in Fig. 12.1, SSA recorded the second lowest amount, averaging $8.7 billion, just above South Asia's $8.31 billion. The region accounted for the least remittances received, totaling $43.4 billion, among the regions in the world for the year under review. In contrast, Europe and Central Asia received the highest, with an average of $175.6 billion, followed by South Asia with $127.4 billion. The reason for the low remittances on the continent is the informal means migrants use to send money to the continent. In effect, most remittances do not pass through the official channel. Most migrants opt for this informal means partly due to the transaction cost.

On average, SSA is the costliest region to send remittances. As of 2023, it costs 7.9% on average to send $200 to SSA (Table 12.1). Nonetheless, from 2019 to 2023, on average, remittance fees declined by 13.4%. This reduction can be attributed to the rise of digital remittances involving electronic money transfers through online platforms, mobile phones, and applications. Digital remittances are faster and reduce dependence on cash-centered remittance services by eliminating intermediaries and informal channels, ultimately lowering transaction costs. Similarly, digital remittances, mobile money, and blockchain-based solutions are crucial in increasing financial inclusion among unbanked and underbanked populations. Research indicates that receiving remittances increases the likelihood of households opening bank accounts and using digital financial services (Biggs, 2016; Abdul Mannan & Farhana, 2023). The World Bank (2021) initially estimated that remittances to Sub-Saharan Africa (SSA) would decline by 23.1% in 2020 due to the COVID-19 pandemic. However, its 2020 report revealed a decline of only 12.5%. Notably, excluding Nigeria, remittances to the region increased by 2.3%, demonstrating resilience during the crisis. This increase in remittance flows at the individual country level was attributed to a rise in the volume of remittances recorded by central banks, as migrants increasingly utilized formal and digital remittance platforms to send money. Travel bans and lockdowns affected the informal means of sending money due to COVID-19. This shift to digital remittances can reduce commission fees and increase the volume sent through formal channels, thereby deepening financial markets.

The UN has designated June 16 of every year as the International Day of Family Remittances

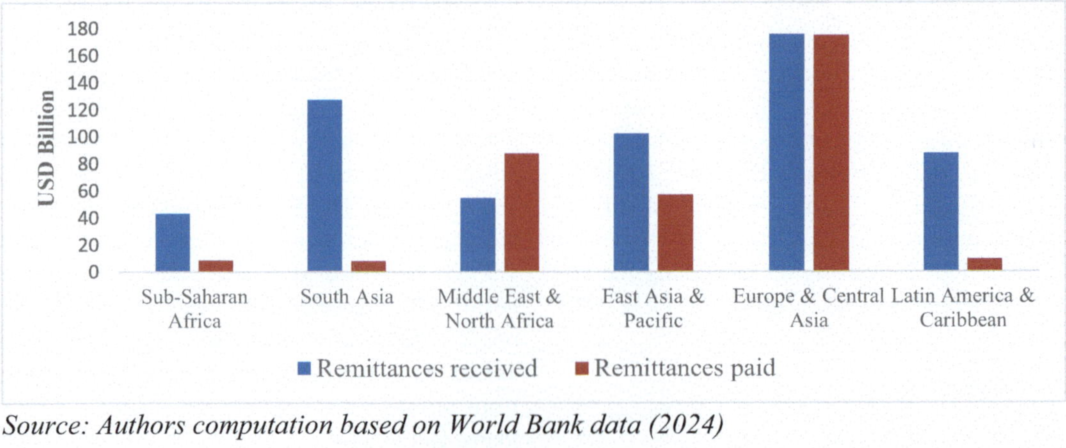

Fig. 12.1 Remittances paid and received, various regions on average (2011–2022). *(Source: Authors computation based on World Bank data (2024))*

Table 12.1 The transaction fee charged for sending $200 regionally (percentage)

Regions	2017	2018	2019	2020	2021	2022	2023
East Asia & Pacific	7.16	7.37	7.16	7.03	6.27	5.75	5.77
Europe & Central Asia	6.71	6.75	6.71	6.37	6.18	6.48	6.47
Latin America & Caribbean	6.11	6.14	6.11	5.83	5.63	5.82	5.88
Middle East & North Africa	6.84	7.06	6.84	7.04	6.38	6.32	6.01
South Asia	5.00	5.29	5.00	5.00	4.46	4.55	4.76
Sub-Saharan Africa	9.11	9.09	9.11	8.51	8.22	8.03	7.89

Source: Authors' computations based on World Bank data (2024)

(IDFR) to highlight the significant contribution of remittances to the SDGs. The theme for the IDFR 2023–2024 campaign is "Digital Remittances for Financial Inclusion and Cost Reduction" (UN, 2023), emphasizing the benefits of digital remittances in reducing transfer costs and promoting financial and digital inclusion for vulnerable senders and recipients.

Despite the benefits of digital remittances in reducing transaction costs, there is a lack of studies focused on digital remittances and sustainable development in SSA. This study aims to bridge that gap by exploring how digital and non-digital remittances are lowering transaction costs and increasing remittance volumes in the region. This work contributes to the growing literature on digital remittances and their role in achieving the UN Sustainable Development Goals. Besides, the study's results could inform policymakers, financial institutions, and development agencies working to enhance financial inclusion while promoting SDGs in SSA. The remainder of the paper is structured as follows: Section 12.2 provides an overview of remittance fees and the growth of digital remittances, Section 12.3 reviews the relevant literature, Section 12.4 details the data and methodology, Section 12.5 presents the empirical research and findings, and Sect. 12.6 concludes with recommendations.

12.2 Remittance Fees and the Rise of Digital Remittances

The cost of remittances, as measured by the World Bank, is calculated as the average fee for sending $200 or $500 through various service providers. The total cost includes the remittance transfer fees and the foreign exchange rate the service providers apply (World Bank, 2021). Global remittances to low- and middle-income countries (LMICs) increased to $647 billion in 2022 (McAuliffe and Oucho, 2024). However, achieving low-cost, safe, and rapid remittance transfers through digital means is still far from universal adoption. As of the third quarter of 2023, average remittance transfer costs remained at 6.39% globally, significantly above SDG 10.c target of 3% (World Bank, 2023).

Hidden fees in remittance transfers are a significant issue. According to Bulck (2023), financial organizations often employ strategies to deceive customers into believing they are receiving a better deal than they are. These hidden fees are particularly frustrating when relatives and friends abroad rely on remittances to buy food, medicine, or pay for education. Bulck further argues that hidden fees are often disguised in inflated exchange rates. For example, in just 1 year, European and American remitters lost $5 billion and $2.2 due to costs hidden in exchange rate mark-ups, while Canadian remitters lost $1 billion (Wise, 2023). Many banks and service providers justify inflated exchange rate practices by citing the necessity to hedge against unforeseen volatility. However, some believe these issues are exaggerated to justify continuous profiteering (Wise, 2023). Due to hidden costs embedded in inflated exchange rates, low-income migrants may be reluctant to use central banks for remittance services. Instead, they often opt for smaller financial institutions, money transfer operators (MTOs), or informal channels such as acquaintances, family members, export-import firms, and transport companies, collectively referred to as traditional methods.

Traditional remittance methods combine formal and informal channels. Formal routes include banks and money transfer operators, while informal methods involve financial transfers through friends, relatives, or courier services (Abdul, 2017). Factors such as accessibility, immigration status, education, and awareness of alternatives influence the choice of the traditional remittance method (Amuedo-Dorantes & Pozo, 2005). Furthermore, Amuedo-Dorantes and Pozo (2005) posit that undocumented migrants prefer non-bank money transfer services over banks, and newly arrived migrants or those remitting to rural areas often favor informal methods. In contrast, digital remittances are electronic money transfers made by migrants to their home countries, facilitated by the internet or mobile apps. Digital remittances have the potential to be more cost-effective, faster, and promote financial inclusion compared to traditional channels (Bersch et al., 2021; Metzger et al., 2019). However, the adoption of digital remittances is hindered by behavioral inertia, limited financial awareness, and legal obstacles (Bersch et al., 2021).

In recent years, digitalizing remittances has created opportunities for cheaper and more transparent transfers through the internet, mobile channels, or a combination of both. Beyond cost savings, digital remittances enhance connections with other digital financial services, fostering long-term consumer financial resilience (UN, 2023). Since 2016, the cost of sending remittances through digital platforms has decreased compared to traditional cash transactions. As of the fourth quarter of 2023, the global average cost for digital remittances was 4.96%, while the global average for non-digital remittances was 7.00% (Fig. 12.2).

Additionally, digital services accounted for 30% of all remittance transaction data collected in the fourth quarter of 2023[1] (World Bank, 2023). The World Bank report further indicates that digital-only money transfer operators (MTOs) have incessantly remained below the global average and MTO index. These trends highlight the importance of digital-only MTOs[2] in lowering transaction costs and achieving the SDG's target of reducing remittance costs to 3%.

In the last quarter of 2022 and 2023, the average global cost of sending $200 in remittances was slightly over 6%, double the UNDP Goal 2030 target of 3%. According to the World Bank (2023) report, banks, mobile operators, and MTOs have seen a general decline in average total cost over time. However, banks have remained above the global average, while MTOs and mobile operators have stayed below it. Figure 12.3 provides an overview of each RSP type in the fourth quarter of 2022 and 2023. Banks are the costliest RSP type, with an average cost of 11.99% as of the fourth quarter of 2023. In contrast, mobile money providers—handling funds sent, stored, and received via mobile phone—were the least costly, at 4.35%. Despite this, mobile operators account for less than 1% of the sample size (World Bank, 2023).

Financial inclusion in SSA has significantly boosted the use of digital remittances in the region. According to Demirgüç-Kunt et al. (2022), 55% of adults in SSA had an account in 2021. The percentage of adults with mobile money accounts increased to 33% between 2014 and 2021, representing a 64% rise (Fig. 12.4). This share is the highest of any region globally, over three times the global average of 10% for mobile money account ownership (Demirgüç-Kunt et al., 2022). Additionally, financial institutions and mobile money accounts for sending and receiving remittances grew by 50% from 2014 to 2021, while traditional methods, such as money transfer operators (MTOs) or cash, declined. This trend highlights the region's growing prevalence and expansion of digital remittances.

[1] A digital remittance must be sent via a payment instrument in an online or self-assisted manner, and received into a transaction account, i.e., bank account, transaction account maintained at a non-bank deposit taking institution (say a post office), mobile money or e-money account.

[2] A digital-only MTO refers to money transfer operators that send remittances predominantly through digital channels. The digital-only MTO index includes five digital-only MTOs, Wise, Remitly, WorldRemit, InstaReM and Xoom. Some of these providers also have physical channels.

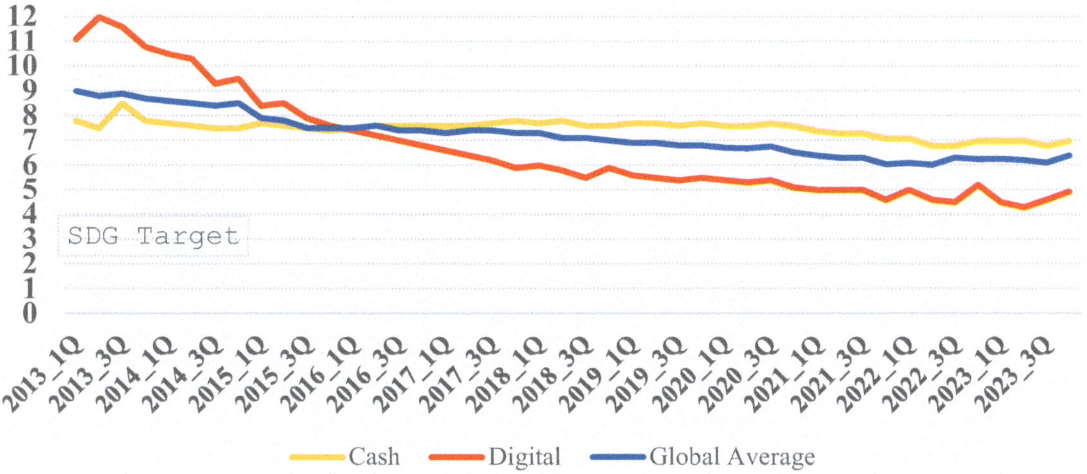

Fig. 12.2 Trends in the global cost of sending $200 in remittances. *(Source: World Bank remittance price worldwide (2023))*

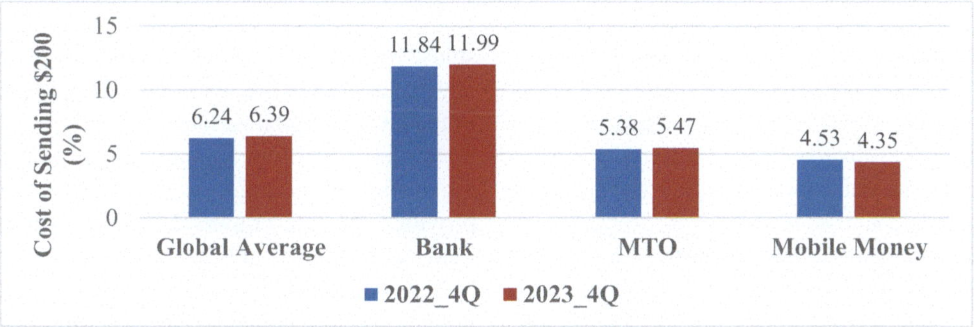

Source: Authors computations based on World Bank data (2024)

Fig. 12.3 The average cost of sending $200 in remittances by RSP type. (*Source: Authors computations based on World Bank data (2024)*)

12.3 Literature Review

The literature on the impact of digital remittances in reducing transaction costs, increasing remittance flows, and contributing to the Sustainable Development Goals (SDGs) is notably limited, particularly in the context of SSA. Research indicates that digital remittances positively impact SDG outcomes in SSA. Remittances have been shown to improve education and health, with a 10% increase in remittances leading to gains in primary education (4.2%), secondary education (8.8%), and health outcomes (1.2%) (Amakom & Iheoma, 2014). Moreover, remittances contribute to economic and social sustainability, though their overall effect on the aggregate SDG index remains ambiguous (Slimani et al., 2024). Furthermore, digital finance innovations have been found to play a vital role in connecting underserved populations to financial services, circumventing the limitations of traditional banking infrastructure (Gapp et al., 2022).

A limited number of studies consistently show that digital remittances can significantly reduce transaction costs and increase remittance volumes (Ahmed et al., 2021; Kpodar & Imam, 2024). This effect is particularly strong in the formal digital remittance sector, where lower costs promote the integration of remittances into the formal economy, enhancing financial inclusion and boosting

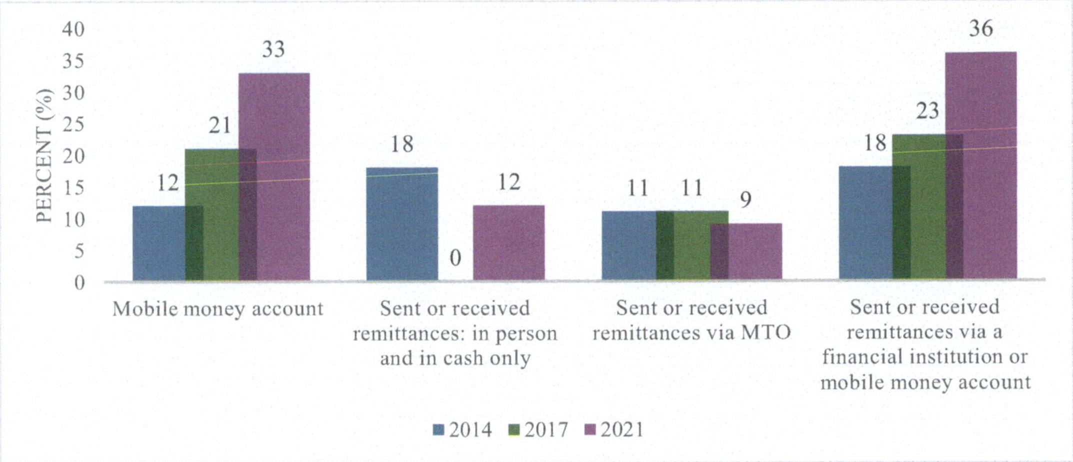

Source: Authors computations based on Global Findex data (2024)

Fig. 12.4 Account ownership and digital remittances growth in Sub-Saharan Africa. *(Source: Authors computations based on Global Findex data (2024))*

the net income of recipient households (Ahmed et al., 2021), more so than in the traditional remittance sector. Additionally, lower remittance fees are linked to increased competition, financial development, and improved digital infrastructure in receiving countries (Carare et al., 2022).

The World Bank has identified that digital remittances were crucial in sustaining formal remittance flows during the COVID-19 pandemic. This digitalization has contributed to a 3.3% reduction in the global average remittance transaction costs, primarily facilitated by mobile money services (World Bank, 2020). Mobile money systems like M-Pesa have significantly reduced remittance costs in Kenya and other African countries, potentially achieving the SDG goal of 3% fees well before 2030 (Metzger et al., 2019). Maloumby-Baka and Kingombe (2016) analyzed technological tools for reducing transaction costs and underscored the potential of digital technologies such as mobile banking and blockchain. Additionally, Bersch et al. (2021) argue that fostering partnerships between traditional financial institutions and fintech firms can further reduce transaction costs associated with digital remittances. Carare et al. (2022) investigated digital money and remittance costs in Central America, Panama, and the Dominican Republic (CAPDR), analyzing factors influencing the variation in digital and non-digital remittance fees over time. Their study reveals an invariable trend of declining remittance fees across these countries, regardless of the level of digitalization.

In the case of lower transaction costs increasing the flow of remittances, Aycinena et al. (2010) provided the first empirical evidence through a field experiment among migrants from El Salvador in Washington. Their study found that reductions in remittance fees led to significant increases in remittances sent to migrants' home countries; a $1 reduction in fees resulted in an additional $25 sent per month. Similarly, Ahmed et al. (2021), using bilateral data on remittance flows and transaction costs across 30 sending and 75 receiving developing countries from 2011 to 2017, observed that reducing the cost of remittances leads to a substantial increase in the volume of remittances. Precisely, a percent decrease in costs corresponded to a 1.6% increase in remittance volumes. In addition, Kpodar and Imam (2024) utilized a panel database on remittances from 71 countries over 2011Q1–2020Q4 to investigate the elasticity of remittances to transaction costs in a high-frequency and dynamic setting. The outcomes revealed that while initial reductions in transaction costs positively impact remittances in the short term, this effect diminishes over time. Yet, microdata from the USA-Mexico corridor confirms that migrants facing higher transaction costs tend to remit less.

Gascón et al. (2023), in their analysis of the impact of digitization on remittance flows in El Salvador using a 2016 household survey, found that digitalization significantly increases the likelihood of households receiving remittances. This suggests that digital remittances play a crucial role in enhancing remittance flows. However, they argue that digitalization does not influence the total remittance amount, as this is driven by migrants' labor conditions and the costs associated with sending money. Similarly, Jemiluyi and Jeke (2024) found that digital technology, including internet usage and mobile cellular subscriptions, positively impacts remittance inflows in Sub-Saharan Africa. The introduction of digital financial services, such as M-Shwari in Kenya, has also shown a direct effect on poverty reduction and increased remittance inflows (Adu & Hartarska, 2022). Conversely, Mookerjee and Roberts (2011), using data from the financial sectors of 64 countries, found that transaction costs do not affect remittance flows.

12.4 Data and Methodology

12.4.1 Data Sample and Hypotheses

The empirical estimation utilizes bilateral remittance flow data and transaction costs for 28 receiving countries in SSA and 8 sending countries from 2011 to 2022. The receiving countries, which are the main focus of our study, were selected based on the availability of reliable and extensive statistics on remittance inflows, thus effectively reflecting typical remittances in Sub-Saharan Africa. The sending countries include the United States of America, the United Kingdom, France, Italy, Germany, Spain, Belgium, and Canada. These countries rank among the top nations that send remittances to SSA and are the most popular destinations for SSA economic migrants. The data depict the fees senders pay for money transfers and the transaction methods used, such as cash, bank account transfers, mobile transfers, and online transactions. The study identified 55 corridors between these sending and receiving countries.

The World Bank defines a digital remittance as those that must be sent via a payment instrument in an online or self-assisted manner and received into a transaction account, that is, a bank account, transaction account maintained at a nonbank deposit-taking institution (say a post office), mobile money or e-money account. The World Bank also posits that remittances sent online from a transaction account and received in cash, over the counter, or sent in cash to a transaction account, especially a mobile money account, are also considered "digital" and are most often designated as "one-leg" digital (World Bank, 2022).

Based on the definition, we sort our data per corridor into two categories in the transaction mode:

- Digital: Remittances are sent and received digitally, including those received in a mobile wallet. The second leg is where remittances are sent non-digitally(cash) and received digitally or sent digitally and received through cash.
- Non-digital: Remittances are sent in a non-digital manner and received non-digital (cash).

We formulate a few hypotheses to assess the role of various determinants of remittance fees in SSA. Our main hypotheses, based on the existing literature, are as follows.

Lower fees are likely associated with:

1. High volume of remittance
2. High competition in the remittance market
3. High digital and financial development
4. High literacy level

12.4.2 Model Description and Methodology

The empirical model for the study is specified as:

$$\begin{aligned}\text{cost}_{srt} &= \alpha_0 \text{cost}_{srt} + \beta_1 \text{rem}_{rt} \\ &+ \beta_2 \text{providers}_{srt} + \beta_3 X_{st} \\ &+ \beta_4 X_{rt} + \beta_5 \theta_{st} + \beta_6 \theta_{rt} \\ &+ \delta_{sr} + \mu_t + \varepsilon_{srt}\end{aligned} \tag{12.1}$$

where "cost" stands for the fees per a US$ 200-dollar remittance expressed as a share of that amount, rem is the amount of remittances expressed in percentage received by a country r in year t, and the variable X stands for country-specific factors. All explanatory variables are lagged except the number of providers. θ is the control variable, δ is the unobserved factors, μ is the time trend; α and β are parameters; s indicates the sending country and r the receiving country and the number of cross-sections (= 1,…, N); t is the number of time series (= 1,…, N) and ε is the error term.

Our analysis utilizes Sys-GMM instead of OLS and the fixed-effect model, which ignores unobservable heterogeneity and endogeneity, likely issues with remittances that could lead to bias estimation. The dynamic panel model is appropriate when unobserved factors affect both the dependent and explanatory variables and when some explanatory variables are significantly linked to past values of the dependent variable, as is likely in our model. Blundell and Bond (1998) introduced the system GMM to address endogeneity by incorporating moment conditions (internal instruments) with lagged variables. Lags are used to ensure non-correlation with the current error term. System GMM also allows for individual fixed effects, heteroskedasticity, and autocorrelation within individuals. Importantly, it addresses problems such as omitted variables, measurement error, endogeneity, and country-specific heterogeneity (Guillaumont & Kpodar, 2006). Internal instruments will be employed for endogenous variables, and their validity determines the system's GMM estimator's precision. Diagnostic tests for these instruments include the Hansen test of over-identifying restrictions and the null hypothesis that the error term exhibits no serial correlation. Accepting the null hypothesis in both tests validates the model (Arellano & Bover, 1995; Blundell & Bond, 1998).

12.4.3 Data Source and Description

The dependent variable, the cost of remitting involved in sending remittances, is extracted from the World Bank remittances price database. One of the primary explanatory variables of interest is remittances (% GDP), sourced from the World Bank, which includes employee transfers and compensation. These transfers encompass current transfers either in cash or in kind received by resident households from non-resident households. We hypothesize that lower transaction fees will positively influence remittance flows to SSA.

Another significant variable is the level of competition in the remittances market, proxied by the number of providers within a corridor over time and aggregated by transaction modes. The degree of competition is a critical determinant of costs in the remittance industry (Afram, 2012; Beck et al., 2022). We anticipate that a higher number of providers will reduce their ability to impose more considerable mark-ups, thereby lowering costs for customers.

We also include data on financial and digital development at the country level. Digital development is proxied by the percentage of individuals using the internet as a proportion of the population, sourced from the International Telecommunication Union (ITU) database. Financial development is proxy by the number of commercial bank branches per 1000 km^2, sourced from the World Bank. High financial and digital development levels will lower fees as more established financial institutions and markets provide more efficient and cost-effective access to financial services.

Literacy is measured by the level of secondary school enrollment obtained from the World Bank. Migrants and their families with lower literacy face challenges in fully utilizing various options for sending money and understanding all the necessary information to select the most cost-effective option. This challenge is particularly evident in situations where price transparency is a concern. Therefore, we anticipate that higher literacy rates will have a beneficial effect on transaction costs. Control variables from the World Bank include the exchange rate and GDP per capita, which control the economic development levels of both sending and receiving countries.

12.5 Empirical Results and Discussion

12.5.1 Correlation Analysis

Table 12.2 displays the correlation analysis, showcasing the relationships between our variables and the diagnostics for multicollinearity in our study. Ratner (2009) proposes that correlation values within the range of 0.7 to 1.0 (or −0.7 to −1.0) indicate a robust positive (negative) association based on a stringent linear standard. We deduce that our variables exhibit low correlation coefficients according to this criterion. Additionally, the correlation coefficients between the variables are modest, indicating the absence of any multicollinearity problems.

The Variance Inflation Factor (VIF) analysis was used to assess the presence of multicollinearity. It is commonly accepted that multicollinearity may be present if the VIF exceeds 5 to 10 or the tolerance is lower than 0.1 to 0.2 (Menard, 1995). As shown in Table 12.2, all our variables meet the rule of thumb for VIF and tolerance levels, indicating no multicollinearity in our data.

12.5.2 Results and Discussion

Table 12.3 displays the empirical estimation for all transaction models, starting with the average, digital, and non-digital derived using the GMM system. Initially, we establish the baseline equation using the average fees and then use digital and non-digital average fee indicators.

The results indicate a strong negative correlation between transaction costs and remittance flows, with statistically significant coefficients observed across all transaction modes. The outcome supports our hypothesis that lower fees are associated with increased remittance flows. Moreover, our findings reveal that digital modes lead to higher remittance volumes than non-digital methods. Specifically, the coefficient in the digital fees suggests that a 1% decrease in the cost of sending USD 200 via digital means corresponds to approximately a 0.64% increase in remittance flows, compared to a 0.32% increase for non-digital means, ceteris paribus. This trend became more evident during the COVID-19 crisis, as formal and digital remittance platforms increased remittance flows to Sub-Saharan Africa by 2.3% (excluding Nigeria), despite expert predictions of a decline. This underscores the importance of digital remittances in enhancing remittance flows to SSA. Our conclusions align with Maloumby-Baka and Kingombe (2016) and Carare et al. (2022). However, they contradict the findings of Mookerjee and Roberts (2011), who argued that transaction costs do not influence remittance flows.

Additionally, the results reveal a negative correlation between the number of providers in each corridor and remittance fees, indicating that increased market competition leads to lower fees. Specifically, in digital transfers, including an additional provider is associated with a decrease of approximately 0.18 percentage points in remittance fees. This conclusion validates the significant role of competition in the money transfer market in determining costs within the remittance industry (Afram, 2012; Beck et al., 2022). This outcome is evident as various participants in Africa's remittance sector are developing solutions to meet the growing demand for affordable and accessible remittance services. For instance, Zepz, a digital remittance provider, has pioneered technology to offer fast, secure, and cost-effective remittance services to over 130 countries, including many in Africa. LemFi, a Nigerian fintech startup, connects mobile money platforms with traditional banks to facilitate cross-border transactions. Chipper Cash allows users to send and receive money across several African countries, while Tanzanian fintech company Nala provides user-friendly remittance services for underserved East African communities (DAI, 2024). Mukuru, a South African remittance and financial services provider, offers banking, insurance, and money transfer services across over 20 African nations. International platforms such as Wise, Xoom, WorldRemit, and Remitly also expand access to affordable digital remittance services. In a less competitive market, remittance service providers may be inclined to raise prices by leveraging their dominant market positions. Conversely, senders

Table 12.2 Correlation analysis

Variables	Cost	REM	COMP	ICTr	ICTs	FDr	FDs	LITr	LITs	GDPr	GDPs	Ex	VIF	Tol
Cost	1													
REM	−0.51	1											1.18	0.88
COMP	−0.39	0.45	1										1.85	0.54
ICTr	0.21	−0.14	−0.29	1									2.58	0.44
ICTs	−0.19	0.04	0.47	−0.03	1								1.91	0.52
FDr	0.12	−0.13	−0.51	0.53	−0.71	1							2.25	0.44
FDs	0.48	−0.14	−0.35	0.22	−0.55	0.41	1						1.64	0.61
LITr	0.29	−0.31	−0.60	0.69	−0.39	0.65	0.56	1					2.63	0.38
LITs	0.52	−0.37	−0.04	0.33	0.16	−0.06	0.24	0.34	1				1.23	0.81
GDPr	0.69	−0.55	−0.38	0.45	−0.13	0.29	0.33	0.51	0.73	1			2.08	0.48
GDPs	−0.34	0.46	0.68	−0.39	0.41	−0.45	−0.57	−0.69	−0.14	−0.34	1		2.58	0.39
Ex	−0.53	−0.01	0.24	−0.29	0.42	−0.37	−0.68	−0.40	−0.33	−0.45	0.14	1	2.62	0.38

Source: Authors' computation

Table 12.3 Empirical results remittance fees across transaction modes

Model	Average	Digital	Non-digital
Variables			
Remittances	−0.415**	−0.640**	−0.320**
	(−2.07)	(−2.10)	(−2.19)
Number of providers	−0.066**	−0.184**	0.059
	(−0.21)	(−0.36)	(0.59)
Internet usage sending country	−0.049**	−0.054***	−0.199*
	(−0.47)	(−0.27)	(−1.75)
Internet usage receiving country	−0.231***	−0.312***	0.091
	(−3.03)	(−2.56)	(1.28)
Financial development sending country	−0.456**	−0.817**	−0.158**
	(−2.06)	(−2.42)	(−2.44)
Financial development receiving country	−0.153**	−0.423**	−0.049*
	(−0.72)	(−1.21)	(−0.43)
Literacy level sending country	−0.101*	−0.178**	−0.091**
	(−1.91)	(−2.12)	(−2.17)
Literacy level receiving country	0.273	−0.408*	0.0407
	(1.67)	(−1.75)	(1.23)
Control variables			
GDP per capita sending country	−0.0002*	−0.0005*	0.0002
	(−0.88)	(−1.80)	(1.52)
GDP per capita receiving country	−0.0006	−0.0009	−0.002
	(−0.45)	(−0.43)	(−2.08)
Exchange rate	0.0004	0.001	2.005
	(0.49)	(0.77)	(0.05)
Tests			
No. of observation	308	308	308
Number of instruments	22	20	21
AR (2)	0.954	0.507	0.777
Hansen test	0.180	0.482	0.313

Source: Author's computations

*Note: t-statistics in parentheses; *** $p < 0.01$, ** $p < 0.05$, * $p < 0.1$ show significance at 1%, 5% and 10%, respectively*

will likely switch to more cost-effective providers in competitive environments, reducing overall costs. Our outcomes are consistent with the work of Kpodar and Imam (2024) and, notably, Olowa (2014), who highlighted that limited competition in the Nigerian market contributes to higher fees.

When considering the role of digital development, it is evident that higher development levels correlate with reduced remittance costs. We found lower transaction costs are associated with high digital development with the digital mode, notably higher internet usage. Specifically, increased internet usage in SSA correlates with reduced transaction fees by 0.312%, which exceeds the overall average (0.231%) mode. However, we observed no significant impact of digital development on non-digital transaction modes. The finding implies that fostering digital remittances and enhancing digital infrastructure contributes to lower transaction costs in SSA. The conclusion aligns with Carare et al. (2022) and Kpodar and Imam (2024), who argue that improved digital infrastructure reduces transaction costs in the remittances market and mitigates information asymmetries. Even in a fully competitive market, insufficient information diminishes price transparency and elevates transaction costs (Zheng, 2020). Therefore, enhanced digital development facilitates greater internet access, enabling remitters to compare costs across different remittance services and select the most competitive option. Consequently, it holds the potential to substantially diminish costs associated with acquiring information regarding hidden fees.

Similarly, the findings reveal that better financial development, measured by bank branch penetration in sending and receiving nations, is generally associated with reduced remittance fees across all regression models, particularly for digital transaction mode. There is a negative correlation between the number of commercial bank branches per 1000 km^2 in sending and receiving nations and remittance fees. For instance, an additional bank branch per 1000 km^2 in SSA is linked to a decrease of 0.423 percentage points in charges for digital transactions and 0.049 percentage points for non-digitized transactions. This outcome was expected, as the primary function of the financial sector is to reduce market frictions caused by information asymmetries and transaction costs. The financial sector expansion allows banks to provide retail remittance services and facilitates cross-border transactions, resulting in lower remittance costs (Mookerjee & Roberts, 2011; Kpodar & Imam, 2024). Furthermore, an improved financial sector enables migrants to send money through official channels and enhances recipients' access to funds through formal banking systems (Mundaca, 2008; Niimi & Ozden, 2006).

The study also found a statistically significant relationship between literacy levels and remittance transaction costs across all transaction models in sending countries, which are primarily highly developed with well-educated populations. In contrast, in SSA, a 1% increase in literacy rates leads to decreased transaction costs in digital transactions mode by 0.408% but does not affect non-digital transactions. A valid inference can be drawn from these results, considering that a certain level of education or knowledge is necessary for digital transactions. Our study is in tandem with the work of Gibson et al. (2014), who argued that financial literacy training enhances migrants' and their families' financial understanding and information-seeking behavior, reducing the likelihood of opting for more expensive remittance alternatives.

Regarding our control variables, the coefficient for GDP per capita, which measures the level of development, consistently shows a negative relationship with transaction costs in the sending countries across all transaction models. This result indicates that a more developed economy is associated with lower remittance fees. In SSA, the correlation between the level of development and remittance fees is negative, although it is not statistically significant in any of the models. The regression results also indicate that the coefficient on the exchange rate is positive but lacks statistical significance. This suggests that there is a possibility that remittance costs may rise as the local currency depreciates.

The validity of our model and the use of instrumental variables are confirmed by diagnostic statistics, including the Hansen test and the second-order serial correlation of the error terms (AR statistics). Based on these diagnostic tests, our model does not suffer from serial correlation or endogeneity, and the instruments are strictly exogenous. In essence, our findings support a robust inference.

12.6 Conclusion and Recommendations

Migrants and remittances play a critical role in attaining the SDGs, and achieving the full benefit of remittances requires reducing remittance fees. Remittance users gain a lot from digitalization, such as lower remittance costs, faster remittances, more assurance regarding payments, security, and tracking, increased service accessibility, an entry point to financial inclusion, and the development of economic resilience. Despite digital remittances reducing transaction costs, SSA still maintains the highest average costs globally for sending remittances. This study investigates the factors influencing the difference in digital and non-digital remittance costs over time in SSA from 2011 to 2022. Using the system GMM modeling framework, we found that lower transaction fees are associated with higher remittance flows into SSA. The findings also indicate that increased competition in the remittance market tends to lower transaction costs, ceteris paribus. Moreover, the results suggest that improved financial and digital development and higher literacy rates are associated with reduced transaction costs in SSA, mainly through digital channels. These outcomes support the idea that digital remittances are crucial in achieving the region's SDG Target 10.c, which aims to reduce remittance costs and, ultimately, SDG 10. Lower transaction costs increase remittance inflows, which are vital in helping low-income families and reducing income inequality. However, to fully realize this goal, policymakers must implement practical and effective policies.

Our findings provide essential practical policy implications. While individuals initiate most remittances, governments can influence these flows by fostering competition among banks and digital money transfer operators (MTOs) through legislation. The legal frameworks and policy environment for digital remittances in SSA are evolving to harness these inflows for development. Remittances have become a key source of foreign exchange for the region, showing resilience even during economic downturns (Mohapatra & Ratha, 2011). However, regulations have not kept pace with their growing importance. Therefore, to maximize the developmental benefits of remittances, banking regulations must improve market access, reduce costs, encourage technological innovation, and prevent illicit flows (Busumtwi-Sam, 2019). Digital technology, particularly internet and mobile cellular subscriptions, has been associated with increased regional remittance inflows. Therefore, partnerships between local banks and internet remittance platforms should be encouraged. Regulating and promoting collaboration between regional central banks and digital platforms will increase competition in the remittance market, lower transaction costs, enhance savings, and combat money laundering, thereby boosting remittance flows through formal channels.

Private and public stakeholders have the potential to educate migrants and households about digital platforms that offer lower transaction costs. Strengthening efforts to improve financial literacy through enhanced educational outcomes is essential. In SSA, access to new digital infrastructures remains uneven, especially in rural areas and certain social groups. This highlights the need for increased investment in education and information and communication technology development. Additionally, the pricing of transfer fees is often complex and lacks transparency, with hidden costs embedded in inflated exchange rates. Governments in SSA and key sending countries can positively influence remittance flows by requiring transfer providers to disclose their rates clearly and provide transparent information. These actions would empower migrants and their families to select the most cost-effective remittance services, lowering overall costs.

Stakeholders and policymakers may also explore using cryptocurrency and artificial intelligence (AI) to transform digital remittances and financial services. Cryptocurrencies and artificial intelligence (AI) are emerging as transformative digital remittances and international payments technologies. AI applications in cryptocurrency have been shown to address issues such as price prediction, fraud detection, and cybersecurity (Sabry et al., 2020). Cryptocurrencies offer potential benefits for remittances, including reduced transaction costs, increased speed, and improved financial inclusion for unbanked populations (Khan et al., 2023). Various AI tools, including market analysis, sentiment analysis, and crypto trading bots, enhance cryptocurrency performance and assist investors (Singh, 2023). Blockchain technology eliminates the need for untrusted third parties in fund transfers, potentially revolutionizing the remittance industry (Kulkarni et al., 2019). However, regulatory challenges and the need for broader adoption remain significant hurdles (Khan et al., 2023). Recent research emphasizes the transformative potential of cryptocurrencies in enhancing remittance efficiency and accessibility, highlighting the need for supportive policies to leverage these benefits effectively (Jegerson & Mertzanis, 2024). As this field evolves, further studies are required to understand the full impact of cryptocurrencies on global remittance markets.

Finally, security and privacy concerns significantly affect the adoption of digital remittance and other financial services. Issues such as identity theft, malware, phishing, and vishing hinder the growth of digital banking (Cele & Kwenda, 2024). Factors like security, privacy, usability, and utility play a role in adopting online platforms. For mobile payment systems, ease of use, perceived advantages, visibility, and security enhance adoption intentions, while privacy concerns reduce security perceptions (Johnson et al., 2018). Therefore, ensuring digital consumer safety is crucial for the uptake of digital remittances and financial inclusion. To encourage adoption, there is a need for awareness campaigns, stronger digital consumer protection laws, and collaborative efforts between mobile network operators, financial institutions, and regulators to fight fraud and cybercrime within the fintech ecosystem (Mutuku, 2020; Bongomin & Ntayi, 2020).

References

Abdul, M. K. (2017). *Remittance model: A study of developing country in Bangladesh.*

Abdul Mannan, K., & Farhana, K. M. (2023). Digital financial inclusion and remittances: An empirical study on Bangladeshi migrant households. *FinTech, 2*(4), 680–697.

Adu, D. T., & Hartarska, V. (2022). Does cellphone banking change lives? Fintech, poverty, and remittances in Kenya. *Fintech, Poverty, and Remittances in Kenya.*

Afram, G. G. (2012). *The remittance market in India: Opportunities, challenges, and policy options*. The World Bank, Washington DC. Available at: https://openknowledge.worldbank.org/server/api/core/bitstreams/5ca2f75a-1ea1-5ce4-9904-b2d8550ffec3/content

Ahmed, J., Mughal, M., & Martínez-Zarzoso, I. (2021). Sending money home: Transaction cost and remittances to developing countries. *The World Economy, 44*(8), 2433–2459.

Amakom, U., & Iheoma, C. G. (2014). Impact of migrant remittances on health and education outcomes in sub-Saharan Africa. *IOSR Journal of Humanities and Social Science, 19*(8), 33–44.

Amuedo-Dorantes, C., & Pozo, S. (2005). On the use of differing money transmission methods by Mexican immigrants. *International Migration Review, 39*(3), 554–576.

Arellano, M., & Bover, O. (1995). Another look at the instrumental variable estimation of error-components models. *Journal of Econometrics, 68*(1), 29–51.

Aycinena, D., Martinez, C. A., & Yang, D. (2010). *The impact of transaction fees on migrant remittances: Evidence from a field experiment among migrants from El Salvador*. University of Michigan.

Beck, T., Janfils, M., & Kpodar, K. (2022). What explains remittance fees? Panel evidence, IMF Working Paper, forthcoming.

Bersch, J., Clevy, J. F., Muhammad, N., Ruiz, M. E. P., & Yakhshilikov, M. Y. (2021). *Fintech potential for remittance transfers: A Central America perspective*. International Monetary Fund.

Biggs, D. C. (2016). How nonbanks are boosting financial inclusion and remittance. In *Banking beyond banks and money: A guide to banking services in the twenty-first century* (pp. 181–196).

Blundell, R., & Bond, S. (1998). Initial conditions and moment restrictions in dynamic panel data models. *Journal of Econometrics, 87*(1), 115–143.

Bongomin, G. O. C., & Ntayi, J. M. (2020). Mobile money adoption and usage and financial inclusion: Mediating effect of digital consumer protection. *Digital Policy, Regulation and Governance, 22*(3), 157–176.

Bulck, M. V. (2023). The cost of remittances is still too high, then fix is easy. In Wise 2023 remittances report.

Busumtwi-Sam, J. (2019). International migrations, diasporas, and remittances. In *The Palgrave Handbook of Contemporary International Political Economy* (pp. 183–199).

Carare, A., Franco, L., Hadzi-Vaskov, M., Lesniak, J., Vasilyev, D., & Yakhshilikov, Y. (2022). Digital money and remittances costs in Central America, Panama, and The Dominican Republic, International Monetary Fund. IMF Working Paper, WP/22/23.

Cele, N. N., & Kwenda, S. (2024). Do cybersecurity threats and risks have an impact on the adoption of digital banking? A systematic literature review. *Journal of Financial Crime*.

DAI. (2024). *Africa's Remittance Market Set to reach $500bn by 2035*. https://www.daimagister.com/resources/remittances/#anchor. Accessed 17 August 2024.

Demirgüç-Kunt, A., Klapper, L., Singer, D., & Ansar, S. (2022). *The Global Findex Database 2021: Financial inclusion, digital payments, and resilience in the age of COVID-19*. World Bank Publications.

Gapp, B., de Fauconberg, Â., Bessis, A., Goble, H., & Obodoekwe, N. (2022). Digital finance in Africa: Accelerating foundations for inclusive and sustainable local innovation. *Environmental Sciences Proceedings, 15*(1), 66.

Gascón, P., Larramona, G., & Salvador, M. (2023). The impact of digitalisation on remittances. Evidence from El Salvador. *Telecommunications Policy, 47*(4), 102500.

Gibson, J., McKenzie, D., & Zia, B. (2014). The impact of financial literacy training for migrants. *The World Bank Economic Review, 28*(1), 130–161.

Global Findex. (2024). *The Global Findex Database (2021): World Bank Publications*.

Guillaumont, S. J., & Kpodar, K. (2006). Financial development, financial instability and economic growth. *Economics and Prevision, 3*(174), 87–111.

Jegerson, D., & Mertzanis, C. (2024). Deciphering the role of cryptocurrencies in global remittances: A comprehensive literature review. *Management & Sustainability: An Arab Review*. (ahead-of-print).

Jemiluyi, O. O., & Jeke, L. (2024). Remittance inflows in a digital economy: A Sub-Saharan African experience. *Journal of Infrastructure, Policy and Development, 8*(4), 2838.

Johnson, V. L., Kiser, A., Washington, R., & Torres, R. (2018). Limitations to the rapid adoption of M-payment services: Understanding the impact of privacy risk on M-Payment services. *Computers in Human Behavior, 79*, 111–122.

Khan, R., Dhaliwal, N., Deorari, R., Anoosha, P., Hussein, A., & Alazzam, M. B. (2023). The way to implement cryptocurrency for international payments and remittances. In *2023 3rd international conference on advance computing and innovative Technologies in engineering (ICACITE)* (pp. 226–229). IEEE.

Kpodar, K., & Imam, P. A. (2024). How do transaction costs influence remittances? *World Development, 177*, 106537.

Kratou, H., & Khlass, N. (2023). Does the cost of remitting matter? Evidence from low-income countries on the effect of remittances on income distribution. *Development Studies Research, 10*(1), 2241641.

Kulkarni, R., Schintler, L., Koizumi, N., Olds, J., & Stough, R. R. (2019). *Cryptocurrency, Stablecoins and Blockchain: Exploring digital money solutions for remittances and inclusive economies*. SSRN Electronic Journal. doi: 10.2139/ssrn.3511139

Maloumby-Baka, R. C., & Kingombe, C. K. (2016). *The quest to lower high remittance costs to Africa: A brief review of the use of mobile banking and bitcoins*. Geneva: Centre for Finance and Development, Graduate Institute of International and Development Studies.

Malpass, D. (2022). *Remittances are a critical economic stabilizer*. https://blogs.worldbank.org/en/voices/remittances-are-critical-economic-stabilizer (accessed 16 August 2024).

McAuliffe, M., & Oucho, L. A. (2024). *World migration report 2024*. International Organization for Migration.

Menard, S. (1995). *Applied logistic regression analysis* (Vol. 106). In Sage university paper series on quantitative application in social sciences, series no. 106 (2nd ed.). Sage

Metzger, M., Riedler, T., & Pédussel Wu, J. (2019). Migrant remittances: Alternative money transfer channels (No. 127/2019). Working Paper.

Mohammed, U., & Karagöl, E. T. (2023). Remittances, institutional quality and investment in Sub-Saharan Africa. *Journal of Business and Socio-economic Development, 3*(4), 322–338.

Mohapatra, S., & Ratha, D. (Eds.). (2011). *Remittance markets in Africa*. World Bank Publications.

Mookerjee, R., & Roberts, J. (2011). Banking services, transaction costs and international remittance flows. *Applied Economics Letters, 18*(3), 199–205.

Mundaca, B. (2008). Remittances, financial market development and economic growth. *Review of Development Economics, 12*, 34–45.

Mutuku, M. K. (2020). A meta-analysis of user perception and use of online digital platforms in Kenya: Evidence from E-citizen platform, Kenya. *International Journal of Management*, 35–42.

Niimi, Y., & Ozden, C. (2006). Migration and remittances: Causes and linkages, World Bank Policy Research Working Paper 4087, World Bank, Washington, DC.

Nurse, K. (2018). Migration, diasporas, remittances and the sustainable development goals in least developed countries. *Journal of Globalization and Development, 9*(2), 20190006.

Olowa, O. W. (2014). Remittances and competition: A policy analysis matrix approach. *Global Journal of Management and Business Research*.

Portes, L. S. V. (2009). Remittances, poverty and inequality. *Journal of Economic Development, 34*(1), 127.

Přívara, A., & Trnovský, K. (2021). The impact of remittances on household savings in the Baltics. *Investment Management and Financial Innovations, 18*(1), 335–345.

Ratner, B. (2009). The correlation coefficient: Its values range between+ 1/− 1, or do they? *Journal of Targeting, Measurement and Analysis for Marketing, 17*(2), 139–142.

Sabry, F., Labda, W., Erbad, A., & Malluhi, Q. (2020). Cryptocurrencies and artificial intelligence: Challenges and opportunities. *IEEE Access, 8*, 175840–175858.

Singh, S. (2023). An analysis of artificial intelligence and its role in the cryptocurrency. *Journal of Research in Information Technology, 4*(1), 1–16.

Slimani, S., Omri, A., & Abbassi, A. (2024). Financing sustainable development goals in Sub-Saharan Africa: Does international capital flows matter? *Sustainable Development.*

United Nations. (2015). Transforming our world: The 2030 agenda for sustainable development. Seventieth session, Agenda items 15 and 116. New York.

United Nations. (2019). *International day for family remittances.* https://www.un.org/en/events/family-remittances-day/un-action.shtml.

United Nations. (2023). *Digital remittances towards financial inclusion and cost reduction.* IDFR.

Wise. (2023). *2023 remittances report: How to sustainably lower costs and achieve the UN goal of 3% by 2030.*

World Bank. (2020). *Migration and remittances, COVID-19 crisis through a migration lens.* https://documents1.worldbank.org/curated/en/989721587512418006/pdf/COVID-19 Crisis-Through-a-Migration-Lens.pdf

World Bank. (2021). *Migration and remittances brief, resilience COVID-19 crisis through a migration lens.* https://www.knomad.org/sites/default/files/202105/Migration%20and%20Development%20Brief%2034_1.pdf

World Bank. (2022). An analysis of trend in remittances service. *Remittances Prices Worldwide Quarterly, 42*(7).

World Bank. (2023). An analysis of the trend in remittances service. *Remittances Prices Worldwide Quarterly.* Issue 48. https://remittanceprices.worldbank.org/sites/default/files/rpw_main_report_and_annex_q423_final.pdf

World Bank. (2024). World Bank development indicator data. World Bank development indicators database. Available at: https://databank.worldbank.org/source/world-development-indicators

Zheng, W. (2020). Information asymmetry, price informativeness and cost of capital. *Price Informativeness and Cost of Capital.*

Charting the Course to SDG 4: Examining the Impact of Remittance Flows on School Enrolment

13

Ibrahim Mohammed Gunu,
Ibrahim Nandom Yakubu, and Siisu Umar

13.1 Introduction

In recent decades, economists worldwide have taken a keen interest in migration and remittances, which have become crucial economic factors. Remittance is defined as the transfer of money or funds from one location or country to another (Mazumder, 2021). It comprises 'household income from foreign economies resulting mostly from temporary or permanent relocation of people to those countries' (IMF, 2009). In most emerging economies, remittances constitute a significant portion of the balance of payments account (Jijin et al., 2022). The flow of remittances is experiencing exponential growth, particularly in low- and middle-income countries (LMICs), with total worldwide remittances surpassing $700 billion in 2019, of which approximately $550 billion was accounted for by LMICs (World Bank, 2021b).

Remittances to Africa have also been gradually growing throughout the years, according to the World Bank. In 2019, for example, remittance inflows to Sub-Saharan Africa amounted to $49.34 billion, reflecting an increase from the previous years. However, in 2020, this value dropped to $42.92 billion. According to the World Bank's (2021a, b) 'Migration and Development Brief 35' report, this decline in remittances to the region was attributed to a reduction in remittances received in Nigeria, which contributes around half of the total remittances in the region. Nevertheless, recent data from the World Bank indicates that in 2021, the region received a total of $45 billion in remittances, which represents an increase from the previous year. There are regional disparities in the pattern of remittances flowing into Africa. Compared to other parts of Africa, the North African nations including Egypt, Tunisia and Morocco are the highest recipients. This is partially attributed to the significant population of North Africans working in Europe and the Middle East, who transfer money to their relatives at home.

Undoubtedly, it is well acknowledged that a steady flow of remittances plays a vital role in economic development, especially in the developing world. For many African countries, remittances are now a vital source of foreign revenue. Empirically, the positive effect of remittances on economic outcomes has been well documented in the literature. For instance, remittances serve as a significant driver of financial development by facilitating growth in financial services and credit markets (Bangake & Eggoh, 2020). Fromentin

I. M. Gunu
Department of Educational Management and Policy Studies, Faculty of Education, University for Development Studies, Tamale, Ghana

I. N. Yakubu (✉) · S. Umar
School of Business, University for Development Studies, Tamale, Ghana
e-mail: ibrahim.yakubu@uds.edu.gh

I. N. Yakubu (ed.), *Strategic Approaches to Banking Business and Sustainable Development Goals*, Sustainable Development Goals Series, https://doi.org/10.1007/978-3-031-80744-2_13

and Leon (2019) postulate that remittances favourably affect both firm and household credit advancement in low- and middle-income economies. Some earlier studies have shown that remittances alleviate poverty, boost access to financing, increase household spending and minimise income distribution inequities (Adams & Cuecuecha, 2013; Inoue & Hamori, 2016; Petrou & Connell, 2017; Abduvaliev & Bustillo, 2020; Musakwa & Odhiambo, 2022; Pal et al., 2022; Tung & Thang, 2023).

The purpose of this study is to investigate the influence of remittance inflow on school enrolment in Africa. Basic and secondary education is essential in equipping young people with the necessary skills for the labour market. Previous research has explored the relationship between education levels and national productivity and growth (see Glewwe et al., 2014; Wang & Liu, 2016; Hassan & Rafaz, 2017; Kotásková et al., 2018; Ramos & Mourelle, 2019; Omodero & Nwangwa, 2020; Maneejuk & Yamaka, 2021; Yakubu & Gunu, 2022).

Recognising the importance of education, it is imperative to devise policies to increase enrolment rates across all levels of education. In order to do so, it is crucial to understand the different factors that influence school enrolment. Remittances have emerged as one of the significant factors impacting enrolment rates in recent years, but the empirical evidence on their impact has been inconsistent. While some studies have found a positive relationship between remittances and enrolment, others have reported a negative association. As a result, further research is needed to clarify the relationship. The goal of this research is to add to the extant literature on how remittances drive enrolment in the case of Africa, where studies on the remittances-education nexus are still in their infancy.

A noteworthy contribution of this study is that we examine how remittances affect enrolment at all levels of education, including primary, secondary and tertiary enrolment. This approach differs from previous studies that focused only on a particular level of education. Aside from this, the study employs the generalised method of moments (GMM) technique to address any endogeneity issues in the model estimation, making the results more reliable.

The remaining sections of the paper are organised as follows: Section 13.2 presents the literature review, while Sect. 13.3 outlines the methodology employed in this study. In Sect. 13.4, we discuss the findings obtained from the analysis, and finally, we conclude in Sect. 13.5.

13.2 Literature Review

Theoretically, this study draws on the human capital theory, family bargaining model and social capital theory. The human capital theory suggests that education is a long-term investment that enhances the economic well-being of individuals and families (Becker, 2009). Remittances, according to the theory, may provide additional income that can be used to support education, resulting in higher enrolment rates. The family bargaining model proposes that remittances can influence the decision-making processes of households, particularly in terms of educational investments (Stark, 1991). This model suggests that remittances can augment the bargaining power of women in the household, resulting in increased investments towards their children's education. The social capital theory, on the other hand, contends that remittances may strengthen recipient households' social networks, which may lead to greater access to information and educational opportunities (Portes, 1998). Remittance flows, according to this notion, may generate social capital that may be leveraged to improve human capital, including education. The social capital theory on the other hand suggests that remittances could enhance the social networks of recipient households, which could increase access to information and opportunities for education (Portes, 1998). This theory argues that remittance flows could create social capital that can be used to enhance human capital, including education.

Empirically, research linking remittances and education is quite scanty. Among the few studies, Zhunio et al. (2012) reported that school attainment (both primary and secondary) is significantly influenced by the growth in remittances. In exploring the impact of remittance on educational attainment in Nepal, Acharya and Leon-Gonzalez (2014) revealed that remittances increase school enrolment and minimise dropout rates. Amakom and Iheoma (2014) investigated how remittances influence education in Sub-Saharan Africa using the Two-stage least squares (2SLS) regression technique. The findings demonstrated that remittances positively impact school enrolment as well as health outcomes in the selected countries in the region. Utilising household survey data, Kalaj (2014) discovered a negative link between remittances and educational attainment in Albania, particularly among males residing in rural areas. According to Jakob (2015), an increase in remittance flow enhances average school enrolment in El Salvador. Using the Probit Model, Ayad and El-Aziz (2018) looked into how remittance flow affects educational attainment in the Egyptian context. The authors established that remittances have a positive influence on the educational attainment of university-aged individuals. Bucheli et al. (2018) used a bivariate probit model to explore the impact of remittances on child education in Ecuador. The study discovered that remittances have both positive and negative consequences based on the scenario of analysis. In analysing the nexus between remittances and educational attainment, Hapsari (2019) concluded that remittances considerably inhibit the likelihood of access to higher education in Indonesia. Arif et al. (2019) studied the effect of remittances on higher education in some selected major remittance inflow nations. The findings from the study reveal that remittances have a strong positive effect on the growth of higher education. Affudo (2020) found that remittances significantly boost educational attainment, with the effect likely being greater at the primary education level. Employing a panel dataset comprising 90 countries from developing economies, Shafiq et al. (2022) revealed that education is positively and significantly driven by remittances in Pakistan. Similarly, Gul et al. (2021) reported that remittance inflow significantly boosts school enrolment in Pakistan. Using the Household Income and Expenditure Survey, Al-Islam et al. (2022) analysed the influence of remittances on school enrolment in the case of Bangladesh. The findings revealed a positive link between remittance inflows and enrolment levels. The study also evidenced that the level of education of parents plays a significant role in increasing school enrolment.

13.3 Methodology

13.3.1 Data and Variables

The study employs panel data from 43 African countries covering from 2000 to 2021. The data is gleaned from the World Bank. School enrolment serves as the dependent variable, whereas our main independent variable is remittances. We control for the effect of government expenditure on education and economic growth. We measure school enrolment using the gross enrolments in primary, secondary and tertiary schools. This allows us to examine how remittances affect enrolment at all educational levels. We gauge remittances inflow by personal remittances received (% of GDP). For the control factors, educational expenditure is the total government expenditure on education (percentage of overall government expenditure). The growth in GDP per capita is employed as an indicator of economic growth.

13.3.2 Model Specification

The baseline models for investigating the link between remittances and enrolment are specified below:

$$\begin{aligned}\mathrm{PSE}_{it} &= \alpha_0 + \beta_1\mathrm{REM}_{it} + \beta_2\mathrm{EDUX}_{it}\\ &+\beta_3\mathrm{ECG}_{it} + \varepsilon_{it}\end{aligned} \quad (13.1)$$

$$\begin{aligned}\mathrm{SSE}_{it} &= \alpha_0 + \beta_1\mathrm{REM}_{it} + \beta_2\mathrm{EDUX}_{it}\\ &+\beta_3\mathrm{ECG}_{it} + \varepsilon_{it}\end{aligned} \quad (13.2)$$

$$\begin{aligned}\mathrm{TER}_{it} &= \alpha_0 + \beta_1\mathrm{REM}_{it} + \beta_2\mathrm{EDUX}_{it}\\ &+\beta_3\mathrm{ECG}_{it} + \varepsilon_{it}\end{aligned} \quad (13.3)$$

From the equations, primary school enrolment (PSE), secondary school enrolment (SSE) and tertiary enrolment (TER) represent primary school enrolment, secondary school enrolment and tertiary enrolment, respectively. Remittances (REM) indicates remittances. Government educational expenditure (EDUX) and economic growth (ECG) are government educational expenditure and economic growth, respectively. The symbol α indicates the constant, and β is the regression coefficient. ε symbolises the error term.

13.3.3 Analytical Technique

The study employs the generalized method of moments (GMM) estimation technique, developed by Arellano and Bond (1991) in the model estimation. This technique is more effective than traditional panel techniques such as least squares, random effects and fixed effects, as it addresses biases caused by unobserved cross-country effects, lagged dependent variables and endogeneity.

The panel model based on the GMM technique has been redefined as follows:

$$Y_{it} = \alpha Y_{it-1} + \beta^{'} X_{it} + \varepsilon_{it} \quad (13.4)$$

In the model (13.4), the dependent variable is represented by Y, while X denotes the independent factors. The period lag of the dependent variable is Y_{t-1}.

13.4 Results and Discussion

13.4.1 Descriptive Statistics

Table 13.1 displays the summary statistics for the variables. The mean primary school enrolment rate is 100.84, and the standard deviation is 21.84. The maximum enrolment rate is 156.45, and the minimum is 32.36. The skewness of primary school enrolment is negative, indicating that the data is slightly left-skewed. The Jarque-Bera test confirms that primary school enrolment is normally distributed with a p-value of 0.39. The descriptive statistics for the other variables follow a similar pattern. Secondary school enrolment has a higher standard deviation and positive skewness. Tertiary school enrolment has a high kurtosis and positive skewness. Remittance has a high kurtosis and a significantly positive skewness. The government educational expenditure has a relatively low skewness, and the economic growth variable is negatively skewed.

13.4.2 Correlation and Variance Inflation Factor (VIF) Analysis

Table 13.2 summarises the outputs of the correlation and variance inflation factor (VIF) analysis. The results indicate that the correlation coefficients between the variables are relatively small, suggesting that there are little concerns for multicollinearity (Kennedy, 2003). Similarly, the results of the VIF confirm that there are no multicollinearity issues. In general, multicollinearity is affirmed by a VIF value surpassing 10 and a tolerance value less than 0.10 (Yakubu, 2019).

13.4.3 Regression Results

The results of the GMM estimation are presented in Table 13.3. The Wald tests' p-values in the diagnostic statistics indicate that the model parameters are significant, demonstrating the models' overall significance. The Arellano-Bond (AR) (2) test evaluates the residuals' autocorrelation, and a p-value greater than 0.05 suggests that the model has no significant autocorrelation. In this case, the p-values for each model exceed 0.05, indicating that there is no significant autocorrelation of the residuals. The Hansen J-statistic test results imply that there is no significant evidence of model misspecification or omitted variable bias. Overall, the diagnostic statistics suggest that the regression models are statistically significant and well-specified, with no significant issues of autocorrelation or omitted variable bias.

The table shows the regression results for three different dependent variables: primary

Table 13.1 Descriptive statistics

	PSE	SSE	TER	REM	EDUX	ECG
Mean	100.8434	45.0927	11.4529	3.2483	16.2416	1.4229
Maximum	156.4451	116.8747	74.2320	53.8262	37.5210	55.5903
Minimum	32.3561	6.1974	0.3520	0.0000	0.8334	−48.3925
Std. dev.	21.8356	24.4067	14.1339	5.7758	5.7471	5.2864
Skewness	−0.1090	0.9580	2.7140	4.5183	0.1850	−0.5494
Kurtosis	2.8886	3.4868	10.7638	30.0297	3.2510	29.1915
Jarque-Bera	1.8863	88.4204	2030.3760	29647.6900	5.6791	26342.8300
Probability	0.3894	0.0000	0.0000	0.0000	0.0585	0.0000

Source: Authors' Computation

Table 13.2 Results of correlation and variance inflation factor (VIF) analysis

	REM	EDUX	ECG	VIF	Tolerance (1/VIF)
REM	1.0000			1.0178	0.9825
EDUX	0.1044	1.0000		1.0271	0.9736
ECG	0.0188	0.1210	1.0000	1.0095	0.9906

Source: Authors' Computation

school enrolment, secondary school enrolment and tertiary school enrolment. The results suggest that the lagged value of each dependent variable is highly significant, with a positive association with the current value of the dependent variable. This implies that past enrolment rates at each level of schooling are strong predictors of current enrolment rates at that same level.

The study's results indicate that remittance inflows have a positive and significant impact on school enrolment across all levels of education in Africa. Specifically, the findings reveal that a percentage increase in remittance inflow leads to a corresponding increase of 0.0866% in primary school enrolment, 0.4175% in secondary school enrolment and 0.0297% in tertiary school enrolment. These results suggest that remittance inflows serve as an additional source of income for recipient households, which, in turn, positively influences their ability to finance education-related expenses for their children. In other words, remittances serve as a significant source of household income and contribute to improving educational opportunities for children in African countries. Our findings support earlier research that found a direct influence of remittances on enrolment (see Arif et al., 2019; Gul et al., 2021; Affudo, 2020; Al-Islam et al., 2022; Shafiq et al., 2022).

Government educational expenditure has a significant positive effect on primary school enrolment but a negative effect on secondary and tertiary school enrolment. This indicates that increased government spending on education has a limited impact on improving enrolment at the higher levels of education, possibly due to issues such as poor infrastructure, inadequate teaching materials or limited opportunities for further education.

The results show that economic growth has a positive and statistically significant effect on primary and tertiary school enrolment rates but a negative impact on secondary school enrolment. The findings suggest that economic growth may contribute to increasing access to education in African countries, particularly at the primary and tertiary levels. However, it may also have negative consequences for the demand for secondary education. The reason for the negative impact of economic growth on secondary school enrolment could be attributed to the possibility of economic growth resulting in increased income inequality, which can restrain the accessibility of secondary education for families with low income.

Table 13.3 GMM estimation results

Variables	Model 1	Model 2	Model 3
Lag_{t-1}	0.7204***	0.8237***	0.8896***
	(0.0027)	(0.0027)	(0.0005)
REM	0.0866**	0.4175***	0.0297***
	(0.0354)	(0.0020)	(0.0051)
EDUX	0.0483***	−0.0673***	−0.0024***
	(0.0060)	(0.0020)	(0.0005)
ECG	0.0500***	−0.0795***	0.0097***
	(0.0089)	(0.0020)	(0.0006)
Diagnostics			
Wald χ^2 (*p-value*)	122278.3 (0.0000)	599056.4 (0.0000)	11465271.0 (0.0000)
AR (2) (*p-value*)	1.193863 (0.2325)	0.008723 (0.9930)	−0.0061 (0.9951)
Hansen J-statistic	35.7215	28.3417	32.5266
Prob(J-statistic)	0.3875	0.6524	0.4905
Number of countries	43	43	43

Source: Authors' Computation
Note: Standard errors in parentheses ***$p < 0.01$ **$p < 0.05$

13.5 Conclusion and Policy Recommendations

Emigration, particularly from underdeveloped to developed countries, has risen in recent years, which has both economic and social benefits. The inflow of remittances is one of the most significant sources of foreign revenue, and its contribution to national growth is well documented in the literature. Even though studies have demonstrated the significance of remittance flow on macroeconomic indicators, little research has been done on its effect on education. In this study, we examined the influence of remittances on school enrolment in 43 African countries using data from the World Development Indicators of the World Bank, which span from 2000 to 2021.

Using the generalized method of moments (GMM) estimator, our study revealed that remittance inflows have a significantly positive effect on school enrolment at all levels of education. We also found that while government expenditure on education increases primary school enrolment, it has little impact on enrolment at the secondary and tertiary levels. Additionally, our results showed that economic growth has a positive and statistically significant influence on primary and tertiary school enrolment rates but a negative impact on secondary school enrolment.

Based on the findings presented, several policy recommendations can be made to improve education outcomes in African countries. First, given the positive effect of remittance inflows on school enrolment at all levels, policies should be put in place to facilitate and encourage remittance flows, such as reducing transaction costs and promoting financial literacy. Remittance-related taxes, in particular, must be abolished or optimally reduced to increase remittance inflows. Additionally, policies should be implemented to ensure that remittance flows are directed towards education-related expenses. Second, the negative effect of government expenditure on secondary and tertiary school enrolment rates suggests the need to re-evaluate current education spending policies. Governments should consider reallocating resources towards secondary and tertiary education to improve access and affordability. Third, the positive effect of economic growth on primary and tertiary school enrolment rates suggests the need to promote sustainable economic growth policies. This could involve investing in infrastructure, promoting foreign direct investment and supporting small and medium-sized enterprises. In future studies, it may be worthwhile to consider the mediating role of government effectiveness, regulatory quality or corruption in analysing the relationship between remittances and education. By examining the mediating effects of these factors, future research could offer a more comprehensive understanding of the link between remittances and education and provide more targeted policy recommendations to optimise the positive impact of remittances on education outcomes.

References

Abduvaliev, M., & Bustillo, R. (2020). Impact of remittances on economic growth and poverty reduction amongst CIS countries. *Post-Communist Economies, 32*(4), 525–546.

Acharya, C. P., & Leon-Gonzalez, R. (2014). How do migration and remittances affect human capital investment? The effects of relaxing information and liquidity constraints. *Journal of Development Studies, 50*(3), 444–460.

Adams, R. H., Jr., & Cuecuecha, A. (2013). The impact of remittances on investment and poverty in Ghana. *World Development, 50*, 24–40.

Affudo, M. (2020). *Remittance inflows and educational outcomes in Kenya*. Masters dissertation, University of Nairobi, Kenya.

Al-Islam, S. M., Prachee, F. S., & Saifullah, M. K. (2022). The impact of remittances on children's education in Bangladesh. *Institutions and Economies*, 87–110.

Amakom, U., & Iheoma, C. G. (2014). Impact of migrant remittances on health and education outcomes in sub-Saharan Africa. *IOSR Journal of Humanities and Social Science, 19*(8), 33–44.

Arellano, M., & Bond, S. (1991). Some tests of specification for panel data: Monte Carlo evidence and an application to employment equations. *The review of economic studies, 58*(2), 277–297.

Arif, I., Raza, S. A., Friemann, A., & Suleman, M. T. (2019). The role of remittances in the development of higher education: Evidence from top remittance receiving countries. *Social Indicators Research, 141*(3), 1233–1243.

Ayad, M. S., & El-Aziz, A. A. (2018). The impact of remittances on children's educational attainment: Evidence from Egypt. *Socioeconomic Challenges, 2*(2), 49–59.

Bangake, C., & Eggoh, J. (2020). Financial development thresholds and the remittances-growth nexus. *Journal of Quantitative Economics, 18*(2), 425–445.

Becker, G. S. (2009). Human capital: A theoretical and empirical analysis, with special reference to education. New York: National Bureau of Economic Research.

Bucheli, J. R., Bohara, A. K., & Fontenla, M. (2018). Mixed effects of remittances on child education. *IZA Journal of Development and Migration, 8*(1), 1–18.

Fromentin, V., & Leon, F. (2019). Remittances and credit in developed and developing countries: A dynamic panel analysis. *Research in International Business and Finance, 48*, 310–320.

Glewwe, P., Maiga, E., & Zheng, H. (2014). The contribution of education to economic growth: A review of the evidence, with special attention and an application to Sub-Saharan Africa. *World Development, 59*, 379–393.

Gul, S., Zeb, A., Ullah, O., & Mingyan, G. (2021). Impact of foreign remittances on school enrolment and educational expenditures in district Peshawar, Pakistan. *Liberal Arts and Social Sciences International Journal (LASSIJ), 5*(2), 209–221.

Hapsari, P. I. (2019). The impact of remittances on children educational outcome in Indonesia. *Jurnal Perencanaan Pembangunan: The Indonesian Journal of Development Planning, 3*(1), 32–46.

Hassan, S. A., & Rafaz, N. (2017). The role of female education in economic growth of Pakistan: A time series analysis from 1990-2016. *International Journal of Innovation and Economic Development, 3*(5), 83–93.

IMF. (2009). *Balance of payments and international investment position manual-sixth edition (BPM6)*. International Monetary Fund.

Inoue, T., & Hamori, S. (2016). Financial access and economic growth: Evidence from sub-Saharan Africa. *Emerging Markets Finance and Trade, 52*(3), 743–753.

Jakob, P. H. (2015). *The impact of migration and remittances on children's education in El Salvador*. Masters Thesis, University of San Francisco.

Jijin, P., Mishra, A. K., & Nithin, M. (2022). Macroeconomic determinants of remittances to India. *Economic Change and Restructuring, 55*(2), 1229–1248.

Kalaj, E. H. (2014). Effects of remittance flows on the school attainment of household members left behind. *Academic Journal of Interdisciplinary Studies, 3*(6), 527–527.

Kennedy, P. (2003). *A guide to econometrics*. MIT Press.

Kotásková, S. K., Procházka, P., Smutka, L., Maitah, M., Kuzmenko, E., Kopecká, M., & Hönig, V. (2018). The impact of education on economic growth: The case of India. *Acta Universitatis Agriculturae Et Silviculturae Mendelianae Brunensis, 66*(1), 253–261.

Maneejuk, P., & Yamaka, W. (2021). The impact of higher education on economic growth in ASEAN-5 countries. *Sustainability, 13*(2), 520.

Mazumder, S. B. A. (2021). Determinants of remittances: A study in Cachar District of Assam. *Global Business Review, 22*(2), 474–484.

Musakwa, M. T., & Odhiambo, N. M. (2022). Remittance inflows and poverty Nexus in Botswana: A multivariate approach. *Journal of Sustainable Finance & Investment, 12*(2), 475–489.

Omodero, C. O., & Nwangwa, K. C. (2020). Higher education and economic growth of Nigeria: Evidence from co-integration and granger causality examination. *International Journal of Higher Education, 9*(3), 173–182.

Pal, S., Villanthenkodath, M. A., Patel, G., & Mahalik, M. K. (2022). The impact of remittance inflows on economic growth, unemployment and income inequality: An international evidence. *International Journal of Economic Policy Studies, 16*(1), 211–235.

Petrou, K., & Connell, J. (2017). Food, morality and identity: Mobility, remittances and the translocal community in Paama, Vanuatu. *Australian Geographer, 48*(2), 219–234.

Portes, A. (1998). Social capital: Its origins and applications in modern sociology. *Annual Review of Sociology, 24*(1), 1–24.

Ramos, L. M., & Mourelle, E. (2019). Education and economic growth: An empirical analysis of nonlinearities. *Revista de economía aplicada, 27*(79), 21–45.

Shafiq, M. N., Yang, X., & Nawaz, M. A. (2022). Do remittances promote education? Empirical evidence from developing countries. *Pakistan Journal of Humanities and Social Sciences, 10*(2), 830–841.

Stark, O. (1991). Migration in LDCs: risk, remittances, and the family. *Finance and Development, 28*(4), 39–41.

Tung, L. T., & Thang, P. N. (2023). Impact of remittances on income inequality: Empirical evidence from emerging economies. In M. Busu (Ed.), *Digital economy and the green revolution: 16th international conference on business excellence, ICBE 2022, Bucharest, Romania, Cham* (pp. 287–297). Springer, Nature Switzerland.

Wang, Y., & Liu, S. (2016). Education, human capital and economic growth: Empirical research on 55 countries and regions (1960-2009). *Theoretical Economics Letters, 6*(02), 347–355.

World Bank. (2021a). *Migration and development brief 34*. World Bank.

World Bank. (2021b). *Defying predictions, remittance flows remain strong during COVID-19 crisis*. Retrieved from: https://www.worldbank.org/en/news/press-release/2021/05/12/defying-predictions-remittance-flows-remain-strong-during-covid-19crisis#:~:text=Officially%20recorded%20remittance%20flows%20to,latest%20Migration%20and%20Development%20Brief

Yakubu, I. N. (2019). Does corruption grease or sand the wheels of bank profitability in Ghana? *Cogent Economics & Finance, 7*(1), 1701909.

Yakubu, I. N., & Gunu, I. M. (2022). The impact of government educational expenditure policy and school enrolment on economic growth in Ghana (1970-2017). *Asian Journal of Research in Education and Social Sciences, 3*(4), 107–117.

Zhunio, M. C., Vishwasrao, S., & Chiang, E. P. (2012). The influence of remittances on education and health outcomes: A cross country study. *Applied Economics, 44*(35), 4605–4616.

14 The Politics of CSR in the Banking Sector: How Does It Align with the SDGs of No Poverty and Quality Education?

Richard Mbayo and Kehinde M. Ige

14.1 Introduction

> A bank is only as strong as the community it serves, and this founding philosophy has not only helped it to provide support and aid during the global health crisis but also forms part of Zenith Bank's continuing commitment to help govern sustainability long into the future. (Onyeagwu & Ovia, 2023)
> Ebenezer Onyeagwu, CEO, Zenith Bank, Nigeria

Corporate social responsibility (CSR) has become integral to contemporary business operations, reflecting an evolving understanding of corporations' roles in addressing societal challenges. In the banking sector, where financial institutions wield significant influence and resources, the alignment of CSR initiatives with the Sustainable Development Goals (SDGs) has emerged as a crucial metric for evaluating their social and ethical commitment (Adeleke, 2014). This study investigates the politics surrounding CSR practices in the banking sector, focusing on how these initiatives intersect with SDGs 1 and 4, No Poverty and Quality Education. Through a comparative case study of Stanbic Bank Uganda and Zenith Bank Nigeria, this chapter aims to uncover the motivations and challenges in aligning CSR efforts with the SDGs.

Stanbic Bank and Zenith Bank are prominent financial institutions within their national financial landscapes. By examining the practices of these two banks, we seek to shed light on the nuances of CSR implementation in diverse socio-economic contexts and regulatory environments. Moreover, the comparative approach allows a deeper understanding of the factors influencing CSR decisions, including stakeholder expectations and regulatory frameworks.

Central to this investigation is an exploration of how CSR initiatives in the banking sector contribute to poverty reduction and educational advancement, as outlined in SDGs 1 and 4. By assessing the nature and impact of CSR projects undertaken by Stanbic Bank and Zenith Bank in these areas, this study evaluates the effectiveness and alignment of their efforts with broader development goals. By examining the experiences of Stanbic Bank and Zenith Bank in Uganda and Nigeria, respectively, this study will generate insights that can inform future CSR strategies, policy interventions, and collaborative initiatives aimed at fostering economic growth and inclusivity such that it contributes to the attainment of the SDGs by 2030. The next sector delves into the overview of CSR in the banking sectors of Uganda and Nigeria.

R. Mbayo (✉) · K. M. Ige
Northern Arizona University, Flagstaff, AZ, USA
e-mail: rm2865@nau.edu

I. N. Yakubu (ed.), *Strategic Approaches to Banking Business and Sustainable Development Goals*, Sustainable Development Goals Series, https://doi.org/10.1007/978-3-031-80744-2_14

14.2 The Overview of CSR in the Banking Sectors of Uganda and Nigeria

Corporate social responsibility (CSR) is becoming increasingly significant in the banking sectors of Uganda and Nigeria, where these banks channel their CSI into education, poverty alleviation programs, healthcare, security, and environmental sustainability programs. Several banks in Uganda assist with educational initiatives by offering scholarships, funding the construction of school facilities, and sponsoring academic programs (Kazibwe, 2023). For instance, Stanbic Bank Uganda has a well-established collaboration with Makerere University, which offers a laptop loan scheme to ease education services for students and other university staff (Wamai, 2022).

Banks contribute to health projects through financial donations, the provision of medical equipment, and support for awareness campaigns. For example, banking institutions have actively participated in efforts focused on raising awareness and preventing HIV/AIDS, with 54 health-related initiatives supported in 2023, covering 17 percent of the annual CSR budget (Centenary Bank, 2023). According to the 2020 CSR report, Zenith Bank Nigeria invested $8.2 million in social investments, representing 1.66 percent of their annual profits, which was $607 million (Onyeagwu & Ovia, 2023).

Many banks have committed to environmental conservation by implementing tree-planting campaigns, trash management initiatives, and assisting with renewable energy projects as contributive efforts toward environmental sustainability (Total Energies Uganda, 2023; *The Guardian*, 2023; SSEBWAMI, 2024). For instance, NCBA Bank Uganda has initiated a collaborative effort with the National Forestry Authority (NFA) to plant 20,000 trees by the year 2024, covering an area of 30 hectares. The trees will be planted at Jubiya Forest Reserve in Masaka as a component of the bank's "*Change the Story*" campaign, which seeks to rehabilitate damaged forest areas (SSEBWAMI, 2024).

Banks engage in several community development initiatives, including establishing clean water facilities and facilitating financial literacy by providing loan access for small and medium companies. The Bank of Africa bolstered the economy by allocating more than US$20 million in financial assistance to small- and medium-sized enterprises (SMEs) operating in the education, trade, and construction industries (Bank of Africa, 2023). In Uganda and Nigeria, CSR operations in the banking sectors have transformed from simple acts of charity to purposeful programs that tackle social, economic, and environmental issues in their localities.

14.3 The Justification of Case Selection

The selection of Stanbic Bank, Uganda, and Zenith Bank, Nigeria, as case studies for examining the alignment of CSR practices in the banking sector with the Sustainable Development Goals (SDGs) of no poverty and quality education is dichotomous, such that both Uganda and Nigeria are developing nations with high levels of poverty and a young population in need of quality education.

Uganda and Nigeria face significant challenges in addressing poverty and access to quality education. As the World Bank Report shows, Nigeria's poverty rate is estimated to have reached 38.9 percent in 2023, with an estimated 87 million Nigerians living below the poverty line (World Bank, 2023a, b). While Uganda's rapid population growth has put a huge portion of the population below the poverty line, human capital and infrastructural shortfalls have stifled growth and social welfare (World Bank, 2023a, b). Hence, the banking sector plays a crucial role in enhancing economic development and has the potential to significantly contribute to the achievement of the SDGs through its CSR initiatives.

Stanbic Bank, Uganda, is a leading financial institution in the country and has been actively involved in various CSR programs focused on education and poverty alleviation (Fitch Group, 2024). The bank's long-standing partnership with Makerere University and Stanbic's National

School Championship provides scholarships and supports entrepreneurship programs that align with SDG 4 (Quality Education) and SDG 1 (No Poverty) by fostering quality education and creating employment opportunities (Kyobutungi, 2024). Similarly, Zenith Bank, Nigeria, has a strong track record of CSR initiatives in the areas of education and poverty reduction. The bank's efforts in promoting financial inclusion, particularly in rural areas, through expanding banking services and mobile banking platforms contribute to SDG 1 (No Poverty) by increasing access to financial services for underserved communities (Onyeagwu & Ovia, 2023).

By examining the CSR practices of these two banks, the case studies provide valuable insights into how the banking sector in developing countries can strategically align its CSR initiatives with the SDGs of no poverty and quality education. The study highlights the nuances aimed at evaluating the impact of CSR programs on sustainable development. Case studies from both Uganda and Nigeria help measure CSR investments and their dealing with poverty and quality education. Overall, the choice of Stanbic Bank, Uganda, and Zenith Bank, Nigeria, as case studies, provides a strong foundation for exploring the politics of CSR in the banking sector and its alignment with the SDGs of no poverty and quality education, contributing to the broader discourse on sustainable development and responsible business practices.

14.4 How CSR Initiatives Align with the SDGs of "No Poverty and Quality Education"

Corporate social responsibility (CSR) plays a critical role in modern business practices, bridging corporate operations and societal needs. Aligning CSR initiatives with the Sustainable Development Goals (SDGs) is particularly important, especially in sectors like banking, which have broad societal impacts. The SDGs provide a global framework for addressing key challenges such as poverty and education, which are crucial areas of focus for many communities. When banks like Stanbic Bank (Uganda) and Zenith Bank (Nigeria) align their CSR activities with these goals, they ensure that their efforts are not only philanthropic but also strategic toward immediate priorities. This alignment helps in effectively channeling resources and efforts to areas where they can create the most significant impact.

Second, such alignment fosters greater accountability and transparency. The SDGs come with specific targets and indicators, which can serve as benchmarks for measuring the impact of CSR activities. By adhering to these benchmarks, banks can provide clearer and more measurable outcomes for their CSR initiatives. This not only builds trust with stakeholders but also demonstrates the bank's commitment to contributing to broader societal goals.

Aligning CSR with the SDGs can enhance the reputation and brand value of banks. In today's socially conscious market, customers and investors are increasingly looking to support businesses that demonstrate a genuine commitment to societal well-being. Banks that align their CSR with the SDGs stand out as responsible and forward-thinking institutions, which can attract more customers and investment. The alignment can contribute to sustainable business practices. By focusing on goals like poverty alleviation and quality education, banks can help build stronger, more educated, and more economically stable communities. This, in turn, can lead to a more stable business environment and potential customer base, fostering long-term business growth. For Stanbic Bank and Zenith Bank, this alignment represents a powerful way to leverage their resources and influence to support meaningful societal progress. The next section focuses on the literature pertaining to CSR.

14.5 Literature Review

The promotion of CSR has been advocated globally through activism aimed at firms, educational programs on sustainable development, and the establishment of a regulatory framework (Wafula & Minja, 2024, 94). Kaplan and Kinderman

(2017) argue that non-market methods, such as cultural beliefs, have significantly influenced the development of corporate social responsibility (CSR) activities (Kaplan & Kinderman, 2017).

Lange and Wyndham (2023) assert that global development agencies, such as the United Nations (UN) and the World Bank, have invited private sector entities to actively participate as development partners (Lange & Wyndham, 2023, 194). Additionally, several of the Sustainable Development Goals (SDGs) are directly connected to the private sector, such as no poverty and quality education, which seem to be inseparable, because we argue that businesses thrive in an environment where people's welfare is stable as education contributes greatly to the development of the middle class, who are the primary consumers of the company's products. Many companies have willingly adopted this new role as a component of their CSR initiatives. Shertans argues that the banking system is vital to economic development (Scholtens, 2009, p. 159).

Businesses, as economic entities in a capitalist system, possess political power as they frequently make decisions that affect different stakeholders, both within and outside the organization (Rotter et al., 2014, 582). This includes decisions related to resources, technological advancements, and product development, which are often made independently of public democratic processes. The CSR initiatives in Nigeria originated from the practices of multinational companies (MNCs) operations in the extraction sectors of the Nigerian economy, especially in the oil sector. Their operations in communities resulted in CSR breaches such as oil spillages, gas flaring, militancy, and community agitations, including the dumping of toxic waste in rivers.

These activities destroyed the sources of income for the communities, which are mainly farming and fishing, leading to widespread poverty and agitation in the communities. Cumulatively, these raised concerns about the role of businesses in Nigerian society (Osemeke et al., 2016). A study by Osemeke (2012) revealed that CSR in developing countries is strategic and linked to companies' corporate philosophies. This study implied that the interests of the managers should be aligned with the interests of the stakeholders to ensure the company's long-term survival and create a win-win situation between the company and the community (Osemeke, 2012).

Also, the corporate social performance theory therefore bolsters the need for more positive impacts of a firm's activities, which are not only about "doing good," as sometimes assumed, but doing what is right. However, constant protests by the host communities attest to their displeasure over the MNC's actions, whose cost outweighs the benefits, hence the demand for more benefits (Enuoh, 2023). To our minds, these realities are evident in the Niger Delta region of Nigeria, wherein the activities of multinational corporations and their oil extraction endeavors have impacted the living conditions of these communities.

Environmental issues such as climate change, life-cycle assessment, and pollution prevention directly and indirectly impact poverty levels and quality education in Nigeria and Uganda, respectively. Community involvement and development, the integration of disabled persons, and consumer interests, including privacy, are all part of the CSR agenda (Ofodile et al., 2012).

14.6 Stakeholder Theory and CSR

In our attempt to dig deeper into the politics of CSR in the banking sector and how it aligns with the SDGs, this chapter used *stakeholder theory,* which is a systematic approach to understanding the politics of CSR, social engagements, and symbiotic relations between agencies, institutions, corporations, and the citizenry (Miles, 2017). Miles continues to argue that the fundamental coalitions and the engagement process regarding corporate social responsibility are an outcome of stakeholder activities. Stakeholder theory is widely accepted, but elementary aspects remain indeterminate as the term "stakeholder" is an essentially contested concept (Miles, 2017). The stakeholder theory deems the CSR of institutions and firms in such a manner that it oscillates between two extremes: one that reduces the firm's commitment to the obtainment of profit for its shareholders, like the firm's obligation to engage with a wide range of actors with a specific interest or "stake" in the firm. As a result, the stakeholder theory is appealing from an ethical standpoint despite lacking a solid foundation that different schools of thought consider tenable (Argandoña, 1998).

CSR highlights that companies pursue their main goals and critical interests relative to maximizing profit and are responsible for contributing to society's overall development and welfare through compelling policy considerations and initiatives such as voluntary efforts. Companies that pursue CSR engagements will gain a competitive advantage due to creating an excellent public image and reputation for their likelihood of generating high revenues and investment returns. However, some authors largely disagree with this perspective. It is imperative to state that CSR is characterized by a dichotomous nature of social development and profit markets, as one factor is not independent of the other (Barauskaite & Streimikiene, 2021).

CSR is based on a set of moral rules that all businesses, organizations, cooperatives, and even individuals must follow to be fair in their dealings with a wide range of stakeholders, such as society, the environment, and the economy. This can be difficult to implement because people must be more ethical. Additionally, the challenges of a regulatory environment may negatively impact the CSR's effectiveness. Existing evidence also proved a mismatch between policy statements, legal documents, statutory provisions, and reality in the practical experience of the implementation of CSR objectives (Maliganya et al., 2023). Using Nigeria and Uganda as examples in this chapter, it is essential to note that CSR is usually limited to charitable giving and, at best, community development projects. In this regard, the stakeholder theory considers the respective coalitions and relations between stakeholders when enacting CSR. However, we argue that the stakeholder theory is limited in scope in its outright definition of the parameters for engagement between stakeholders, seeing there are no concise frameworks, rules, or guidelines for engagement (Amodu, 2020). The next section explains the politics of CSR and sustainable development.

14.7 The Politics of CSR and Sustainable Development

The politics of CSR and sustainable development seek to inquire about the policy engagement and legal criteria between state institutions, bureaucracies, financial institutions, and the masses in the implementation of CSR. Pragmatic necessities now serve as drivers that define ethical and tenable approaches within the confines of the implementation of corporate social responsibility. The politics of CSR and sustainable development are intertwined as they deal with the various engagements between governmental and non-governmental stakeholders. Some institutional shareholders seek to leverage politics by influencing corporate policies and practices. (Chen et al., 2020).

These business-driven initiatives rely on enacted policies and regulations alongside NGOs and advocacy groups. Several reasons could motivate institutional shareholders to improve firms' social outcomes. Capital providers now consider environmental and social impacts when making investment decisions. Prior studies have demonstrated that, in addition to financial returns, individual investors may profit by making investments based on their social preferences for social responsibility within the confines of the legal frameworks and requirements set forth for such engagements (Chen et al., 2020).

Many misconceptions about corporate sustainable development are that it fundamentally depends on stakeholders and their engagements with the community. Still, principally, the government determines the regulatory conditions, the rules in use, and the actions in use that either directly or indirectly create equity amidst the numerous CSR implementation processes. Likewise, firms proffer effective managerial and organizational changes to help them better fulfill their proposed social responsibility instead of window-dressing their activities. The integration of CSR activities with the economic operations of the business is imperative (Dahiya & Singh, 2020). Hence, on this backdrop, we establish our methodology based on reliability and validity in our assessment of the CSR endeavors of Stanbic Bank and Zenith Bank, respectively, which serves as the premise for our methodology.

14.8 Methodology

This study is part of a larger research project titled *Strategic Approaches to Banking Business and Sustainable Development Goals*. The study used ratio analysis from Uganda's Stanbic Bank and Nigeria's Zenith Bank by reviewing 5-year annual financial and sustainability reports spanning from 2019 to 2023 that highlight the annual profits made at the end of the financial year and how much of that profit was invested in CSR projects that align with the Sustainable Development Goals of poverty alleviation and quality education. Stanbic Bank, Uganda, is the leading financial institution in the country and has been actively involved in various CSR programs focused on education and poverty alleviation, which explains why we selected it as a case study from a Ugandan perspective (Fitch Group, 2024).

The bank's long-standing partnership with educational institutions and Stanbic's National School Championship (NSC), providing scholarships and supporting entrepreneurship programs, aligns with SDG 4 (Quality Education) and SDG 1 (No Poverty) by promoting access to education and creating employment opportunities (Kyobutungi, 2024). Similarly, Zenith Bank, Nigeria, has a strong track record of CSR initiatives in the areas of education and poverty reduction. The bank's efforts in promoting financial inclusion, particularly in rural areas, through the expansion of banking services and mobile banking platforms contribute to SDG 1 (No Poverty) by increasing access to financial services for underserved communities (Onyeagwu & Ovia, 2023).

Besides our main objective of understanding the politics of CSR in the banking sector, we seek to understand whether these financial institutions in question align with the UN Sustainable Development Goals of No Poverty and Quality Education. Using ratio analysis, it helps to answer those questions by reviewing the financial and sustainability reports. It also helps to find out whether these banks dedicate a reasonable amount of their annual profits to achieve these SDGs of no poverty and quality education, as we believe that financial institutions have a great role to play, especially by providing capital, which is very important in fighting poverty as well as investing in education.

The empirical data were analyzed using ratio analysis to get the exact percentage of profit allocated to corporate social responsibility (CSR) initiatives. We focused on Sustainable Development Goals 1 and 4, which are No Poverty and Quality Education, to be attained by 2030. Although these banks invested in other sectors of the economy, such as healthcare and environment protection, Zenith Bank and Stanbic Bank have partnered with local NGOs and government agencies to plant trees and provide healthcare services.

Our focus is on poverty alleviation and quality education because reducing poverty can help improve education and the quality of life. Quality education can help reduce poverty, assuming that those skills can be invested in income-generating activities. Ratio analysis evaluates the financial revenues and record statements of corporations to evaluate their profitability. These financial statements serve as the basis of our analysis of the expenditure of Stanbic Bank, Uganda, and Zenith Bank, Nigeria, on corporate social responsibility vis-à-vis the UN SDGs of No Poverty and Quality Education.

14.9 Stanbic and Zenith Banks' Role in Poverty Alleviation and Quality Education

(a) *Stanbic Bank Uganda*

In our attempt to understand the politics of CSR in the banking sector and how it aligns with the SDGs of poverty alleviation and quality education, we present a case study of Uganda's Stanbic Bank and Nigeria's Zenith Bank. We analyzed these banks' corporate social investments in the last 5 years, spanning from 2019 to 2023. Based on the profit margins of these annual financial and sustainability reports, Stanbic Bank has been investing less than 2 percent of its annual profit after taxes in CSI activities aimed at improving the lives of communities, as seen in Table 14.1.

In the year 2023, Stanbic Bank Uganda (SBU) made a profit after taxes of 412 billion Ugandan shillings and spent less than *1 percent* on CSR projects, according to its financial annual report for 2023. This 1 percent was greatly invested in education, where UGX 1 billion was channeled to the National School Championship (NSC). Here, over 60,000 students benefited from the initiative. Stanbic Bank's NSC attracted 158 schools across the country, and the bank also offered 70 scholarships through Stanbic Busitema University Scholarships, which were awarded to students from underprivileged families. UGX20 million was invested in the Women in Technology initiative, 10 graduated, and 800 million was invested in the health sector through Stanbic *Corporate Society for Safe Motherhood*, which helped 44 percent of maternal deaths and 200,000 mothers benefit from this program across 52 health centers (Stanbic Bank Uganda, 2023, 46).

In 2022, SBU invested 1.2 percent of its annual profit into CSR projects impacting the lives of 331,684 Ugandans in the sector of education, where 1 billion Ugandan shillings were injected into the NSC supporting 60,000 students across 100 schools, 200 business ideas were developed, and 800 million were donated to the health sector, of which 78,0000 mothers benefited from the Stanbic Maternal Health Program and 74,000 trees were planted through Stanbic's Running Out of Trees (ROOT) Campaign (Stanbic Bank Uganda Report to Society, 2022, 45). To fight poverty, Stanbic Business Incubator trained 1074 SMEs and supported 260,000 SACCO members, and 197,000 women groups were financed through Stanbic for Her Campaign.

Stanbic Bank Uganda invested UGX3.5 billion in CSR projects in 2021, which is 1.3 percent of the total profits made by the bank in the same year, which is UGX350 billion (Stanbic Bank Sustainability Report, 2021, 98). Bank invested 800 million Ugandan shillings in the National School Championship (NSC), where 60,000 high school students participated across 100 schools, 400 teachers, and 2300 students acquired skills. While in the health sector, UGX230 million was invested through the Stanbic *Every Mother Counts* program, and 4000 expectant mothers benefited from the initiative. In the environment sector, Stanbic's *Taasa Obutonde* Campaign, together with other corporations, donated 30 million Ugshs to support waste recycling management initiatives, and 19,279 trees were planted in the calendar year. Through Stanbic's Employee Community Initiative (ECI), over UGX447 million was inserted into community projects aimed at fighting poverty, supporting youth projects, and women's initiatives, of which over 73,045 individuals benefited.

Table 14.1 Stanbic Bank's (Uganda) corporate social investments (CSI) in the past 5 years

Year	Annual profits after tax (UGX)	Annual CSI (UGX)	CSI (Percentage)	Annual beneficiaries (Persons)	Sectors invested
2023	412 billion	4 billion	0.97%	3,532,376	Education, health, environment, and poverty alleviation projects
2022	357 billion	4.2 billion	1.2%	331,684	Education, health, environment, and poverty alleviation projects
2021	269 billion	3.5 billion	1.3%	331,600	Education, health, environment, and poverty alleviation projects
2020	241 billion	3.9 billion	1.6%	459,138	Education, health, environment, and poverty alleviation projects
2019	259 billion	2.9 billion	1.1%	374,169	Education, health, environment, and poverty alleviation projects

Source: Author's illustration

In 2020, when the world was battling with COVID-19, SBU continued with its commitment to education, as UGX1 billion was invested in education, which benefited over 60,000 students. In health, the bank provided UGX260 million in the form of PPE items that were given to the employees of the Ministry of Health. The SBU, through ECI, collected UGX100 million, and over 30 staff members participated in community activities aimed at providing services. To fight poverty, Stanbic Business Incubator trained 2392 individuals and 1216 companies. Overall, 459,138 individuals nationwide benefited from the SBU's CSI projects in 2020 (Stanbic Uganda Holdings Limited Annual Report, 2019, 107).

In the year 2019, SBU recorded a profit of UGX 259 billion, and it invested 1.1 percent of that profit in CSR. These CSR programs have impacted the lives of 374,169 individuals. In total, 1.85 billion Ugandan shillings were invested in education through Stanbic's NSC, where 72 schools, 72 teachers, and 21,600 students participated, and 432 new business ideas were cropped out. SBU also offered 75 million Ugandan shillings worth of scholarships to students from underprivileged families to study at Uganda Christian University. Together with the USAID/RTI program, Stanbic invested 20 million Ugandan shillings in the *Reading Card Project*, which benefited 25,025 primary students and allowed them to read and write.

Over 500 million Ugandan shillings were invested in poverty alleviation projects, such as offering training to 660 SMEs and 1746 individuals. In total, 65 million Ugandan shillings were channeled to leadership training through the Young Emerging Leaders Project (YELP). Hence, 80 million Ugandan shillings were invested in the solar project through the *Stanbic-Mkopa project*, where 22 primary schools received solar lights. SBU donated 250 million Ugandan shillings to MTN Marathon, which helped equip health facilities in the country and benefited 42,000 individuals. 30 million were contributed to the *Rotary Cancer Run* that supported cancer patients countrywide, and through ECI and other philanthropic commitments to humanitarian endeavors, over 26 million Ugandan shillings were donated to health, education, and poverty alleviation projects for over 39,112 people. The next section analyzes the practice of CSR from a Nigerian perspective.

(b) *Zenith Bank Nigeria*

Zenith Bank has been and remains an integral part of the sustainability endeavor in Nigeria and the entirety of the continent of Africa. The said contributions of Zenith Bank cut across every facet of socio-economic development. This section of this chapter highlights the significant CSR contribution of Zenith Bank from 2019 to 2023. Much of our analysis and assessment of these CSR contributions is derived from the annual sustainability reports from Zenith within the respective years of 2019 up until 2023, as highlighted in Table 14.2.

According to the Zenith Bank (Nigeria) Financial and Sustainability Report of 2023, we see that the bank earned a profit margin of N676 billion in annual profit after tax. They then went ahead to allocate the sum of N5.67 billion naira for their annual corporate social investment, which made up a CSI percentage of 0.83 percent. The sectors to which this investment was allocated are security, sports, health, and education (Zenith Bank Sustainability Report, 2023a, b).

Likewise, in the year 2022, Zenith Bank's record profit after tax was N294 billion, with an annual CSI of N2.6 billion, making it a 0.88 CSI percent. The sectors invested in were education, health, environment, sports development, government, public infrastructure, women's empowerment, and youth empowerment (Zenith Bank Sustainability Report, 2022).

In 2021, Zenith Bank's annual profit after tax was N233.13 billion, with an annual CSI allocation of N4.372 billion, making up 1.88 percent of its annual profit after tax. The sectors targeted were education, health, and the environment (Zenith Bank Sustainability Report, 2021). For 2020, the annual profit after tax was N294 billion with an annual CSI of N3.285 billion, which made up a CSI percentage of 1.66 percent. The sectors for investments were environment, education, health, and social welfare. Key sectors for

Table 14.2 Zenith Bank (Nigeria) corporate social investments in the past 5 years (2019–2023)

Year	Annual profit after tax (₦)	Annual CSI (₦)	CSI percentage	Sectors invested
2023	N676 billion	N5.67 billion	0.83%	Security, sports, health, and education
2022	N294 billion	N2.6 billion	0.88%	Education, health environment, sports development, government and public infrastructure, and women and youth empowerment
2021	N233.13 billion	N4.372 billion	1.88%	Education, health, and environment
2020	N294 billion	N3.285 billion	1.66%	Education, health, and social welfare
2019	N178 billion	N2.73 billion	1.53%	Education, healthcare, and youth projects

Source: Authors' illustration

investments were education, health, and social welfare. The sustainability report from 2020 indicates that about 8 of the SDGs were addressed in its sustainability endeavors for the year. Zenith Bank has always played a critical role in communities by providing critical infrastructure and support via its social investment efforts.

Though the COVID-19 pandemic hindered investment in some of the targeted priority areas, Zenith Bank's response to the health crisis was robust. Its allocation to health was precisely 1 billion, 495 million naira, 85 million naira were allocated to education, and 87 million naira were made available for sports, youth, and women's development. Zenith Bank likewise provided 508 million naira toward government infrastructure and policy, which largely involved road construction, security trust funds, car purchases, government CSR, and program sponsorship in Lagos State (Abule-Ado explosion emergency relief fund). The total sum of 110 million naira was allocated for other projects, which include the Lagos State Government and other projects (Zenith Bank Sustainability Report, 2020).

For the year 2019, Zenith Bank recorded an annual profit after tax of N178 billion, with an annual CSI of N2.73 billion, which made up 1.53 percent of its CSI allocation. According to Zenith Bank's 2019 sustainability report, Zenith Bank's contributions to corporate social responsibility cut across the areas of healthcare, education, sports, youth development, women's development, government, public infrastructure, and policy, among others. The report also provided a detailed overview of the financial allocation to these endeavors. For instance, the report indicated that the sum of 345 million naira was allocated specifically to the area of healthcare, wherein donations were made to the private sector health alliance of Nigeria, the O'Five charity initiative, and provisions of medical supplies to Iga-Idunganran Primary Healthcare Center. Additionally, within the area of education, 210 million naira were allocated (Zenith Bank Sustainability Report, 2019).

Here, Zenith Bank engaged in the construction of a multi-purpose hall at Maryland Comprehensive Secondary School and upgraded the facilities at St. Francis Catholic Secondary School, Idimu; donations of laptops to Gateway Polytechnic, Ogun State; the construction of a new hostel at St. Finbarr's College, Akoka; sponsorship of the CFA Institute Research Challenge, as well as sponsoring the Zenith Bank Academic Excellence Award for best-graduating students in some federal universities in the country. Further, in sports/youth and women's development, 238 million naira were allocated.

Much of these funds went into the title sponsorship of the Delta State principal cup and provisional support for the Nigerian Football Federation and the Zenith Bank National Women's Basketball League in partnership with the Nigerian Basketball Federation (NBBF). Other areas, such as government/public infrastructure and policy, had an allocation of 1 billion, 420 million naira, while other projects had an allocation of 516 million naira (Zenith Bank sustainability report, 2019).

14.10 Conclusion

In conclusion, the chapter sheds light on the politics of CSR in the banking sector and how it aligns with the SDGs of no poverty and quality education. While CSR plays a key role in the Global North, its influence in the Global South is still minimal, as seen in the analysis that Africa's giants in the banking sector (Stanbic & Zenith) invest less than 2 percent of their annual profits. Still within CSI investments, a limited amount of money is dedicated to fighting poverty compared to other sectors such as education, health, and environment. The chapter also reveals that banks, especially Stanbic Bank Uganda, conduct their own research that enables them to make decisions pertaining to the SDGs the bank aligns with the most, as they depend on the community needs in allocating the CSI funds, which is through the Employee Community Initiative (ECI). While Zenith Bank also invests much of its CSR resources in education, healthcare, and the environment, Zenith Bank also invests in the security sector, which is not the same as SBU.

14.11 Recommendations

In our attempt to understand the politics of CSR, we found out that Stanbic Bank Uganda conducts its independent research to identify sectors that need the most support. The bank's CSR projects are mainly focused on education, health, and the environment. It is evident that soft loans and capacity building are offered to individuals, SMEs, and SACCOs through *Stanbic Business Incubator* and *Stanbic for Her*, which support women in financial literacy skills and support poverty alleviation programs.

Furthermore, SBU, through its Employee Community Initiative (ECI), contributes funds to support specific issues in the communities where employees have identified in their respective communities of operations. On the other hand, Zenith Bank, Nigeria, also invests most of its CSR monies in education, women's groups, the environment, and partly in poverty alleviation projects. This implies that both banks allocated a few resources aimed at fighting poverty, directly through their CSR programs. However, these banks in question provide affordable loans and offer financial literacy training to individuals, SMEs, and women groups, something that, in the long run, can help fight poverty.

Our research also confirms that both Stanbic and Zenith banks invest less than 1.5 percent of their annual total profits after taxes in CSR projects. For instance, Stanbic Bank Uganda reported UGX 412 billion in profit and only invested UGX 4 billion in CSR projects in 2023, which is less than 1 percent of its total annual profit. This is not a different story from Zenith Bank of Nigeria, which is the leading financial institution, and it also invested less than 2 percent of its annual profits, as seen above. Although we acknowledge that the banking sector contributes toward the SDGs of no poverty and quality education, the small amount of money allocated to CSR programs makes financial institutions' role minimal to achieve the SDGs by 2030.

For instance, we reviewed five-year financial and sustainability reports from 2019 to 2023, which indicate on average that Stanbic Bank Uganda impacts the lives of 300,000 people in a year with just 1 percent investment of its profits in CSR projects. Except in 2023, when the bank was reported to have impacted the lives of over 3 million people through its CSR projects by 1 percent. This seems strange, but that was the number of people reported, yet in 2022, a similar amount of money was invested in CSR but impacted only 330,000 people. Therefore, we recommend that banks and other corporations like MTN, Coca-Cola, and those in the construction and mining sectors increase their investment in CSR to only 5 percent of their annual profit. This will have a great impact on a great number of beneficiaries, ranging from 1.5 to 3 million people per year. This sounds beautiful because it can make a fundamental impact on the SDGs examined in this chapter.

Banks should prioritize poverty alleviation as they do with other sectors such as education, healthcare, and the environment. Based on a five-year financial and sustainability report of both banks, a limited amount of CSR monies is allocated to poverty alleviation, and in most cases, what is reported are soft loans given to SACCOs, women groups and financial literacy.

We join other scholars like Ediagbonya (2023) to recommend that governments across the African continent introduce laws that encourage corporations to commit at least 5 percent of their annual profits to investment in CSR programs to achieve the SDGs by 2030 (Ediagbonya, 2023). Imagine the circumstance where Stanbic Bank invests 5 percent of its profits; this means that a bank alone can support 1.5 to 3 million people a year and other corporations will follow suit. This can make these SDGs achievable by 2030, especially in the Global South, where most of the corporations are from the Global North. This means all profits are repatriated instead of invested again in the economy, and the only way for developing countries to benefit is by introducing a law barring corporations from investing less than 5 percent of their annual profits in CSR programs.

14.12 Limitations and Future Research Directions

The key limitation of our research was the inability to interview the CSR managers of Zenith Bank and Stanbic Bank and possibly provide an accountability report in our research as to the rationale behind the CSR practices adopted by each bank. This would have provided us with insights into the legal implications of these banks and the regulatory framework for ethical and transparent CSR engagement.

Focusing solely on Stanbic Bank in Uganda and Zenith Bank in Nigeria limits the generalizability of our findings. The results may not apply to other banks or regions. Also, the study may not account for all external factors influencing CSR activities, such as regional economic conditions, regulatory environments, and cultural differences. The study examines CSR initiatives within a specific time frame, which might not capture the long-term impact and sustainability of these programs. Changes in CSR strategies over time due to shifts in corporate policy, management, or external pressures are not fully explored.

We suggest conducting longitudinal studies to track the long-term impact of CSR initiatives on poverty alleviation and educational quality. This would provide deeper insights into the effectiveness of these programs. Comparative studies involving a broader sample can highlight best practices and common challenges in CSR.

References

Adeleke, C. J. (2014). *Corporate social responsibility in the Nigerian banking sector* (Doctoral dissertation, Walden University). Walden University. https://www.proquest.com/docview/1642439680?pq-origsite=gscholar&fromopenview=true&sourcetype=Dissertations%20&%20Theses

Amodu, N. A. (2020). CSR in Nigeria. In *Corporate social responsibility and law in Africa*. Routledge.

Argandoña, A. (1998). The stakeholder theory and the common good. *Journal of Business Ethics, 17*(9), 1093–1102. https://doi.org/10.1023/A1006075517423

Bank of Africa. (2023). *Community - Bank of Africa Uganda*. Bank of Africa Uganda. Retrieved May 22, 2024, from https://boauganda.com/brand/community/

Barauskaite, G., & Streimikiene, D. (2021). Corporate social responsibility and financial performance of companies: The puzzle of concepts, definitions and assessment methods. *Corporate Social Responsibility and Environmental Management, 28*(1), 278–287. https://doi.org/10.1002/csr.2048

Centenary Bank. (2023). *Health & livelihood*. Centenary Bank. Retrieved May 22, 2024, from https://www.centenarybank.co.ug/csr/health

Chen, T., Dong, H., & Lin, C. (2020). Institutional shareholders and corporate social responsibility. *Journal of Financial Economics, 135*(2), 483–504. https://doi.org/10.1016/j.jfineco.2019.06.007

Dahiya, M., & Singh, S. (2020). The linkage between CSR and cost of equity: An Indian perspective. *Sustainability Accounting, Management and Policy Journal, 12*(3), 499–521. https://doi.org/10.1108/SAMPJ-10-2019-0379

Ediagbonya, V. (2023). Towards a Mandatory Corporate Social Responsibility for banks in challenging institutional contexts: A case study of Nigeria Crowther, D., & Seifi, S. (Ed.), *Innovation, social responsibility and sustainability* (Developments in Corporate Governance and Responsibility, Vol. 22, pp. 3–24), Emerald Publishing Limited. https://doi.org/10.1108/S2043-052320230000022001

Enuoh, R. O. (2023). *Evaluating corporate social responsibility using corporate social performance: A case of the Niger Delta, Nigeria*. Retrieved June 18, 2024, from https://core.ac.uk/download/pdf/234648742.pdf

Fitch Group. (2024, April 16). *Stanbic Bank Uganda Limited*. Fitch Ratings. Retrieved May 23, 2024, from https://www.fitchratings.com/research/banks/stanbic-bank-uganda-limited-16-04-2024

Kaplan, R., & Kinderman, D. (2017). The business-led globalization of CSR: Channels of diffusion from the United States into Venezuela and Britain, 1962–1981. *Business & Society, 59*(3), 439–488. https://doi.org/10.1177/0007650317717958

Kazibwe, K. (2023, August 6). *Centenary Bank donates 20 computers to Kampala school to boost e-learning*. Nile Post. Retrieved May 22, 2024, from https://nilepost.co.ug/business/167298/centenary-bank-%C2%A0donates-20-computers-to-kampala-school-to-boost-e-learning

Kyobutungi, R. (2024). The ninth Stanbic Entrepreneurship title run assembles Uganda's Educational Titans. *The New Vision*. https://www.newvision.co.ug/category/education/9th-stanbic-entrepreneurship-title-run-assemb-NV_186779

Lange, S., & Wyndham, V. (2023). Gender, regulation, and corporate social responsibility: The case of equinor's social investments in Tanzania. In *Corporate social responsibility and the paradoxes of state capitalism: Ethnographies of Norwegian energy and extraction businesses abroad* (pp. 193–218). Berghahn Books. https://www.jstor.org/stable/jj.5590567.12

Maliganya, W., Bengesi, K. M. K., Mbota, M. M., & Bulengela, G. (2023). Corporate social responsibility and the challenges of the regulatory environment in the Tanzanian mining sector. In S. K. Msosa, S. Mugova, & C. Mlambo (Eds.), *Corporate social responsibility in developing countries: Challenges in the extractive industry* (pp. 157–173). Springer International Publishing. https://doi.org/10.1007/978-3-031-27512-8_10

Miles, S. (2017). Stakeholder theory classification: A theoretical and empirical evaluation of definitions. *Journal of Business Ethics, 142*(3), 437–459. https://doi.org/10.1007/s10551-015-2741-y

Ofodile, U. E., Altschuller, S. A., Dolize, A., & Fessler, M. (2012). Corporate social responsibility. *The International Lawyer, 46*(1), 181–197.

Onyeagwu, E., & Ovia, J. (2023). *Zenith Bank is balancing the scorecard with ESG policies*. World Finance. Retrieved May 22, 2024, from https://www.worldfinance.com/banking/zenith-bank-is-balancing-the-scorecard-with-esg

Osemeke, L. (2012). *The effects of different institutional investors and board of director characteristics on corporate social responsibility of public listed companies: The case of Nigeria* [PhD Thesis, University of Greenwich]. http://gala.gre.ac.uk/id/eprint/9405/

Osemeke, L., Adegbite, S. A., & Adegbite, E. (2016). Corporate social responsibility initiatives in Nigeria. *SSRN Electronic Journal*. https://doi.org/10.2139/ssrn.3512804

Rotter, J. P., Airike, P.-E., & Mark-Herbert, C. (2014). Exploring political corporate social responsibility in global Supply chains. *Journal of Business Ethics, 125*(4), 581–599. https://www.jstor.org/stable/24702316

Scholtens, B. (2009). Corporate social responsibility in the international banking industry. *Journal of Business Ethics, 86*(2) (May, 2009), 159–175. https://www.jstor.org/stable/40294882

Ssebwami, J. (2024, May 7). *NCBA Bank Uganda, NFA to plant 20,000 trees*. PML Daily. Retrieved May 22, 2024, from https://www.pmldaily.com/news/2024/05/ncba-bank-uganda-nfa-to-plant-20000-trees.html

Stanbic Bank Sustainability Report. (2021). *Sustainability Report*. Stanbic Bank Uganda. https://www.stanbic.co.ug/static_file/Uganda/Downloadable%20files/Sustainability%20Report%202022.pdf

Stanbic Bank Uganda. (2023). *Report to society*. Stanbic Bank, Uganda. https://www.stanbicbank.co.ug/static_file/Uganda%20Holdings/Downloadable%20files/Annual%20Reports/

Stanbic Bank Uganda Report to Society. (2022). *Stanbic Bank, Uganda*. Report to Society. https://www.stanbicbank.co.ug/static_file/Uganda%20Holdings/Downloadable%20files/Annual%20Reports

Stanbic Uganda Holdings Limited Annual Report. (2019). *Stanbic Uganda Holdings Limited Annual Report*. Stanbic Bank, Uganda. https://www.stanbic.co.ug/static_file/Uganda%20Holdings/Downloadable%20files/SUHL%20ANNUAL%20REPORT%202019_low%20res.pdf

The Guardian. (2023, October 30). FirstBank begins 50,000 tree planting campaign. *The Guardian*. Retrieved May 22, 2024, from https://guardian.ng/business-services/firstbank-begins-50000-tree-planting-campaign/

Total Energies Uganda. (2023, October 6). *Uganda is running out of trees: Partners target to grow 40 million trees this year*. TotalEnergies Uganda. Retrieved May 22, 2024, from https://totalenergies.ug/uganda-running-out-trees-partners-target-grow-40-million-trees-year

Wafula, G. M., & Minja, D. (2024). Corporate social responsibility: Evolution, controversies, and practical implications for Sino-Africa relations. *International Journal of Applied Science and Research, 7*(3)., SSN: 2581–7876. https://doi.org/10.56293/IJASR.2024.5906

Wamai, M. (2022, February 16). *Mak Signs Laptop Loan Scheme Agreement with Stanbic Bank*. Makerere University News. Retrieved May 22, 2024, from https://news.mak.ac.ug/2022/02/mak-signs-laptop-loan-scheme-agreement-with-stanbic-bank/

World Bank. (2023a). *Nigeria Overview: Development news, research, data*. World Bank. Retrieved May 23, 2024, from https://www.worldbank.org/en/country/nigeria/overview

World Bank. (2023b). *Uganda Overview: Development news, research, data*. World Bank. Retrieved May 23, 2024, from https://www.worldbank.org/en/country/uganda/overview

Zenith Bank. (2020) *Annual Sustainability Report*. African Financials. Retrieved May 29, 2024, from https://africanfinancials.com/document/ng-zenith-2020-ar-00/

Zenith Bank. (2022). *Annual Sustainability Report*. Retrieved May 29, 2024, from https://www.zenithbank.com/sustainability_report_2022/index.html#p=44

Zenith Bank. (2023a). *Annual Sustainability Report 2023*. Retrieved from May 29, 2024, https://www.zenithbank.com

Zenith Bank. (2021). *Annual Sustainability Report*. African Financials. Retrieved May 29, 2024, from https://africanfinancials.com/document/ng-zenith-2021-ar-00/

Zenith Bank. (2023b). *Annual Sustainability Report*. African Financials. Retrieved May 29, 2024, from https://africanfinancials.com/document/ng-zenith-2023-ar-00/

Part V

Macroeconomic Policy, Financial Stability, and Sustainable Growth

15 Examining Sustainable Economic Growth, Unemployment and Inflation: The Roles of Financial Development and Legal Systems

Mubarik Salifu, James Atta Peprah, and William Godfred Cantah

15.1 Introduction

Growth and inflation in sub-Saharan Africa (SSA) are significant issues for policymakers and academics. The region faces challenges like poor financial systems, weak property rights and legal frameworks, extreme poverty, inequality and regular shocks. These factors impact people's well-being and economic success in various ways.

A key concern is how growth, unemployment and inflation in Sub-Saharan Africa are affected by property rights, legal frameworks and financial development. Better financial intermediation quality and effectiveness can boost investment, innovation and output. Financial development promotes increased production, innovation and investment by enhancing the amount of financial intermediation. In order to lower transaction costs, build trust and promote economic activity, legal systems and property rights refer to the organisations that defend the rights of creditors, investors and business owners as well as the enforcement of agreements and rules.

Many researchers studied how financial growth affects economies in Africa south of the Sahara Desert. However, they disagree on the results. Some studies (e.g. Ibrahim et al., 2022) showed that financial development has a negative or negligible impact on growth, yet others (like Ustarz & Fanta, 2021) proved the opposite is true. Similarly, evidence exists that property rights and legal systems impact African economies positively (Acemoglu et al., 2001), negatively (Easterly andamp; Levine, 2003) or not at all (Rodrik et al., 2004).

This nexus also has a complicated and unclear inflationary role. Growth, unemployment and the evolution of the financial system can all be impacted by inflation, directly or indirectly. Moderate inflation, on the one hand, can boost gross domestic product (GDP) through raising profits, promoting investment and lowering real wages. On the other hand, excessive inflation can impede growth by reducing purchasing power, warping relative prices, causing ambiguity and discouraging saving. The influence of inflation on labour supply and demand, pay negotiations

M. Salifu (✉)
Department of Economics, School of Economics, University for Development Studies, Tamale, Ghana
e-mail: smubarik@uds.edu.gh

J. A. Peprah
Department of Applied Economics, School of Economics, University of Cape Coast, Cape Coast, Ghana
e-mail: jpeprah@ucc.edu.gh

W. G. Cantah
Department of Data Science and Economic Policy, School of Economics, University of Cape Coast, Cape Coast, Ghana
e-mail: william.cantah@ucc.edu.gh

I. N. Yakubu (ed.), *Strategic Approaches to Banking Business and Sustainable Development Goals*, Sustainable Development Goals Series, https://doi.org/10.1007/978-3-031-80744-2_15

and labour market institutions (LMIs) can also have an effect on unemployment. The supply and demand for financial services, the profitability of financial intermediaries and the effectiveness of financial markets are other ways that inflation can have an impact on financial development.

Empirical research on how inflation impacts Sub-Saharan Africa's economy, unemployment rate and financial development is sparse and unclear. Khan and Senhadji (2001) found a negative impact whilst Bittencourt et al. (2015) indicated negligible or adverse effects on unemployment, yet Ouedraogo et al. (2023) observed favourable influences. Regarding financial development, some literature (Boyd et al., 2001) suggested inflation hindered progress, yet (Ang & McKibbin, 2007) noted positive or negligible consequences.

Thus, the study sought to examine the impact of financial development on economic growth, unemployment and inflation in sub-Saharan Africa and also investigated the moderating effects of property rights and legal systems on these outcome variables.

15.2 Literature Review

Many theoretical and empirical studies have outlined the role financial development plays in influencing the various sectors of economies, especially its role in economic growth, unemployment, stability and inflation. Various theories, measurements and methods have been used to study these relationships. These sections present some of these studies and their arguments.

15.2.1 Financial Development and Economic Growth

The relationship between financial development and economic growth is a complex subject that has multiple aspects. There are many pathways through which financial institutions and markets influence the efficiency and productivity of an economy (Rousseau & Wachtel, 2000). This paper aims to bring together different research pieces on how far there is a cause-effect link between financial development and economic growth. According to Levine et al. (2000), financial services may lead to growth by providing ex-ante information about potential investment opportunities; observation of transactions for better corporate governance; risk management and diversification; raising savings (mobilisation) and pooling of them; facilitating trade in goods and services.

There exist diverging perspectives regarding the correlation between financial development and economic growth. Four arguments have been outlined by Chuah and Thai (2004). The initial argument suggests that financial development could stimulate economic growth by supporting innovative initiatives (King & Levine, 1993; Bencivenga & Smith, 1991). The second argument posits that economic growth precedes financial development, with the expansion of an economy giving rise to a financial sector (Robinson, 1952). The third argument asserts that both the first and second arguments are valid, highlighting a bi-directional relationship (Greenwood & Jovanovic, 1990). Lastly, the final argument implies that the impact of financial development on economic growth is often exaggerated, as the actual effect is minimal (Lucas, 1988).

Financial development essentially reduces informational, transactional and monitoring costs whilst optimising the distribution of money and risk throughout the economy. This leads to higher investment and higher quality investment, which in turn promotes economic growth (Machado et al., 2021).

Abbas et al. (2022) in a study comparing lower middle-income countries and upper middle-income countries find how economic growth and income distribution are affected by the level of financial inclusion. For the economic growth model, they use 44 middle-income countries, whilst for the inequality model, they use 42 middle-income countries. Using an autoregressive distributed lag (ARDL) model, they find that financial development affects both groups of economies in the long run. They, however, find that the effect is more noticeable amongst the upper middle-income countries than the lower middle-income countries. Using a Granger causality test, they further find that there existed a two-way relationship between economic growth and financial development in both lower and middle-income countries.

Alananga and Mutasa (2022) using a sample of countries from the East Africa Community and the Southern African Development Community examined the finance-growth nexus. They examined how economies of smaller countries could utilise monetary policy to stimulate the rate of their economic growth with a particular focus on proximity. Using a panel error correction model, their findings suggest that the ability of relatively small countries to pursue monetary expansionary policies is threatened by their proximity to trading giants. They also find that economic growth in the earmarked economies was influenced by the level of development of their financial systems.

Ductor and Grechyna (2015) evaluate the interdependence linking the progress of the financial sector and the production of the tangible sector and their influence on the advancement of the economy, utilising a data panel comprising 101 nations during the period of 1970–2010. The findings of this study demonstrate that when a state of equilibrium is achieved between the growth rates of the financial sector and the tangible sector, a beneficial impact of financial development on economic growth is observed. Conversely, if the growth rates of these two sectors are disproportionate, the positive impact is diminished or potentially even reversed. Similarly, Puşcaşu (2022), using annual data between the period 1990–2020, examined how economic growth is affected by financial systems in European member countries. The results of their research suggest that the advancement of financial systems, by means of the operations conducted by banks and capital markets, has a beneficial impact on the distribution of resources, the collection of savings and the effective handling of risks. Consequently, this leads to economic growth, as there exists a connection between the amount of funds invested and the output generated by the real sector.

Chen (2023) using data from 50 counties between 1997 and 2014 examined how economic volatility and economic growth are affected by financial development. Their findings demonstrate that, in both developed and developing nations, the amount of financial development inhibits economic progress, but the quality of financial development can stimulate it. Financial development affects economies more in emerging nations than in developed nations, but it has less of an effect on overall economic volatility. This helps to explain why economic volatility is higher in emerging nations than in developed nations.

Marwa and Zhanje (2015), in a review of theories that explain the finance-growth nexus, attempt to examine how development fits into the nexus. The main emerging tendency is that the majority of modern theories that attempt to explain the relationship between finance and growth have only been concentrating on conventional finance as a whole. Understanding the role of development finance on the finance-growth nexus has received little attention. It concludes that the role of development finance should be integrated into the theory of the finance-growth nexus for a more thorough understanding of the relationship. The study shows that the traditional model of the relationship between finance and growth is more likely to underestimate the extent to which finance influences economic growth, particularly in less developed nations.

Ustarz and Fanta (2021) carried out an examination regarding the influence of financial development on the performances of various sectors within sub-Saharan African economies. Their discoveries propose that financial development has a positive impact on the agriculture and service sectors, albeit necessitating a certain level of advancement to favourably influence the growth of the industrial sector. The outcomes of their investigation exhibit consistency even when alternative estimation techniques are taken into account.

Guru and Yadav (2019) carried out an investigation focusing on the association between financial development and economic growth in five emerging economies—Brazil, Russia, India,

China and South Africa—during the time frame spanning from 1993 to 2014. Evident differences were detected in the primary indicators of financial development and macroeconomic variables amongst these economies. Their analysis revealed a significant and positive correlation between economic growth and parameters such as the value of shares traded, credit-to-deposit ratio and domestic credit to the private sector. In a similar vein, Bist (2018) examined the persistent correlation between financial development and economic growth from 1995 to 2014 in 16 low-income nations. The outcomes exposed a continual co-integration between financial development and economic growth, with financial progressions playing a noteworthy and favourable role in economic advancement, in conjunction with global connections.

Ikhsan and Satrianto (2023) delved into the impact of financial development on economic growth in high-income nations. By employing the fixed effect model for their analysis, the scholars established that the level of financial development indeed affects economic growth in high-income countries. Their research accentuates the importance of financial development as a fundamental prerequisite for economic advancement, especially in nations concentrating on enhancing their financial sectors through efficient capital mobilisation approaches. It is imperative for financial development to act as the bedrock of policymaking, particularly concerning financial institutions and markets, to effectively guide the distribution and absorption of financial resources and significantly contribute to propelling economic progress. Additionally, investments and savings play pivotal roles in fulfilling the demands for financial resources and should be nurtured to bolster economic growth.

The body of literature concerning the association between finance and economic growth has illustrated a significant relationship between financial development and economic well-being. Recent investigations have further substantiated this concept by demonstrating that the progression of financial systems plays a critical role in nurturing economic growth, aligning with the contentions posited by Levine et al. (2000).

15.2.2 Financial Development and Inflation Rate

Economists and policymakers have exhibited significant interest in the intricate nexus between financial development and inflation, a topic that has been extensively studied empirically (Rajan, 2019; Cecchetti & Kharroubi, 2015; Fry, 1980). The influence of financial systems on macroeconomic variables like inflation grows as these systems evolve and become more sophisticated (Arcand et al., 2015; Levine, 1997; Beck & Levine, 2004). The importance of financial markets and institutions' depth and efficiency, referred to as financial development, is widely recognised as a fundamental factor shaping inflation dynamics (Claessens et al., 2018; Demetriades & Luintel, 1996). Recent scholarly works delve into the complex interconnections amongst financial innovation, credit accessibility and the transmission mechanisms of monetary policy, enriching our comprehension of how financial development impacts inflation across various global settings (Cecchetti & Kharroubi, 2012; Dées et al., 2019).

Scholars have established a link between financial development and inflation, proposing that inflation may lead to changes in financial development (Mahawiya et al., 2020; Seo et al., 2019; Seo & Shin, 2016; Abbey, 2012; Bittencourt, 2011). This group of researchers aims to argue that inflation can exert either a threshold effect or a linear effect on financial development, suggesting a causal relationship between inflation and financial development.

In their comparative analysis, Mahawiya et al. (2020) examined the impact of inflation on financial development in the economic community of west african states (ECOWAS) and southern african development community (SADC) regions from 1980 to 2011, identifying a distinct inflation threshold for both areas. The study revealed threshold levels of 17.9% for ECOWAS and 14.5% for SADC, indicating that exceeding these thresholds could have detrimental effects on financial development. Consequently, the researchers emphasise the importance of prioritising price stability policies within an inflation-targeting framework to enhance financial development in these regions. In a separate study, Odhiambo (2012) investigated the relationship between inflation and financial development in Zambia from 1980 to 2011 using the ARDL-bounds testing approach, finding an inverse association between inflation and financial development, with private sector lending serving as a proxy for financial development.

Nguyen (2019) utilised a vector autoregressive model to examine the interactions amongst financial development, GDP growth and inflation in five emerging Southeast Asian countries from 1997 to 2016. The study uncovered several key relationships, including a positive correlation between GDP per capita growth and money supply, a positive impact of domestic credit on GDP per capita growth but a negative effect on inflation, a negative correlation between stock market capitalisation and inflation, and positive associations between GDP per capita growth and money supply, as well as directional relationships between GDP per capita and inflation, and inflation and money supply, and domestic lending to the private sector.

Keho (2010) delved into the causal and enduring connection between the performance of the banking sector and inflation across seven West African Economic and Monetary Union (UEMOA) nations. The study methodology was based on the Granger causality test and the bound testing strategy proposed by Pesaran et al. (2001). The study outcomes revealed a long-term partnership in one country but not in the remaining six. Additionally, in five nations, financial development emerged as a driver of inflation, whereas in two countries, the inflation rate acted as a catalyst for financial development.

Almalki and Batayneh (2015) examined the long-term correlation between the inflation rate and the development of the financial sector through the application of an autoregressive distributed lag (ARDL) -bound testing methodology as suggested by Pesaran et al. (2001). The findings substantiated the significant negative link, both in the short and long term, between financial development and inflation. Furthermore, the results confirmed that the current development of the financial sector is positively influenced by the policies implemented in the previous period in a statistically significant manner.

Fuinhas et al. (2021) used Granger causality assessments based on a panel vector autoregressive framework, as well as mechanism impulse response functions, to examine the relationship between economic growth, inflation, stock market development and banking sector development in sixteen high-income countries from 2001 to 2016. Their findings indicate a two-way relationship between financial progress and inflation. Furthermore, they find a negative association between these variables, meaning that financial system improvement is expected to reduce inflation in the studied economies.

Theoretical literature has tended to overlook Boyd et al. (2001), nevertheless demonstrating an inverse relationship between inflation and financial market performance. Furthermore, distinct balances lead to pronounced inflation under specific conditions and moderate inflation under others, particularly when there is a sizable government deficit. The repercussions of financial progress on growth hinge predominantly on the initial state of equilibrium. Succinctly stated, financial advancement curtails the equilibrium inflation rate and nurtures economic expansion in nations commencing with relatively low inflation rates.

Effiong et al. (2020) investigate the impact of financial development on the efficacy of monetary policy concerning output and inflation in Africa. Their analysis encompasses annual data spanning 39 African countries from 1990 to 2015, employing conventional panel data methodologies. Their findings unveil a restricted correlation between financial development and the effectiveness of monetary policy in Africa. Whilst the association with output growth lacks statistical backing, the inflation scenario displays a negative relationship—but solely at contemporaneous levels. Additionally, as evidenced by Ciccarelli et al. (2017), the credit channel intensifies the repercussions of monetary policy shocks on production and inflation, heightening its potency amidst financial frictions.

In accordance with Ma and Lin (2016), there exists an adverse correlation between monetary policy effectiveness and financial development, as heightened levels of financial development exert a mitigating impact on output and inflation via monetary policy. Furthermore, due to financial expansion diminishing output and inflation in emerging economies and advanced economies (AEs), respectively, its influence on the efficacy of monetary policy is disparate.

15.2.3 Financial Development and Unemployment

Understanding the complex relationship between financial development and unemployment across different economies has been the focus of an increasing amount of empirical research in recent years.

In a sample of 19 emerging market economies, Raifu and Afolabi (2023) examined how financial development affected unemployment. The participants' age groups and gender dichotomy were explicitly taken into account by the researchers. According to the study's conclusions, financial development has a conditional mean-reducing impact on unemployment. Furthermore, a reducing effect on the distribution of unemployment was identified in relation to financial development.

Tsaurai (2022) examined the relationship between unemployment in North African countries, the expansion of the financial industry and the usage of renewable energy. The complementarities between (1) financial development (FIN) and renewable energy consumption and (2) financial development and economic growth, along with their effects on unemployment, were also studied within the context of North African states. According to the results of their study, financial development significantly reduces unemployment in North African economies.

Chen et al. (2021) investigated the nonlinearities in the relationship between financial structure and development and unemployment. They discovered that the impact varies depending on the kind of financing and degree of labour market flexibility in a nation, using a system generalized method of moments (GMM) estimation using a panel of 97 organisation for economic co-operation and development (OECD) and non-OECD countries for the years 1991–2015. In particular, (i) countries with more rigid labour markets experience lower unemployment due to excessive financial development; (ii) countries with more flexible labour markets experience higher unemployment due to bank-centred or underdeveloped financial systems and (iii) countries with more rigid labour markets experience higher unemployment due to excessive credit to private enterprises, whilst higher unemployment rates are associated with under credit to households in flexible labour markets.

Ajide (2020) investigates how financial development has an uneven impact on unemployment in Nigeria. Data from 1980 to 2017 were analysed using the nonlinear autoregressive distributed lag (NARDL) technique, and the findings indicate that there is a long-term equilibrium between the variables. The Wald test verifies the unbalanced relationship between short- and long-term unemployment and financial progress. The results verify that the beneficial aspects of financial development have a greater impact on unemployment than the negative aspects.

Raifu (2019) looked at the relationship between financial development and unemployment over the short and long terms whilst accounting for various financial development metrics. They discovered that only domestic deposit to GDP has the ability to lower the unemployment rate over the long and short terms. They arrived at this conclusion by utilising ARDL as an estimating technique and a variety of financial development indicators. The unemployment rate is only temporarily decreased by other financial variables including lending to the private sector, liquidity, financial efficiency and financial stability. Additionally, they discovered a co-integration between the unemployment rate and financial development, which includes the inflation rate and real GDP. Comparisons between Nigeria's financial sector development and those of emerging and developed countries are said to have provided the results.

Epstein and Finkelstein Shapiro (2019) investigated the relationship between sectoral dynamics, unemployment volatility and financial development. In developing and emerging economies (DEMEs), they show a negative and significant association between the volatility of unemployment and domestic financial development; in advanced economies (AEs, this relationship is absent. These facts can be mathematically rationalised using a business cycle labour search model that includes company heterogeneity, collateral limitations and interfirm input credit capital. Increased bank credit, increased capital accumulation and decreased use of input credit capital are all linked to higher financial development and increase a firm's resilience to financial shocks. The enhanced shock resilience of firms stabilises hiring decisions, which eventually results in less fluctuating unemployment. Then, under conditions of increased financial development, interfirm input credit plays a significant role in promoting reduced volatility amongst enterprises by creating clear connections between them. Therefore, in economies with higher average bank credit-to-GDP ratios and lower input credit usage—AEs—instead of economies with lower average bank credit-to-GDP ratios and higher input credit usage—DEMEs—increases in financial development have less of an effect on stabilising unemployment.

Olowu et al. (2019) conducted an empirical analysis to investigate the effects of agricultural sector value added and financial development on unemployment. For the study, eleven developing nations from the Southern African Development Community were chosen. To conduct the empirical analysis, second-generation econometric techniques were employed. The findings of the regression analysis showed that financial development and agricultural value added are significant factors influencing unemployment in the area. The findings also demonstrate that there is a long-term negative correlation between unemployment and both financial efficiency and depth. Panel bootstrap causality tests were used to examine the relationship between agricultural value-added financial development and unemployment in more detail. The findings of the causality test showed that there is a substantial one-way causal relationship between the region's unemployment rate and financial depth and agricultural value addition.

15.2.4 Institutions and Economic Growth, Inflation and Unemployment

Conventional growth theory (see Lucas, 1988; Barro, 1990; Romer, 1990) places a strong emphasis on the role that human capital, technical diffusion and public infrastructure or innovation play in explaining variations in cross-national growth. Nonetheless, the importance of institutions and governance in economic growth has gained more and more emphasis in recent years. Earlier proponents of the role of institutions argue that institutions act as the sources of variations for growth between countries (Coatsworth, 1975).

Recently, the literature has paid close attention to the role of institutions in achieving various economic outcomes. For instance, Kharisma et al. (2022) argue that in Asian countries, economic institutions affect economic growth and play a major role in controlling the activities of markets, whilst political institutions do not affect economic growth. Sabir et al. (2019) also find that economic growth is significantly improved by institutional development in various economies. Uberti and Knutsen (2021) indicate that both property rights institutions and human capital development positively affect economic growth, although human capitals tend to be more robust when country-fixed effects are accounted for. Uddin et al. (2021) also argue that when institutions are dysfunctional and weak, additions to human capital negatively affect economic growth due to the fact that new stock tends to be employed in socially unproductive activities. Hartmann and Spruk (2021) argue that income and growth are negatively affected by institutional instability and acknowledge that de jure instability is less important than de facto institutional instability. Hussen and Çokgezen (2021) in SSA found a unidirectional effect of institutional quality on economic growth in low-income economies and a bi-directional effect in lower income economies.

The literature on the relationship between institutional quality and inflation is extant. Various studies have examined either the moderating or mediating role of institutions in the effect of inflation on other macroeconomic outcomes. For instance, Law and Soon (2020) in a study on the effect of inflation on income inequality indicate that inflation has a detrimental effect on income inequality but find that the effect is mitigated by the role of institutional quality. Khan and Hanif (2020) also found in their study that there exists a minimum level of institutional quality above which there is a negative inflation-growth relationship. However, below that threshold, the relationship is non-existent. Fazio et al. (2018) indicate in their results that inflation targeting does not significantly influence stability in economies with low institutional quality, but the effect is significant in economies with average levels of institutional quality. Louati and Boujelbene (2020) demonstrate a significant positive correlation between inflation targeting and institutional quality, based on data from developing economies.

Unemployment rates are greatly influenced by institutional factors, such as economic institutions (EIs), and labour market institutions (LMIs). The minimum wage, immigration inflows, wage bargaining coordination and centralisation, employment protection laws and other LMIs are inversely correlated with overall unemployment rates (Perone, 2022). Conversely, Kilishi et al. (2021) found that there is a positive association between unemployment rates and union density, tax wedge and product market regulation (PMR). In a similar vein, long-term unemployment is reduced by EIs associated with the monetary system, trade flows, government expenditure and fiscal process (Kim & Kim, 2017). Nevertheless, there is no discernible correlation between labour taxes and the level of job security and employment or unemployment rates (Seiler & Roncone, 2019). Sachs and Smolny (2015) in a study on OECD countries found that LMIs such as powerful unions and employment protection laws could have a positive or negative effect on unemployment, with the elderly being less affected than the young. These findings suggest that LMIs and economic institutions play a crucial role in shaping unemployment dynamics.

15.3 Methodology and Data

15.3.1 Data

The current investigation aims to assess the moderating influence of legal frameworks and enforcement of property rights on the efficacy of financial advancement in SSA. Consequently, we evaluate the effectiveness of legal frameworks and property rights enforcement in moderating the impact of financial development on various outcome variables such as economic growth, lending rates, trade, foreign direct investment, unemployment and inflation. A panel dataset comprising 41 countries in SSA is utilised due to data availability spanning from 2000 to 2020. Information on financial development, economic growth rates, exchange rates, inflation, gross fixed capital formation and unemployment is extracted from the World Development Indicators database. Data concerning legal systems and property rights originates from the Fraser Institute's Economic Freedom database.

15.3.2 Theoretical and Estimation Techniques

A panel quantile regression model is used to investigate how financial development affects inflation, unemployment and economic growth. Quantile regression is a statistical methodology that involves minimising weighted absolute deviations, sometimes referred to as the L_1 method. The estimate of conditional quantile functions, also known as percentile functions, is done using this technique. This approach has been greatly influenced by the groundbreaking work of Koenker and Bassett (1978) as well as Koenker and Hallock (2001). Whereas asymmetric weights are applied to all other quantiles (e.g. 0.1..., 0.9), symmetric weights are only used on the median (quantile = 0.5). In contrast, conditional mean functions are estimated using traditional ordinary least squares regression, which is based on the L_2 approach. Quantile regression stands out from ordinary least squares (OLS) as it is not confined to explaining the mean of the dependent variable, offering an opportunity to elucidate the determinants of the dependent variable at any given point across the distribution.

For the outcome variables (economic growth, unemployment and inflation), the quantile regression estimation makes it possible to examine how the independent variables could affect the outcome variables across different distributions.

Theoretically, we could specify the quantile regression equation as given below:

$$Q_{yi}\left(\phi / x_i\right) = x_i^{'}\alpha_\phi \tag{15.1}$$

where ϕ denotes the quantile and the equation denotes a conditional quantile y_i in every given x_i. When utilising the technique of quantile regression in panel data analysis, it is imperative to account for unobserved heterogeneity. This necessitates the utilisation of a panel quantile regression model with a fixed effect, which grants the ability to regulate individual heterogeneity that has gone unnoticed. The aforementioned model is referred to as the panel quantile regression model with a fixed effect.

$$Q_{yit}\left(\phi_k / \varphi_i, x_{it}\right) = \varphi_i + x_{it}^{'}\alpha\left(\phi_k\right) \tag{15.2}$$

Here, φ_i captures the fixed effect that also brings the incidental parameter problem (Lancaster, 2000). As the number of cross-sectional units approaches infinity, the estimator becomes inconsistent when fixed time-series observations are present for each unit (Galvao & Kato, 2016). Consequently, conventional linear approaches are not suitable for employment in the panel quantile regression model. To address this issue in panel quantile regression, Koenker (2004) devises the shrinkage method. This approach introduces a penalty term that eliminates the unobserved fixed effects. The parameters of the model are estimated accordingly.

$$\left(\hat{\alpha}\left(\phi_k,\eta\right),\left\{\varphi_i,(\eta)\right\}_{i=1}^{N}\right) = \arg\min \sum_{k}^{K}\sum_{t}^{T}\sum_{n}^{N}\Omega_k\rho_{\phi k}\left(y_{it} - \varphi_i - x_{it}^{'}\alpha\left(\phi_k\right)\right) + \eta\sum_{i}^{N} I\varphi_i I \tag{15.3}$$

i and t denote the country and time periods, respectively. $\rho_{\phi k}$ is the quantile loss function, and k is the quantile. Ω_k represents the weight assigned to the k-th quantile and also captures the contributions of the various distributions. η represents the tuning parameter that is used to plunge the individual effects to 0 so that better slope coefficients can be estimated. As used by Zhu et al. (2018), this value is set to $\eta = 1$.

15.3.3 Empirical Model Specification

Three (3) main financial development outcome variables are our main focus when assessing the efficacy of financial development. Based on the assumption that financial development aims to attain economic expansion, generate job opportunities and maintain stability via stable prices, we calculate the impact of financial development on economic growth, unemployment and inflation. Additionally, we model how property rights and legal frameworks moderate the relationship between financial development and its outcome variables. Thus, one could model this relationship as

$$QY_{it}\left(\phi_k / \varphi_i, x_{it}\right) = f\left(X_{it}, Z_{it}\right) \tag{15.4}$$

where Y_{it} represents the outcome variables earlier indicated (economic growth, unemployment and inflation). X_{it} represents financial development, which is country and time variant. Z_{it} also represents some country and time-variant control variables, which could be included in the models. All other parameters are as defined earlier. For the economic growth model, we estimate the following model:

$$Q_{\mathrm{ECG}_{it}}\left(\phi_k/\phi_i, x_{it}\right)=\alpha_i+\beta_{1\phi k}\mathrm{FD}_{it}+\beta_{2\phi k}\mathrm{TRD}_{it}+\beta_{3\phi k}\mathrm{INF}_{it}+\beta_{4\phi k}\mathrm{GFCF}_{it}+\beta_{5\phi k}\mathrm{FDI}_{it}+ \beta_{6\phi k}\mathrm{EXC}_{it}+\beta_{7\phi k}\mathrm{UNE}_{it}+\beta_{8\phi k}\mathrm{LR}_{it}+\mu_{it} \tag{15.5}$$

$$Q_{\mathrm{ECG}_{it}}\left(\phi_k/\phi_i, x_{it}\right)=\alpha_i+\beta_{1\phi k}\mathrm{LSY}_{it}+\beta_{2\phi k}\mathrm{TRD}_{it}+\beta_{3\phi k}\mathrm{INF}_{it}+\beta_{4\phi k}\mathrm{GFCF}_{it}+\beta_{5\phi k}\mathrm{FDI}_{it}+ \beta_{6\phi k}\mathrm{EXC}_{it}+\beta_{7\phi k}\mathrm{UNE}_{it}+\beta_{8\phi k}\mathrm{LR}_{it}+\mu_{it} \tag{15.6}$$

$$Q_{\mathrm{ECG}_{it}}\left(\phi_k/\phi_i, x_{it}\right)=\alpha_i+\beta_{1\phi k}\mathrm{FD}_{it}+\beta_{2\phi k}\mathrm{LSY}_{it}+\beta_{3\phi k}\mathrm{FD}_{it}\times\mathrm{LSY}_{it} +\beta_{4\phi k}\mathrm{TRD}_{it}+\beta_{5\phi k}\mathrm{INF}_{it}+\beta_{6\phi k}\mathrm{GFCF}_{it}+ \beta_{7\phi k}\mathrm{FDI}_{it}+\beta_{8\phi k}\mathrm{EXC}_{it}+\beta_{9\phi k}\mathrm{UNE}_{it}+\beta_{10\phi k}\mathrm{LR}_{it}+\mu_{it} \tag{15.7}$$

For the unemployment model, the following quantile regressions are estimated:

$$Q_{\mathrm{UNE}_{it}}\left(\phi_k/\phi_i, x_{it}\right)=\alpha_i+\beta_{1\phi k}\mathrm{FD}_{it}+\beta_{2\phi k}\mathrm{TRD}_{it}+\beta_{3\phi k}\mathrm{INF}_{it}+\beta_{4\phi k}\mathrm{GFCF}_{it}+\beta_{5\phi k}\mathrm{FDI}_{it}+ \beta_{6\phi k}\mathrm{EXC}_{it}+\beta_{7\phi k}\mathrm{ECG}_{it}+\beta_{8\phi k}\mathrm{LR}_{it}+\mu_{it} \tag{15.8}$$

$$Q_{\mathrm{UNE}_{it}}\left(\phi_k/\phi_i, x_{it}\right)=\alpha_i+\beta_{1\phi k}\mathrm{LSY}_{it}+\beta_{2\phi k}\mathrm{TRD}_{it}+\beta_{3\phi k}\mathrm{INF}_{it}+\beta_{4\phi k}\mathrm{GFCF}_{it}+\beta_{5\phi k}\mathrm{FDI}_{it}+ \beta_{6\phi k}\mathrm{EXC}_{it}+\beta_{7\phi k}\mathrm{ECG}_{it}+\beta_{8\phi k}\mathrm{LR}_{it}+\mu_{it} \tag{15.9}$$

$$Q_{\mathrm{UNE}_{it}}\left(\phi_k/\phi_i, x_{it}\right)=\alpha_i+\beta_{1\phi k}\mathrm{FD}_{it}+\beta_{2\phi k}\mathrm{LSY}_{it}+\beta_{3\phi k}\mathrm{FD}_{it}\times\mathrm{LSY}_{it}+\beta_{4\phi k}\mathrm{TRD}_{it}+\beta_{5\phi k}\mathrm{INF}_{it}+ \beta_{6\phi k}\mathrm{GFCF}_{it}+\beta_{7\phi k}\mathrm{FDI}_{it}+\beta_{8\phi k}\mathrm{EXC}_{it}+\beta_{9\phi k}\mathrm{ECG}_{it}+\beta_{10\phi k}\mathrm{LR}_{it}+\mu_{it} \tag{15.10}$$

For the inflation model, we estimate the following quantile regressions:

$$Q_{\mathrm{INF}_{it}}\left(\phi_k/\phi_i, x_{it}\right)=\alpha_i+\beta_{1\phi k}\mathrm{FD}_{it}+\beta_{2\phi k}\mathrm{TRD}_{it}+\beta_{3\phi k}\mathrm{ECG}_{it}+\beta_{4\phi k}\mathrm{GFCF}_{it}+\beta_{5\phi k}\mathrm{FDI}_{it}+ \beta_{6\phi k}\mathrm{EXC}_{it}+\beta_{7\phi k}\mathrm{UNE}_{it}+\beta_{8\phi k}\mathrm{LR}_{it}+\mu_{it} \tag{15.11}$$

$$Q_{\mathrm{INF}_{it}}\left(\phi_k/\phi_i, x_{it}\right)=\alpha_i+\beta_{1\phi k}\mathrm{LSY}_{it}+\beta_{2\phi k}\mathrm{TRD}_{it}+\beta_{3\phi k}\mathrm{ECG}_{it}+\beta_{4\phi k}\mathrm{GFCF}_{it}+\beta_{5\phi k}\mathrm{FDI}_{it}+ \beta_{6\phi k}\mathrm{EXC}_{it}+\beta_{7\phi k}\mathrm{UNE}_{it}+\beta_{8\phi k}\mathrm{LR}_{it}+\mu_{it} \tag{15.12}$$

$$Q_{\mathrm{INF}_{it}}\left(\phi_k/\phi_i, x_{it}\right)=\alpha_i+\beta_{1\phi k}\mathrm{FD}_{it}+\beta_{2\phi k}\mathrm{LSY}_{it}+\beta_{3\phi k}\mathrm{FD}_{it}\times\mathrm{LSY}_{it}+\beta_{4\phi k}\mathrm{TRD}_{it}+\beta_{5\phi k}\mathrm{ECG}_{it}+ \beta_{6\phi k}\mathrm{GFCF}_{it}+\beta_{7\phi k}\mathrm{FDI}_{it}+\beta_{8\phi k}\mathrm{EXC}_{it}+\beta_{9\phi k}\mathrm{UNE}_{it}+\beta_{10\phi k}\mathrm{LR}_{it}+\mu_{it} \tag{15.13}$$

15.3.4 Definition and Measurement of Variables

Financial Development (FD) The proportion of GDP derived from domestic credit supplied to the private sector is known as financial development. In additional studies, it has served as a stand-in for the level of financial development. The expectation is that the rate of economic growth will favourably correlate with financial development, whilst the rates of unemployment and inflation will adversely connect with it.

Economic Growth (ECG) GDP is a crucial indicator of a nation's investment activity. We use the GDP growth rate annually in the current study. A dependable measure of a nation's economic activity, income and demand is its GDP. It is anticipated that there will be a negative correlation between the growth rate and the unemployment and inflation rates.

Inflation (INF) The consumer price index, which measures inflation, tracks yearly variations in the average cost of goods and services in an economy. It is anticipated that unemployment and economic growth will be negatively correlated with inflation.

Unemployment (UEN) This is the official unemployment statistics for each country in the panel. It is measured as the total number of unemployed as a percentage of the total labour force. It is expected that unemployment will have a negative relationship with the level of economic growth and inflation.

Legal System (LSY) This refers to how successfully the legal systems and property rights are enforced in the various research countries. It assesses the extent to which governments uphold the rule of law and legal systems and regimes that protect private property rights. It also examines whether private property can be expropriated, the impartiality of the judiciary, judicial corruption and the economic feasibility of contract enforcement for both private and corporate parties. Higher values indicate better legal protection for property rights.

Trade (TRD) This measures the volume of trade activities ongoing in an economy. It is measured as the ratio of net trade to GDP. It is expected that trade will have a positive relationship with economic growth and a negative relationship with inflation and the unemployment rate.

Foreign Direct Investment (FDI) Foreign Direct Investment is the inflow of capital from foreign funds for investment, trade or business operations. It is measured as the net inflow of FDI to GDP. FDI is expected to positively affect economic growth and negatively affect unemployment, but the effect on inflation is indeterminate.

Exchange Rate (EXC) This is measured as the rate at which the Ghana cedi is exchanged for 1 dollar at every period of time. This data is sourced from the World Development Indicators.

Lending Rate (LR) The average interest rate at which formal financial institutions lend money to the private sector is known as the lending rate. Greater values suggest higher borrowing costs, which may deter private investment.

Gross Fixed Capital Formation (GFCF) GDP to private investment as a percentage of gross fixed capital creation is available from the World Development Indicators database. It includes gains in the value of non-manufactured assets (such as subsurface assets or significant changes in the quantity, prominence or efficiency of land) as well as the total value of a manufacturer's acquisitions less the sales of durable assets over the course of the monetary cycle.

15.4 Results and Discussions

15.4.1 Descriptive Statistics

The study's variables' descriptive statistics are shown in Table 15.1. The average annual rate of inflation was 8.983 per cent, with a minimum value of approximately −9.616 per cent. During the same period, some countries recorded annual rates of inflation as high as 513.907 per cent. The average growth rate of the GDP during this period is 4.086. The maximum annual economic growth rate is approximately 33.629 per cent, whilst the lowest recorded GDP growth rate is −36.392 per cent. The sub-region's average unemployment rate (8.318) has been comparatively low, with the percentage of the labour force unemployed ranging from 0.32 per cent at minimum to 35.457 per cent at maximum. There is a minimum of roughly 0.1 per cent of GDP allocated to the private sector and a maximum of approximately 142.21 per cent. In SSA, the private sector accounts for roughly 20.145 per cent of GDP on average. With a minimum value of roughly 1.66 and a maximum value of roughly 6.94, the legal system and property rights variable has a mean value of roughly 4.086. The result is that SSA has comparatively low levels of institutional development.

Table 15.1 Descriptive statistics

Variable	Obs	Mean	Std. Dev.	Min	Max
Inflation	790	8.983	27.201	−9.616	513.907
Private sector credit	740	20.145	23.112	0.1	142.21
Legal system propert~s	861	4.086	1.21	1.66	6.94
Trade	861	69.655	3.825	62.676	75.903
Unemploy	840	8.318	7.708	0.32	35.457
Log_gfcf	731	3.03	0.42	0.092	4.395
Log_foreign	816	19.826	1.303	15.573	23.16
Log_exch	819	4.965	2.133	−0.607	9.193
Growth	860	4.086	4.677	−36.392	33.629
Lending rate	861	17.756	3.817	14.518	25.795

Source: Authors' Computation

Table 15.2 shows the Pearson correlation matrix between the explanatory variables used in the study. This is necessary to detect possible issues of collinearity amongst the independent variables used. The results of the correlation matrix indicate that the correlation between the independent variables is not very high to create issues of collinearity in the model.

15.4.2 Effect of Financial Development on Economic Growth in Sub-Saharan Africa

Using a quantile regression model, we estimate the effect of financial development on economic growth rate at the 25th, 50th and 75th quantiles of economic growth. For each quantile, we test the effect of financial development, legal systems and property rights and also the moderating role of legal systems and property rights in the relationship between financial development and economic growth, respectively. The results are shown in Table 15.3. The results indicate that financial development positively affects economic growth in the 25th quantile of economic growth and the 75th quantile of economic growth but does not affect economic growth at the 50th quantile. A developed financial system facilitates the effective mobilisation of savings by households, businesses and governments. These savings may then be channelled into productive investments, such as infrastructure projects, research and development, and business expansion. The allocation of capital to its most efficient and effective uses results in a boost to economic growth. A percentage point increase in the level of financial development will result in 0.155 and 0.105 percentage point increases in the economic growth in the 25th and 75th quantiles of economic growth, respectively. The outcomes also demonstrate that the beneficial impact of financial advancement on economic expansion diminishes at higher levels of economic growth, suggesting that at higher levels of economic growth, the influence of financial advancement decreases. This outcome is in line with the discoveries of Abbas (2022), Puşcaşu (2022) and Ustarz and Fanta (2021), who all observe that financial development can have a positive impact on economic growth in developing nations.

The implementation of property rights and legal systems is crucial in shaping economic growth in SSA. The results reveal that at the 25th and 50th percentiles of economic growth, legal systems and property rights protection boost economic growth. Based on the results, at the 25th and 50th percentiles, respectively, a one-percentage-point enhancement in the quality of legal systems and property rights enforcement will lead to a 0.483 and 0.279 percentage point rise in the economic growth rate. Additionally, the findings indicate that at higher levels of economic growth, the influence of legal systems and property rights enforcement on growth diminishes. Strong and well-defined property rights are a fundamental requirement for economic progress. When individuals and businesses have con-

Table 15.2 Correlation matrix

Variables	(1)	(2)	(3)	(4)	(5)	(6)	(7)	(8)	(9)	(10)
(1) Inflation	1.000									
(2) FD	0.014	1.000								
	(0.700)									
(3) Legal System Pro~s	−0.118*	−0.086*	1.000							
	(0.001)	(0.015)								
(4) Trade	−0.089*	0.136*	0.069*	1.000						
	(0.012)	(0.000)	(0.044)							
(5) Unemploy	−0.039	0.190*	0.242*	−0.026	1.000					
	(0.283)	(0.000)	(0.000)	(0.451)						
(6) Log_gfcf	−0.020	0.263*	0.212*	0.216*	0.149*	1.000				
	(0.591)	(0.000)	(0.000)	(0.000)	(0.000)					
(7) Log_foreign	0.031	−0.071	−0.008	0.216*	−0.435*	0.127*	1.000			
	(0.393)	(0.051)	(0.820)	(0.000)	(0.000)	(0.001)				
(8) Log_exch	−0.112*	−0.078*	−0.430*	0.015	−0.545*	−0.173*	0.184*	1.000		
	(0.002)	(0.032)	(0.000)	(0.660)	(0.000)	(0.000)	(0.000)			
(9) Growth	−0.098*	0.118*	0.070*	0.193*	−0.201*	0.117*	0.228*	0.013	1.000	
	(0.006)	(0.001)	(0.040)	(0.000)	(0.000)	(0.002)	(0.000)	(0.714)		
(10) Lending rate	0.140*	0.024	−0.178*	−0.564*	0.018	−0.275*	−0.278*	−0.072*	−0.004	1.
	(0.000)	(0.501)	(0.000)	(0.000)	(0.595)	(0.000)	(0.000)	(0.038)	(0.902)	

Source: Authors' Computation
*** $p < 0.01$, ** $p < 0.05$, * $p < 0.1$

Table 15.3 Financial development and economic growth

Variable	25th			50th			75th		
	Coefficient	Coefficient	Coefficient	Coefficient	Coefficient	Coefficient	Coefficient	Coefficient	Coefficient
Private sector credit	0.155* (0.094)		0.258*** (0.0920)	0.038 (0.059)		0.107 (0.0658)	0.105** (0.047)		0.136** (0.0599)
Legal		0.483*** (0.148)	0.998*** (0.195)		0.279** (0.141)	0.683*** (0.158)		0.140 (0.169)	0.426*** (0.149)
Interaction			−0.931* (0.475)			−1.083** (0.451)			−0.825*** (0.308)
Trade	0.122** (0.048)	0.123*** (0.0429)	0.0797 (0.0541)	0.128*** (0.049)	0.109*** (0.0355)	0.109*** (0.0403)	0.141*** (0.054)	0.115** (0.0503)	0.131** (0.0510)
Log_inflation	−0.338* (0.178)	−0.344* (0.183)	−0.196 (0.201)	−0.142 (0.106)	−0.162 (0.116)	−0.0987 (0.169)	0.165 (0.107)	0.00862 (0.119)	0.148 (0.127)
Log_gfcf	1.265** (0.59)	1.246** (0.587)	0.151 (0.421)	1.574*** (0.419)	1.461*** (0.555)	1.022* (0.525)	1.727*** (0.4)	1.261** (0.491)	1.547*** (0.508)
Log_foreign	0.841*** (0.204)	0.714*** (0.154)	0.783*** (0.172)	0.488*** (0.112)	0.445*** (0.115)	0.365*** (0.137)	0.423*** (0.142)	0.338*** (0.130)	0.363** (0.151)
Log_exch	−0.352*** (0.076)	−0.151 (0.0959)	−0.142* (0.0763)	−0.064 (0.084)	−0.0826 (0.0807)	−0.138 (0.107)	−0.132 (0.127)	−0.181** (0.0867)	−0.208* (0.122)
Unemploy	−0.152*** (0.023)	−0.15*** (0.0288)	−0.12*** (0.0407)	−0.081** (0.034)	−0.088*** (0.0278)	−0.10*** (0.0210)	−0.104*** (0.035)	−0.11*** (0.0284)	−0.099*** (0.0271)
Lending rate	0.057 (0.094)	0.116 (0.0709)	0.0656 (0.0689)	0.065 (0.052)	0.0785 (0.0576)	0.0716 (0.0443)	0.108** (0.047)	0.127** (0.0535)	0.0822* (0.0452)
Constant	−25.09*** (8.217)	−25.4*** (6.620)	−23.5*** (6.227)	−19.03*** (5.388)	−17.44*** (3.760)	−15.9*** (4.039)	−18.79*** (5.617)	−13.5*** (4.100)	−16.97*** (6.110)
Observations	604	644	604	604	644	604	604	644	604
Pseudo R Square	0.1218	0.1128	0.1461	0.1005	0.0923	0.1191	0.0962	0.0818	0.1049

Source: Authors' Computation

fidence that their assets and property will be safeguarded by the legal system, they are more likely to invest, innovate and engage in economic activities. Clear property rights promote long-term planning and investment, resulting in increased productivity and economic growth. These results align with the studies of Kharisma et al. (2022), Uberti and Knutsen (2021) and Sabir et al. (2019).

We analyse the moderating impact of property rights enforcement and legal systems on the connection between financial development and economic growth rate in the third column for each percentile, as displayed in Table 15.3. Legal frameworks and financial development have a considerable positive impact on the economic growth rate at the 25th quantile of growth. Legal and financial systems, respectively, have substantial positive effects on economic growth at the 75th quantile. In the 25th, 50th and 75th quantiles of economic growth, the relationship between financial development and economic growth is negatively impacted by legal systems and the enforcement of property rights. The results indicate that the moderating effect on economic growth is 0.931, 1.083 and 0.825 percentage points, respectively. Whilst financial development can provide positive contributions to economic growth, its efficacy can be hindered by insufficient institutional structures. In scenarios where the rule of law is weak, political instability is rampant, regulation is inadequate, corruption is prevalent, governance is inefficient, investor protection is lacking, informal sectors dominate, political interference is present, transparency deficits occur, financial power is concentrated and more, the potential benefits of financial development may be restricted or even reversed. To ensure that the mechanisms of financial development are translated into sustained and inclusive economic growth by fostering trust, stability and a conducive environment for investment, entrepreneurship and resource allocation, effective institutions are imperative. Uddin et al. (2020) and Hartmann and Spruk (2020) also found similar results.

In the 25th quantile of economic growth, trade (0.122), private investment activities (1.265) and foreign direct investment (0.841) have significant positive effects on the level of economic growth, whilst inflation (0.338), exchange rate (0.352) and unemployment rate (0.152) are negatively related to the growth of GDP in SSA. In the 50th quantile, trade (0.128), private investment (1.574) and foreign direct investment (0.488) positively influence economic growth, whilst unemployment (0.081) dampens the economic growth rate. In the 75th quantile, trade (0.141), private investment (1.727), foreign direct investment (0.423) and lending rate (0.108) positively influence economic growth, whilst the lending rate (0.104) has a negative relationship with the rate of economic growth. One thing, which is noticeable, is that the effect of explanatory variables on economic growth tends to reduce at higher quantiles of economic growth, implying that the effect of these variables is higher at lower quantiles of economic growth rate.

15.4.3 Effect of Financial Development on Unemployment in Sub-Saharan Africa

Table 15.4 reports the results of quantile regression describing the effect of financial development on unemployment in SSA at the selected quantiles of unemployment (0.25, 0.5 and 0.75). Further in columns 2, 5 and 8, we estimate the effect of legal systems and property rights enforcement on unemployment in SSA. In columns 3, 6 and 9, we find the moderating role of legal systems and property rights enforcement on the relationship between financial development and unemployment.

The results indicate that financial development is a significant determinant of unemployment at the 50th and 75th quantiles of unemployment but does not significantly affect unemployment at the 25th quantile. At the 50th and 75th quantiles, a percentage point increase in financial development could potentially lead to an increase in unemployment by about 0.223 and 0.673 percentage points, respectively. This could in part be due to the fact that many economies in

Table 15.4 Effect of financial development on unemployment

	25th			50th			75th		
Variable	Coefficient	Coefficient	Coefficient	Coefficient	Coefficient	Coefficient	Coefficient	Coefficient	Coefficient
Private sector credit	0.0467 (0.0545)		0.0537 (0.0414)	0.223** (0.107)		0.213** (0.0884)	0.672*** (0.110)		0.364*** (0.112)
Legal		−0.211* (0.120)	−0.569*** (0.156)		0.0974 (0.214)	−0.398 (0.273)		0.165 (0.320)	−1.406*** (0.437)
Interaction			−4.335*** (1.287)			−5.274*** (0.837)			−7.555*** (0.619)
Trade	−0.0589 (0.0508)	−0.0251 (0.0387)	−0.0257 (0.0486)	−0.00708 (0.0665)	0.0002 (0.0595)	−0.0281 (0.0637)	−0.0246 (0.0955)	0.0310 (0.146)	0.000983 (0.0899)
Log_inflation	−0.0863 (0.106)	−0.109 (0.120)	−0.142 (0.142)	−0.338 (0.278)	−0.388** (0.178)	−0.348 (0.259)	−0.326 (0.225)	−0.514* (0.305)	−0.341 (0.293)
Log_gfcf	0.927** (0.426)	1.443*** (0.496)	0.515 (0.394)	2.221*** (0.503)	2.850*** (0.627)	1.831*** (0.523)	1.586* (0.910)	2.132*** (0.787)	3.352*** (1.105)
Log_foreign	−1.372*** (0.227)	−1.456*** (0.147)	−0.893*** (0.288)	−1.779*** (0.266)	−1.732*** (0.316)	−1.735*** (0.166)	−2.619*** (0.397)	2.132*** (0.787)	−2.434*** (0.335)
Log_exch	−0.894*** (0.0734)	−0.990*** (0.103)	−0.719*** (0.0900)	−1.054*** (0.103)	−1.258*** (0.174)	−0.788*** (0.0888)	−1.873*** (0.242)	−1.895*** (0.118)	−0.843*** (0.154)
Growth	−0.172*** (0.0585)	−0.183*** (0.0565)	−0.161*** (0.0385)	−0.216*** (0.0674)	−0.252*** (0.0831)	−0.184*** (0.0488)	−0.436*** (0.121)	−0.290* (0.154)	−0.216* (0.113)
Lending rate	−0.254*** (0.0615)	−0.206*** (0.0537)	−0.179** (0.0753)	−0.189** (0.0822)	−0.115 (0.0876)	−0.142*** (0.0515)	−0.398*** (0.136)	−0.305*** (0.117)	−0.166* (0.0927)
Constant	42.28*** (6.242)	41.11*** (4.949)	29.60*** (8.813)	44.09*** (9.828)	42.36*** (5.534)	42.97*** (5.476)	74.25*** (13.71)	78.05*** (13.27)	55.98*** (9.416)
Observations	604	644	604	604	644	604	604	644	604
Pseudo R Square	0.1677	0.1853	0.1954	0.2424	0.2595	0.3248	0.3663	0.3926	0.4376

Source: Authors' Computation

the sub-region are largely informal such that an increasing financial development could result in the formalisation of the sectors which could lead to disruptions in employment as businesses struggle to meet job demands. Also, financial development can lead to shifts in the demand for labour. With the increasing availability of financial services, individuals may be inclined to shift away from traditional industries like agriculture or small-scale manufacturing. This shift has the potential to result in job displacement and a rise in unemployment within those sectors. These results, however, are inconsistent with the findings of Bayar (2016), Epstein and Finkelstein Shapiro (2019) and Raifu and Afolabi (2023), who all find that financial development could reduce unemployment.

Legal systems and property rights enforcement only significantly affect unemployment at the 25th quantile of unemployment but have no effect on the 50th and 75th quantiles of unemployment. At the 25th quantile, the relationship between financial development and the unemployment rate is negative. An increase in the level of development of legal institutions and property rights enforcement could result in a reduction in unemployment by about 0.211 percentage points. Ragmoun (2022) argued that institutional quality ensures the proper allocation of property rights and therefore increases the ease of doing business and self-employment, which could in the long run result in the reduction of unemployment. The establishment of precise and rigorously enforced property rights cultivates trust in individuals and corporations that their investments will be protected. Such assurance stimulates the development of entrepreneurial ventures and investment in novel business opportunities, thereby engendering the emergence of fresh commercial enterprises and employment prospects.

In the interaction equations in columns 3, 6 and 9, we find that the effect of financial development on unemployment is positive in both the 50th (0.213) and 75th (0.364) quantiles, and also noticeable is that the effect of financial development on unemployment increases at higher distributions of the rate of unemployment. Legal systems and property rights also negatively influence unemployment at the 25th (0.569) and 75th (1.406) quantiles, also indicating an increasing effect at higher quantiles. The results also indicate that legal systems and property rights enforcement negatively moderate the effect of financial development on unemployment. The coefficient of the interaction at the 25th quantile (4.335), 50th quantile (5.274) and 75th quantile (7.555) indicates a reduction in unemployment in percentage points when the effect of financial development on unemployment is considered in the context of good regulations and institutional frameworks. This, thus, implies that for financial development to effectively reduce unemployment, there need to be sound and working institutional frameworks which ensure the rule of law and property enforcement of property rights. The implementation of efficient property rights enforcement measures can incentivise enterprises operating within the informal sector to formalise their operations. This shift towards formality has the potential to enhance their access to financial services, technology and markets, thereby bolstering business expansion and facilitating employment opportunities. This result is consistent with the findings of Kilishi et al. (2020) and Perone (2022).

Inflation has a negative relationship with unemployment in the 50th quantile of unemployment (0.388) and the 75th quantile (0.514). Investment also has a positive relationship with unemployment at the 25th quantile (0.927), 50th quantile (2.221) and 75th quantile (1.586). The results are also true for columns 2, 3, 5, 6, 8 and 9. Implying that investment is positively related to unemployment at the various quantiles of unemployment. Foreign direct investment is negatively related to the rate of unemployment at 25th quantile (1.372), 50th quantile (1.779) and 75th quantile (2.619). It is realised that the effect of foreign direct investment on unemployment increases at higher quantiles of unemployment. The same result is confirmed in columns 2, 3, 4, 5 and 9. The rate of exchange also significantly influences the rate of unemployment in SSA. Increasing the exchange rate reduces the unemployment rate, as indicated in the results in Table 15.4. It is realised that the effect of exchange rate increases on unemployment tends to increase at higher quantiles of unemployment. The results also show that the growth rate of GDP is inversely related to the level of unemployment. Thus, higher growth rates of GDP reduce the rate of unemployment, and the coefficients increase at higher quantiles of unemployment. An increase in the rate of growth of GDP will result in a decrease in unemployment by 0.172 per cent at the 25th quantile, 0.216 per cent at the 50th quantile and 0.436 at the 75th quantile. The same trend is observed in columns 2, 3, 4, 6, 8 and 9, respectively.

15.4.4 Effect of Financial Development on Inflation in Sub-Saharan Africa

Table 15.5 presents the results of the quantile regression of financial development on inflation in SSA at the 25th, 50th and 75th quantiles of the inflation rate. In columns 2, 5 and 8, we estimate the effect of legal systems and property rights enforcement on inflation in SSA. In columns 3, 6 and 9, we find the moderating role of legal systems and property rights enforcement on the relationship between financial development and inflation.

We find that financial development dampens inflation in the 25th, 50th and 75th quantiles of inflation. Specifically, a percentage point increase in financial development could result in a reduction in inflation by about 0.184 percentage points in the 25th quantile of inflation, 0.305 percentage points in the 50th quantile and 0.315 percentage points in the 75th quantile, respectively. This result is consistent with the findings of Bandura and Leshoro (2022), who also find a negative relationship between financial development and inflation rate. Financial development could regulate inflationary pressures through an enhanced transmission mechanism. A well-established financial system plays a crucial role in transmitting monetary policy to the wider economy. Efficient financial markets increase the likelihood of policy rate adjustments being passed on to borrowers and savers, thereby influencing spending and investment patterns. In turn, this contributes towards stabilising demand and supply dynamics, potentially mitigating inflationary pressures. The result is also consistent with the findings of Nguyen (2019).

Legal systems and property rights enforcement significantly affect inflation only at higher distributions of inflation. Thus, at the 25th and 50th quantile, legal systems and property rights enforcement have no significant effect on inflation. However, at the 75th quantile, increasing the level of legal systems reduces inflation by about 1.689 percentage points. A legal framework that safeguards property rights serves as an impetus for investment in both physical and human capital. The confidence engendered by such a system prompts businesses to readily invest in technology, equipment and workforce training with the assurance that their investments will be protected. The resultant outcome of such investment is an increase in production efficiency, thereby contributing to price stability. Legal systems and property rights enforcement moderate the relationship between financial development and inflation at the 25th quantile of inflation but do not affect the relationship at the 50th and 75th quantiles. It is realised that the combined effect of legal systems and property rights enforcement and financial development on inflation (0.701) is higher than the direct effect of financial development (0.184) on inflation. The implication, thus, is that legal systems and property rights enforcement re-enforces the effect of financial development on inflation in SSA. These findings are consistent with those of Law and Soon (2020) and Khan and Hanif (2018), who all find that institutions can help mitigate the detrimental effects of inflation.

In the 25th quantile of inflation, trade (0.130), foreign direct investment (0.437) and lending rate (0.214) have a positive significant relationship with the rate of inflation, whilst unemployment (0.0446) and exchange rate (1.007) have a significant negative relationship with inflation in SSA. In the 50th quantile of inflation, foreign direct investment (0.503) and lending rate (0.175) have positive relationships with inflation, whilst the unemployment rate (0.0518) and exchange rate (0.802) have a significant negative relationship with inflation. In the 75th quantile, foreign direct investment (1.433) and lending rate (0.330) have positive significant relationships with inflation, whilst the unemployment rate (0.168) and exchange rate (0.971) negatively affect the rate of inflation.

Table 15.5 Effect of financial development on inflation

	25th			50th			75th		
Variable	Coefficient	Coefficient	Coefficient	Coefficient	Coefficient	Coefficient	Coefficient	Coefficient	Coefficient
Private sector credit	−0.184** (0.0879)		−0.163 (0.106)	−0.305*** (0.0719)		−0.274*** (0.105)	−0.315* (0.185)		−0.359* (0.201)
Legal		−0.226 (0.171)	0.207 (0.256)		−0.348 (0.240)	−0.0228 (0.431)		−1.689*** (0.435)	−1.208** (0.524)
Interaction			−0.701** (0.304)			−0.655 (0.418)			−0.327 (0.551)
Trade	0.130** (0.0518)	0.112*** (0.0384)	0.111** (0.0438)	0.0799 (0.0819)	0.113* (0.0587)	0.0540 (0.0633)	0.158 (0.118)	−0.0300 (0.104)	−0.00132 (0.178)
Unemployment	−0.0446* (0.0253)	−0.0424** (0.0194)	−0.0113 (0.0213)	−0.0518* (0.0294)	−0.089*** (0.0209)	−0.0465 (0.0409)	−0.168*** (0.0549)	−0.207*** (0.0475)	−0.161** (0.0641)
Log_gfcf	0.261 (0.628)	−0.322 (0.491)	−0.119 (0.621)	0.707* (0.393)	0.570 (0.686)	0.462 (0.573)	0.770 (0.699)	1.271 (1.208)	1.390** (0.685)
Log_foreign	0.437*** (0.0968)	0.528*** (0.191)	0.533*** (0.162)	0.503*** (0.109)	0.639*** (0.148)	0.660*** (0.251)	1.433*** (0.326)	1.286*** (0.213)	1.332*** (0.308)
Log_exch	−1.007*** (0.111)	−1.121*** (0.137)	−1.017*** (0.137)	−0.802*** (0.138)	−1.050*** (0.168)	−0.993*** (0.229)	−0.971*** (0.182)	−1.397*** (0.293)	−1.351*** (0.233)
Growth	−0.0228 (0.0517)	−0.0132 (0.0572)	−0.0296 (0.0492)	0.0419 (0.0600)	−0.00421 (0.0826)	0.0141 (0.0621)	−0.0415 (0.119)	−0.0448 (0.126)	0.0235 (0.125)
Lending rate	0.214*** (0.0664)	0.134*** (0.0467)	0.170*** (0.0596)	0.175*** (0.0570)	0.131*** (0.0473)	0.168*** (0.0532)	0.330** (0.161)	0.0594 (0.109)	0.135 (0.110)
Constant	−12.73** (5.899)	−10.14* (5.468)	−12.15** (5.093)	-9.519 (7.175)	−12.13** (5.636)	-8.571 (6.803)	−29.71** (13.29)	−3.050 (11.48)	-7.901 (12.53)
Observations	651	692	651	651	692	651	651	692	651
Pseudo R Square	0.0695	0.0701	0.0731	0.0501	0.0483	0.0537	0.0493	0.0596	0.0583

Source: Authors' Computation

15.5 Conclusions and Recommendations

As the literature has argued, financial development is crucial for achieving unemployment, economic growth and price stability in general. The study emphasised how important financial development is to SSA's efforts to achieve economic growth, lower unemployment and preserve stable price levels. This is in line with previous research that highlights the critical role financial markets and institutions play in fostering economic growth. In this study, the impact of financial development on three outcome variables—economic growth, inflation and unemployment—in SSA is examined, along with the potential moderating effects of legal frameworks and property rights. A sequential quantile regression model is estimated using a panel of 41 sub-Saharan African nations from 2000 to 2020. In particular, we calculate the impact of financial development at the 25th, 50th and 75th quantiles on inflation, unemployment and economic growth. We also examine the potential moderating effects of legal frameworks and property rights enforcement on the relationships at the 25th, 50th and 75th quantiles. The models also account for trade, investment, foreign direct investment, exchange rate and lending rate amongst other macroeconomic variables.

The outcomes of the research indicate that although not situated at the 50th percentile, financial advancement exerts a favourable influence on economic progress at the 25th and 75th percentiles. Nonetheless, the correlation between property rights, legal structures and financial development negatively affects economic growth. These findings imply that despite the beneficial impact of financial development on growth, property rights and legal frameworks do not enhance the association. Moreover, financial development positively affects unemployment at the 25th, 50th and 75th percentiles. Nevertheless, the correlation weakens regarding property rights and legal systems. Consequently, property rights and efficient legal systems become imperative in supporting the progression of financial development to mitigate unemployment. In addition, financial development diminishes inflation at the 25th, 50th and 75th percentiles. Particularly at the 25th percentile, the connection is further reinforced by the existence of robust legal systems and the enforcement of property rights. The conclusions of the study underscore the importance of upholding property rights and the rule of law to fully leverage the potential benefits of financial development concerning inflation, unemployment and economic growth. Furthermore, the study's conclusions reveal that macroeconomic elements such as trade, investment, foreign direct investment, exchange rates and lending rates significantly influence inflation, unemployment and economic growth in SSA. According to the study, although financial development alone may positively impact inflation, growth and unemployment, these effects can be either magnified or mitigated by the presence of sturdy legal frameworks and the preservation of property rights.

The study, therefore, recommends the following: policymakers in the SSA region should enhance the enforcement of property rights and legal frameworks to establish a conducive environment for financial development. This involves reducing corruption, safeguarding property rights and improving the efficiency, predictability and transparency of legal processes. Additionally, recognising the substantial role played by financial development in economic growth, unemployment and inflation, it is crucial for financial sector stakeholders in the SSA region to initiate targeted financial development schemes by promoting financial inclusion, fostering the growth of capital markets and creating and broadening financial instruments in SSA.

Whilst the study furnishes valuable insights into the connections between financial development and economic growth, unemployment and inflation, it does not address the potential policy challenges that could impede the advancement of financial sectors and the obstacles to institutional progress. Therefore, further investigations should explore the probable impediments that might obstruct the execution of financial sector reforms aimed at fostering financial development. Subsequent studies could also conduct subsample scrutinise to ascertain the regional diversities.

Appendix A: Quantile Plots of the Dependent Variables

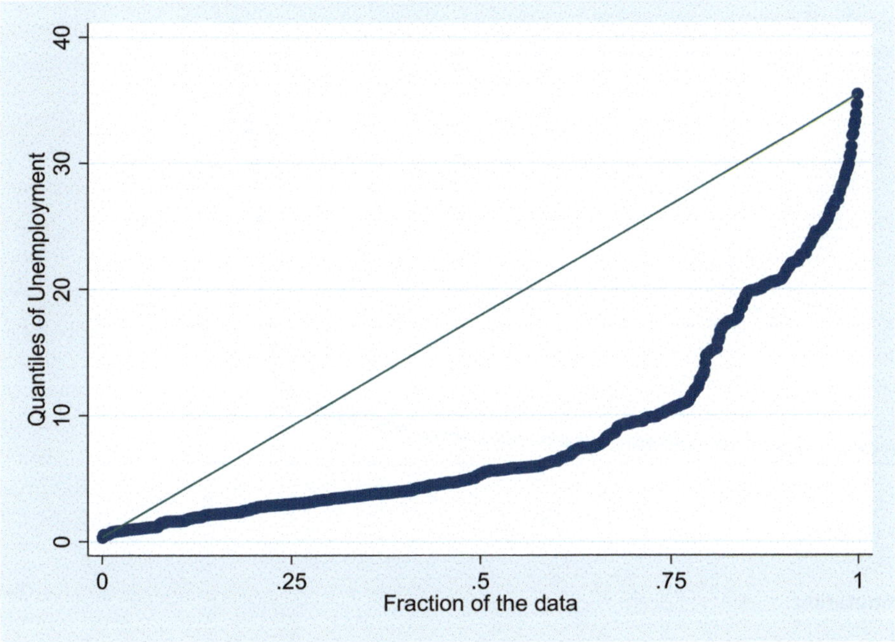

Source: **Authors' Computation**

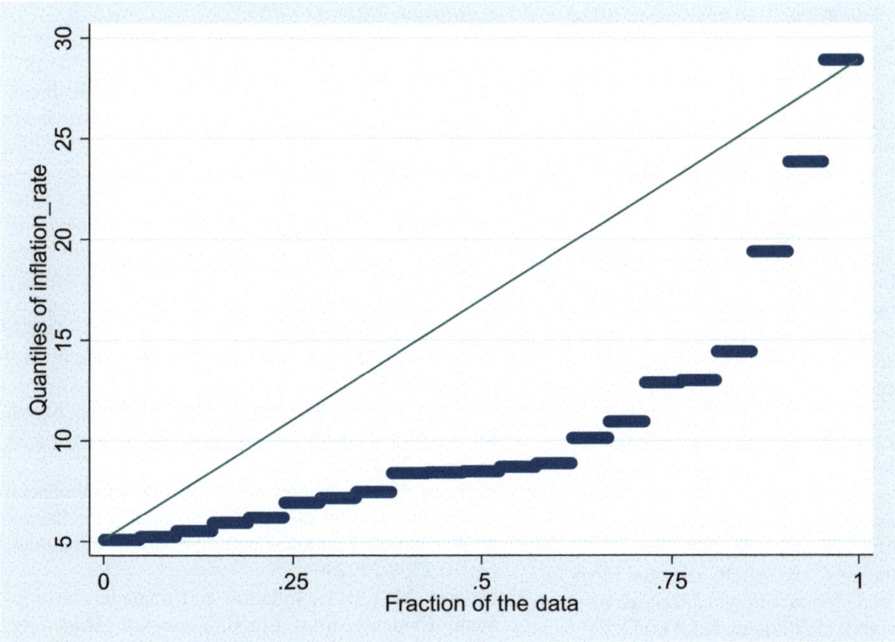

Source: **Authors' Computation**

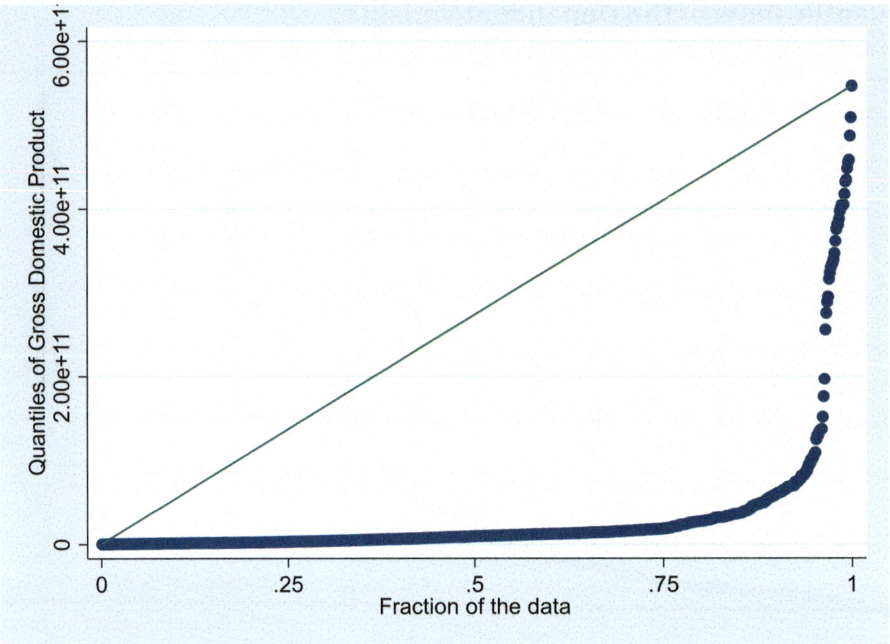

Source: **Authors' Computation**

References

Abbas, U. (2022). Inflation and Unemployment Effect Economic Growth. *Pakistan Journal of Economic Studies (PJES), 5*(2), 427–437.

Abbas, Z., Afshan, G., & Mustifa, G. (2022). The effect of financial development on economic growth and income distribution: Empirical evidence from lower-middle and upper- middle-income countries. *Development Studies Research, 9*(1), 117–128. https://doi.org/10.1080/21665095.2022.2065325

Abbey, E. N. (2012). Inflation and financial development: Evidence. *American Journal of Economics and Business Administration, 4*(4), 227–236.

Acemoglu, D., Johnson, S., & Robinson, J. A. (2001). The colonial origins of comparative development: An empirical investigation. *American Economic Review, 91*(5), 1369–1401.

Ajide, F. (2020). Asymmetric influence of financial development on unemployment in Nigeria. *Ilorin Journal of Economic Policy, 7*(2), 39–52.

Alananga, S., & Mutasa, F. (2022). Does proximity matter? The effect of financial sector development on economic growth in the EAC and SADC regional blocs. *Journal of International Economics and Management, 22*(3). https://doi.org/10.38203/jiem.022.3.0052

Almalki, A. M., & Batayneh, K. I. (2015). The relationship between inflation and financial development in Saudi Arabia. *The Journal of Developing Areas, 49*(6), 321–332. https://doi.org/10.1353/jda.2015.0088

Ang, J. B., & McKibbin, W. J. (2007). Financial liberalization, financial sector development and growth: Evidence from Malaysia. *Journal of Development Economics, 84*(1), 215–233.

Arcand, J. L., Berkes, E., & Panizza, U. (2015). Too much finance? *Journal of Economic Growth, 20*, 105–148.

Bandura, W., & Leshoro, T. (2022). Inflation and financial development in sub-Saharan Africa. *Ekonomski Anali, 67*(233), 85–112. https://doi.org/10.2298/eka2233085b

Barro, R. J. (1990). Government spending in a simple model of endogenous growth. *Journal of Political Economy, 98*(2), 103–125. https://doi.org/10.1086/26172

Bayar, Y. (2016). Financial development and unemployment in emerging market economies. *Scientific Annals of Economics and Business, 63*(2), 237–245. https://doi.org/10.1515/saeb-2016-0119

Beck, T., & Levine, R. (2004). Stock markets, banks, and growth: Panel evidence. *Journal of Banking & Finance, 28*(3), 423–442.

Bencivenga, V. R., & Smith, B. D. (1991). Financial intermediation and endogenous growth. *The Review of Economic Studies, 58*(2), 195–209. https://doi.org/10.2307/2297964

Bittencourt, M. (2011). Inflation and financial development: Evidence from Brazil. *Economic Modelling, 28*(1-2), 91–99.

Bittencourt, M., Van Eyden, R., & Seleteng, M. (2015). Inflation and Economic Growth: Evidence from the Southern A frican D evelopment C ommunity. *South African Journal of Economics, 83*(3), 411–424.

Bist, J. P. (2018). Financial development and economic growth: Evidence from a panel of 16 African and non-African low-income countries. *Cogent Economics and Finance, 6*(1), 1449780. https://doi.org/10.1080/23322039.2018.1449780

Boyd, J. H., Levine, R., & Smith, B. D. (2001). The impact of inflation on financial sector performance. *Journal of Monetary Economics, 47*(2), 221–248. https://doi.org/10.1016/s0304-3932(01)00049-6

Cecchetti, S., & Kharroubi, E. (2012). *Reassessing the Impact of Finance on Growth. Production*.

Cecchetti, S., Kharroubi, E., Beck, T., Arcand, J. L., Berkes, E., & Panizza, U. (2015). *Finance and Growth–Beware the Measurement*.

Chen, W. (2023). The effect of financial development on economic growth and economic volatility. *BCP Business and Management, 38*, 717–734. https://doi.org/10.54691/bcpbm.v38i.3765

Chen, T. C., Kim, D. H., & Lin, S. C. (2021). Nonlinearity in the effects of financial development and financial structure on unemployment. *Economic Systems, 45*(1), 100766. https://doi.org/10.1016/j.ecosys.2020.100766

Chuah, H. L., & Thai, W. (2004). *Financial development and economic growth: Evidence from causality tests for the GCC countries* (Vol. 4, No. 20). IMF Working Paper.

Ciccarelli, M., Osbat, C., Bobeica, E., Jardet, C., Jarocinski, M., Mendicino, C., et al. (2017). Low inflation in the euro area: Causes and consequences. *ECB Occasional Paper, 181*.

Claessens, S., Frost, J., Turner, G., & Zhu, F. (2018). Fintech credit markets around the world: size, drivers and policy issues. *BIS Quarterly Review September*.

Coatsworth, J. H. (1975). *Douglas C. North y Robert Paul Thomas. The rise of the Western World: A new economic history*. Cambridge University Press, 1973. 171 p. *Estudios Demográficos Y Urbanos, 9*(01), 118. https://doi.org/10.24201/edu.v9i01.320

Dées, S., & Zimic, S. (2019). Animal spirits, fundamental factors and business cycle fluctuations. *Journal of Macroeconomics, 61*, 103123.

Demetriades, P. O., & Luintel, K. B. (1996). Financial development, economic growth and banking sector controls: evidence from India. *The Economic Journal, 106*(435), 359–374.

Ductor, L., & Grechyna, D. (2015). Financial development, real sector, and economic growth. *International Review of Economics and Finance, 37*, 393–405. https://doi.org/10.1016/j.iref.2015.01.001

Effiong, E. L., Esu, G. E., & Chuku, C. (2020). Financial development and monetary policy effectiveness in Africa. *Journal of Social and Economic Development, 22*(1), 160–181. https://doi.org/10.1007/s40847-020-00098-x

Epstein, B., & Finkelstein Shapiro, A. (2019). Financial development, unemployment volatility, and sectoral dynamics. *Journal of Economic Dynamics and Control, 99*, 82–102. https://doi.org/10.1016/j.jedc.2018.11.004

Fazio, D. M., Silva, T. C., Tabak, B. M., & Cajueiro, D. O. (2018). Inflation targeting and financial stability: Does the quality of institutions matter? *Economic Modelling, 71*, 1–15. https://doi.org/10.1016/j.econmod.2017.09.011

Fry, M. J. (1980). Money, interest, inflation and growth in Turkey. *Journal of Monetary Economics, 6*(4), 535–545.

Fuinhas, J. A., Koengkan, M., & Belucio, M. (2021). Exploring the causality between economic growth, financial development and inflation in sixteen high-income countries. *Revista De Estudos Sociais, 22*(45). https://doi.org/10.19093/res11094

Galvao, A. F., & Kato, K. (2016). Smoothed quantile regression for panel data. *Journal of Econometrics, 193*(1), 92–112.

Greenwood, J., & Jovanovic, B. (1990). Financial development, growth, and the distribution of income. *Journal of Political Economy, 98*(5, Part 1), 1076–1107. https://doi.org/10.1086/261720

Guru, B. K., & Yadav, I. S. (2019). Financial development and economic growth: Panel evidence from BRICS. *Journal of Economics, Finance and Administrative Science, 24*(47), 113–126. https://doi.org/10.1108/jefas-12-2017-0125

Hartmann, S., & Spruk, R. (2020). Long-term effects of institutional instability. *Empirical Economics, 61*(4), 2073–2112. https://doi.org/10.1007/s00181-020-01934-z

Hartmann, S., & Spruk, R. (2021). Long-term effects of institutional instability. *Empirical Economics, 61*(4), 2073–2112.

Hussen, M. S., & Çokgezen, M. (2021). The institution-growth nexus in Sub-Saharan Africa: New evidence from heterogeneous panel causality approach. *Evolutionary and Institutional Economics Review, 19*(1), 141–157. https://doi.org/10.1007/s40844-021-00225-1

Ibrahim, M., Aluko, O. A., & Vo, X. V. (2022). The role of inflation in financial development– Economic growth link in sub-Saharan Africa. *Cogent Economics and Finance, 10*(1). https://doi.org/10.1080/23322039.2022.2093430

Ikhsan, A., & Satrianto, A. (2023). The effect of financial development on economic growth in high- income countries. *Asian Economic and Financial Review, 13*(3), 202–215. https://doi.org/10.55493/5002.v13i3.4743

Keho, Y. (2010). Effect of financial development on economic growth: Does inflation matter? Time series evidence from the UEMOA countries. *International Economic Journal, 24*(3), 343–355. https://doi.org/10.1080/10168730903502416

Khan, M., & Hanif, W. (2018). Institutional quality and the relationship between inflation and economic growth. *Empirical Economics, 58*(2), 627–649. https://doi.org/10.1007/s00181018-1479-7

Khan, M., & Hanif, W. (2020). Institutional quality and the relationship between inflation and economic growth. *Empirical Economics, 58*, 627–649.

Khan, M. S., Senhadji, A. S., & Smith, B. D. (2001). IMF Working Paper.

Kharisma, B., Wardhana, A., & Sofyan, M. N. F. (2022). Do institutions cause growth? Evidence from Asian countries. *Jurnal Ekonomi and Studi Pembangunan, 23*(1), 51–65. https://doi.org/10.18196/jesp.v23i1.13897

Kilishi, A. A., Adebowale, H. A., & Oladipupo, S. A. (2020). The nexus between economic institutions and unemployment: Evidence from sub-Saharan Africa. *International Journal of Development Issues, 20*(1), 74–94. https://doi.org/10.1108/ijdi-03-2020-0048

Kilishi, A. A., Adebowale, H. A., & Oladipupo, S. A. (2021). The nexus between economic institutions and unemployment: evidence from sub-Saharan Africa. *International Journal of Development Issues, 20*(1), 74–94.

Kim, Y., & Kim, T. B. (2017). The effects of institutions on the labour market outcomes: Cross- country analysis. *KDI Journal of Economic Policy, 39*(4), 69–94.

King, R. G., & Levine, R. (1993). Finance, entrepreneurship and growth. *Journal of Monetary Economics, 32*(3), 513–542. https://doi.org/10.1016/0304-3932(93)90028-e

Koenker, R. (2004). Quantile regression for longitudinal data. *Journal of Multivariate Analysis, 91*(1), 74–89.

Koenker, R., & Bassett, G., Jr. (1978). Regression quantiles. *Econometrica: journal of the Econometric Society*, 33–50.

Koenker, R., & Hallock, K. F. (2001). *Quantile regression. Journal of Economic Perspectives, 15*(4), 143–156.

Lancaster, C. (2000). *Transforming foreign aid: United States assistance in the 21st century*. Peterson Institute.

Law, C. H., & Soon, S. V. (2020). The impact of inflation on income inequality: The role of institutional quality. *Applied Economics Letters, 27*(21), 1735–1738. https://doi.org/10.1080/13504851.2020.1717425

Levine, R. (1997). Financial development and economic growth: views and agenda. *Journal of economic literature, 35*(2), 688–726.

Levine, R., Loayza, N., & Beck, T. (2000). Financial intermediation and growth: Causality and causes. *Journal of Monetary Economics, 46*(1), 31–77. https://doi.org/10.1016/s0304-3932(00)00017-9

Louati, S., & Boujelbene, Y. (2020). Inflation targeting and bank risk: The interacting effect of institutional quality. *Cogent Business and Management, 7*(1), 1847889. https://doi.org/10.1080/23311975.2020.1847889

Lucas, R. E. (1988). On the mechanics of economic development. *Journal of Monetary Economics, 22*(1), 3–42. https://doi.org/10.1016/0304-3932(88)90168-7

Ma, Y., & Lin, X. (2016). Financial development and the effectiveness of monetary policy. *Journal of Banking & Finance, 68*, 1–11.

Machado, C. M., Saraiva, A. F., & Vieira, P. D. (2021). Finance-growth nexus in sub-Saharan Africa. *South African Journal of Economic and Management Sciences, 24*(1), 1–11. https://doi.org/10.4102/sajems.v24i1.3435

Mahawiya, S., Haim, A., & Eric Fosu, O. A. (2020). In search of inflation limits for financial sector development in ECOWAS and SADC regions: A panel smooth transition analysis. *Cogent Economics and Finance, 8*(1), 1722306. https://doi.org/10.1080/23322039.2020.1722306

Marwa, N., & Zhanje, S. (2015). A review of finance–growth nexus theories: How does development finance fits in? *Studies in Business and Economics, 10*(1), 83–91. https://doi.org/10.1515/sbe-2015-0007

Nguyen, A. T. (2019). The relationship between financial development, economic growth, and inflation: Evidence from Southeast Asia. *International Journal of Business, Economics and Management, 6*(4), 201–214. https://doi.org/10.18488/journal.62.2019.64.201.214.

Odhiambo, N. M. (2012). The impact of inflation on financial sector development: Experience from Zambia. *Journal of Applied Business Research, 28*(6), 1497.

Olowu, G., Olasehinde-Williams, G., & Bein, M. (2019). Does financial and agriculture sector development reduce unemployment rates? Evidence from Southern African countries. *Agricultural Economics (Zemědělská Ekonomika), 65*(5), 223–231. https://doi.org/10.17221/263/2018-agricecon

Ouedraogo, R., Camara, I., & Sy, A. (2023). *Unbearable costs: When is inflation impeding job creation? Evidence from Sub-Saharan Africa.* IMF Working Papers, 2023(046), 1. https://doi.org/10.5089/9798400234347.001

Perone, G. (2022). The effect of labor market institutions and macroeconomic variables on aggregate unemployment in 1990–2019: Evidence from 22 European countries. *Industrial and Corporate Change, 31*(2), 500–551. https://doi.org/10.1093/icc/dtab074

Pesaran, M. H., Shin, Y., & Smith, R. J. (2001). Bounds testing approaches to the analysis of level relationships. *Journal of Applied Econometrics, 16*(3), 289–326. https://doi.org/10.1002/jae.616

Puşcaşu, E. A. (2022). The impact of financial systems on economic growth in European Union member countries. *Proceedings of the International Conference on Business Excellence, 16*(1), 722–731. https://doi.org/10.2478/picbe-2022-0068

Ragmoun, W. (2022). Institutional quality, unemployment, economic growth and entrepreneurial activity in developed countries: A dynamic and sustainable approach. *Review of International Business and Strategy, 33*(3), 345–370. https://doi.org/10.1108/ribs-10-2021-0136

Raifu, I. A. (2019). The effect of financial development on unemployment in Nigeria: do measures of financial development matter?. *DBN Journal of Economics and Sustainable Growth, 2*(2), 1–35.

Raifu, I. A., & Afolabi, J. A. (2023). The effect of financial development on unemployment in emerging market countries. *Global Journal of Emerging Market Economies, 15*(3), 354–384. https://doi.org/10.1177/09749101221116715

Rajan, R. (2019). *The third pillar: How markets and the state leave the community behind*. Penguin Press. .

Robinson, J. (1952). The Generalization of the General Theory. In *The Rate of Interest and other essays*. Macmillan Publisher.

Rodrik, D., Subramanian, A., & Trebbi, F. (2004). Institutions rule: the primacy of institutions over geography and integration in economic development. *Journal of Economic Growth, 9*, 131–165.

Romer, P. M. (1990). Human capital and growth: Theory and evidence. *Carnegie-Rochester Conference Series on Public Policy, 32*, 251–286. https://doi.org/10.1016/0167-2231(90)90028-j

Rousseau, P., & Wachtel, P. (2000). Equity markets and growth: Cross-country evidence on timing and outcomes, 1980–1995. *Journal of Banking and Finance, 24*(12), 1933–1957. https://doi.org/10.1016/s0378-4266(99)00123-5

Sabir, S., Latif, R., Qayyum, U., & Abass, K. (2019). Financial development, technology and economic development: The role of institutions in developing countries. *Annals of Financial Economics, 14*(03), 1950012. https://doi.org/10.1142/s201049521950012x

Sachs, A., and Smolny, W. (2015, August 1). Youth unemployment in the OECD: The role of institutions. *Jahrbücher Für Nationalökonomie Und Statistik, 235*(4–5), 403–417. https://doi.org/10.1515/jbnst-2015-4-505

Seiler, V., & Roncone, F. (2019). Business cycles, labour market institutions and unemployment—Further evidence from OECD countries. *SSRN Electronic Journal*. https://doi.org/10.2139/ssrn.3426852

Seo, M. H., & Shin, Y. (2016). *Dynamic panels with threshold effect and endogeneity. Journal of Econometrics, 195*(2), 169–186.

Seo, M. H., Kim, S., & Kim, Y. J. (2019). Estimation of dynamic panel threshold model using Stata.*T. he Stata Journal, 19*(3), 685–697.

Tsaurai, K. (2022). Financial development, renewable energy and unemployment in North Africa. *Journal of Accounting and Finance in Emerging Economies, 8*(3), 413–424. https://doi.org/10.26710/jafee.v8i3.2395

Uberti, L. J., & Knutsen, C. H. (2021). Institutions, human capital, and economic growth. *The Palgrave Handbook of Comparative Economics*, 461–492.

Uddin, M. A., Ali, M. H., & Masih, M. (2020). Institutions, human capital and economic growth in developing countries. *Studies in Economics and Finance, 38*(2), 361–383. https://doi.org/10.1108/sef-10-2019-0407

Uddin, M., Chowdhury, A., Anderson, K., & Chaudhuri, K. (2021). The effect of COVID–19 pandemic on global stock market volatility: Can economic strength help to manage the uncertainty? *Journal of Business Research, 128*, 31–44.

Ustarz, Y., & Fanta, A. B. (2021). Financial development and economic growth in sub-Saharan Africa: A sectoral perspective. *Cogent Economics and Finance, 9*(1), 1934976. https://doi.org/10.1080/23322039.2021.1934976

Zhu, Y., Yang, F., & Ye, W. (2018). Financial contagion behavior analysis based on complex network approach. *Annals of Operations Research, 268*, 93–111.

Monetary Policy and Inclusive Growth in the Era of Economic Uncertainty: Navigating Challenges and Opportunities

16

Khadijah Iddrisu, Joshua Yindenaba Abor, Osman Issah, and Felix Owusu Gyebi

16.1 Introduction

Monetary policy, a critical tool in the hands of central banks, aims to control the money supply and interest rates to ensure price stability and foster economic growth (Abor et al., 2024). Traditionally, its primary focus has been on macroeconomic objectives such as controlling inflation, managing employment levels and ensuring overall economic stability. For instance, the Bank of Ghana (BoG) uses an Inflation-Targeting (IT) framework, employing the Monetary Policy Rate (MPR) to set the monetary policy stance and manage inflation expectations in the economy.[1] In the same way, some Asian countries used monetary policy to control inflation and improve income levels (Cobham, 2023). Major central banks have used monetary policy to mitigate the negative effects of global crises like the 2008 financial crisis and coronavirus disease 2019 (COVID-19), geopolitical risks[2] such as the Russia-Ukraine War and other political pressures such as those that occurred in Africa. For example, the Federal Reserve implemented quantitative easing (QE) during the 2008 financial crisis to boost liquidity and support economic recovery.[3] Similarly, the Reserve Bank of India (RBI) adjusted the repo rate, cutting interest rates multiple times in 2020 to counter the economic impact of the COVID-19 pandemic. During the European debt crisis, the European Central Bank (ECB) stabilised affected economies by purchasing government bonds.[4] In response to inflation driven by surging global commodity prices dur-

[1] https://www.bog.gov.gh/wp-content/uploads/2019/08/Monetary-Policy-Report-March-2017.pdf. Accessed 11 July 2024

[2] Global crises and geopolitical risks influence macroeconomic stability leading to economic uncertainty.

[3] https://www.federalreserve.gov/newsevents/speech/kohn20100513a.htm. Accessed 11 July 2024

[4] https://www.ecb.europa.eu/press/key/date/2024/html/ecb.sp240528~a4f151497d.en.html. Accessed 11 July 2024

K. Iddrisu (✉)
School of Business, Simon Diedong Dombo University of Business and Integrated Development Studies, Wa, Ghana
e-mail: khadijahiddris@iug.edu.gh

J. Y. Abor
Department of Finance, University of Ghana Business School, Legon, Ghana
e-mail: joshabor@ug.edu.gh

O. Issah
Department of Accounting, School of Business, University for Development Studies, Tamale, Ghana
e-mail: oissah@uds.edu.gh

F. O. Gyebi
Department of Accountancy, Faculty of Business And Management Studies, Koforidua Technical University, Koforidua, Ghana
e-mail: felixo.gyebi@ktu.edu.gh

I. N. Yakubu (ed.), *Strategic Approaches to Banking Business and Sustainable Development Goals*, Sustainable Development Goals Series, https://doi.org/10.1007/978-3-031-80744-2_16

ing the Russia-Ukraine war, the Central Bank of Kenya raised the Central Bank Rate (CBR) from 7.00% to 7.50% in March 2022 to stabilise prices and manage inflation.[5] Amid conflicts in Africa, including those in Niger, Gabon and Burkina Faso, the Central Bank of Mauritius (BoM) implemented monetary policy measures by closely monitoring and managing the Mauritian rupee's exchange rate.[6] This approach aimed to mitigate currency volatility, maintain stability and address the economic disruptions affecting Mauritius.

In recent years, the scope of monetary policy has expanded to include more inclusive growth objectives, recognising that sustainable and economic growth must benefit all segments of society. For example, the RBI has implemented schemes like the *Pradhan Mantri Jan Dhan Yojana* to increase access to banking services for underserved populations, ensuring broader economic benefits (Khera, 2023). Similarly, the ECB's targeted longer term refinancing operations (TLTROs) aim to boost lending to households and small and medium-sized enterprises (SMEs), which are vital for inclusive economic development. As previously highlighted, central banks utilise monetary policy to mitigate economic uncertainty. Given this context, it is essential to discuss how monetary policies can influence inclusive growth in the presence of economic uncertainty. This chapter investigates how monetary policy has evolved to support inclusive growth amid ongoing economic uncertainties. It also highlights the challenges and opportunities that arise in aligning monetary policy with inclusive growth objectives in the current volatile environment. The discussion is especially pertinent given recent global crises, such as the COVID-19 pandemic, the Russia-Ukraine war and various regional conflicts in Africa, all of which have intensified the challenges of achieving equitable economic development.

The rest of the chapter is structured as follows: Sect. 16.2 captures the meaning and tools of monetary policy. How central bank independence (CBI) promotes monetary policy is discussed in Sect. 16.3. In Sect. 16.4, we discussed the challenges central banks or financial regulators face when implementing monetary policy. We provide an overview of inclusive growth which is shown in Sect. 16.5. Section 16.6 discusses the role of monetary policy in inclusive growth. Challenges, opportunities and future prospects of monetary policy on inclusive growth during economic uncertainty are discussed in Sect. 16.7. We end this chapter by providing the concluding remark in Sect. 16.8.

16.2 Monetary Policy: Meaning and Tools

16.2.1 What Is Monetary Policy?

Monetary policy involves regulating the money supply in an economy and the mechanisms through which new money is introduced (Bordo & Siklos, 2017; IMF, 2023). It can also be explained as the actions undertaken by a central bank or other regulatory authority to manage the money supply and interest rates to achieve macroeconomic objectives such as controlling inflation, managing employment levels, stabilising the currency and fostering economic growth. Central banks or financial regulatory authorities responsible for implementing monetary policy can choose between contractionary and expansionary approaches. A contractionary monetary policy is used by central banks to counter rising inflation and stabilise the economy. This involves reducing monetary expansion to curb inflation. On the other side, expansionary monetary policy is a strategy used by central banks to stimulate economic growth, typically during periods of recession or low economic activity. The main objectives are to increase the money supply, lower unemployment and boost consumer and business spending. Monetary policy can be con-

[5] https://www.centralbank.go.ke/uploads/monetary_policy_reports/628277515_29th%20Monetary%20Policy%20Committee%20Report.pdf. Accessed 22 August 2024

[6] https://www.elibrary.imf.org/view/journals/002/2024/139/article-A001-en.xml. Accessed 22 August 2024

tractionary or expansionary depending on the situation at hand. For instance, the U.S. Federal Reserve used contractionary monetary policy (i.e., raised interest rates) in 1980 under Paul Volcker to combat double-digit inflation, and Brazil and Argentina implemented similar measures during hyperinflation to restore currency confidence (Bordo & Siklos, 2017; Rocheteau et al., 2018).

Monetary policy originates from the foundational principles of central banking. The need to regulate the money supply and facilitate monetary exchanges was the driving force behind the establishment of central banks (Ajayi, 1999; Goodhart, 1988, 2011). Consequently, monetary policy has undergone substantial evolution across various monetary regimes (Bordo, 2008). Until 1914, the dominant monetary regime was the gold standard, where the value of a country's currency was directly tied to a specific amount of gold (Bordo, 2008;Goodhart, 1988; Thornton, 1802). Under the classical gold standard, the primary goal of monetary policy was to maintain gold convertibility (Goodhart, 1988; Thornton, 1802). This meant that the currency could be exchanged for a fixed amount of gold, ensuring stability and confidence in the currency's value. Consequently, there was limited focus on managing the domestic economy, such as controlling inflation or unemployment. The stability of the currency and international trade facilitation were prioritised over domestic economic conditions (Bagehot, 1873; Thornton, 1802). Therefore, Bagehot (1873) proposed the concept of the lender of last resort to incorporate domestic economic conditions into monetary policy. Thus, during the eighteenth and nineteenth centuries, central banks functioned as the ultimate last resort to both the banking and payment systems. The Bank of England pioneered this concept as the first central bank to adopt it (Bordo & Siklos, 2017). Subsequently, other central banks followed suit. This idea significantly contributed to maintaining financial stability by averting panic and enhancing confidence within the financial sector (Goodhart, 2011).

The lender of last resort was expanded to address economic depression and manage inflation and deflation by regulating asset prices (Meltzer, 2011). This occurred when the US Federal Reserve adopted the real bills doctrine in its early years. The doctrine advocated that a central bank should pre-emptively curb asset price booms to prevent inflation, subsequent depression and deflation (Bordo & Siklos, 2017; Meltzer, 2011). As more central banks emerged, they gained full authority to issue currency and engage in various activities to stabilise the macroeconomic environment. During the twentieth century, monetary policy evolved significantly to prioritise financial stability, a goal now recognised as crucial across the entire financial sector, unlike in the eighteenth and nineteenth centuries. The focus of monetary policy shifted towards financial stability due to rising inflationary pressures in the 1960s and 1970s, leading to the implementation of stricter monetary measures (Ajayi, 1999; Ocampo, 2017). These pressures contributed to the collapse of the Bretton Woods system, an international monetary framework established after World War II (Ocampo, 2017). In response to rising inflation, the focus of monetary policy shifted. Central banks began prioritising the maintenance of the domestic value of the currency, aiming to control inflation and stabilise the economic environment through tighter monetary measures (Ajayi, 1999; Bordo & Siklos, 2017). Some of these policies include the adoption of IT frameworks, where central banks set explicit inflation rate targets and use interest rates and other tools to keep inflation within the desired range. The use of interest rates became a primary tool for influencing economic activity and managing inflation expectations (Ocampo, 2017). Central banks increased focus on communication and transparency, offering clearer guidance on policy intentions to anchor inflation expectations and build public confidence. These strategies helped navigate post-Bretton Woods challenges, balancing economic growth and price stability.

Since the 2008/2009 financial crises, central banks have implemented diverse monetary policies to ensure financial stability by addressing financial imbalances and mitigating asset price booms (Bordo & Siklos, 2017). These include unconventional measures like QE, where banks buy securities to boost the money supply and lower interest rates. Negative interest rates have been used in some regions to spur lending and discourage cash hoarding. Macroprudential policies, such as countercyclical capital buffers and stress testing, manage systemic risks by ensuring banks hold adequate capital. Forward guidance has also guided market expectations, enhancing stability. These strategies collectively prevent financial imbalances and moderate asset price cycles, crucial for post-crisis financial stability. In the twenty-first century, there has been a notable shift in focus towards promoting inclusive growth and achieving sustainable economic and environmental development (IMF, 2023). In response to these evolving priorities, monetary policies have adapted to support this growth trajectory. Central banks and monetary authorities are increasingly integrating considerations of social equity and environmental sustainability into their policy frameworks. This includes promoting financial inclusion, supporting SMEs and fostering green financing initiatives. Moreover, monetary policies now often incorporate measures aimed at achieving long-term economic stability while addressing environmental challenges such as climate change and resource depletion. This holistic approach aims to ensure that economic growth is not only robust but also inclusive and environmentally sustainable in the modern era.

16.2.2 Monetary Policy Versus Fiscal Policy

Monetary policy and fiscal policy are two important tools used by governments and central banks to influence the economy. Monetary policy involves measures taken by the central bank, such as setting interest rates and controlling the money supply (IMF, 2023). These measures aim to achieve macroeconomic objectives such as price stability, full employment and economic growth (Bordo & Siklos, 2017). For example, lowering interest rates can stimulate borrowing and spending, thereby increasing economic activity. Conversely, raising interest rates will curb inflation but may slow economic growth. Fiscal policy, on the other hand, refers to decisions about fiscal spending and taxation.[7] This directly affects aggregate demand and economic activity. Governments use fiscal policy to manage deficits and surpluses, boost growth during economic downturns or curb inflation during overheating. For example, during a recession, governments may increase spending on infrastructure projects or cut taxes to encourage consumer spending and business investment. Conversely, in times of high inflation, governments can raise taxes or cut spending to calm the economy.

In a nutshell, monetary policy focuses on regulating the money supply and interest rates to influence credit, spending and inflation, while fiscal policy involves government spending and taxation to achieve broader goals such as economic growth and stability. Both policies are solutions for policymakers, but they differ in their mechanism and direct impact on the economy.

16.2.3 Tools of Monetary Policy

Monetary policy tools are instruments used by central banks to manage the money supply, interest rates and overall economic activity. These tools are crucial for achieving macroeconomic objectives such as price stability, full employment and economic growth. Here are some key tools of monetary policy:

16.2.3.1 Open Market Operations

Open market operations (OMOs) involve buying and selling government securities including bonds and treasury bills on the open market (Rocheteau et al., 2018). When a central bank buys securities, it injects money into the banking system, increasing the money supply (i.e. expansionary monetary policy). Conversely, selling

[7] https://www.imf.org/en/Publications/fandd/issues/Series/Back-to-Basics/Fiscal-Policy#:~:text=Fiscal%20policy%20is%20the%20use,sustainable%20growth%20and%20reduce%20poverty. Accessed July 11 2024

securities reduces the money supply (i.e. contractionary monetary policy). OMOs are flexible and effective in influencing short-term interest rates and controlling liquidity in the economy. Many central banks have used this tool to enhance monetary policy. For example, the People's Bank of China used OMO to curb rising inflation in China (see Box 16.1).

Box 16.1: People's Bank of China Used Open Market Operations for Monetary Policy

Open Market Operations (OMOs) were introduced as a monetary tool in 1993. Since 1998, following efforts by the People's Bank of China (PBOC) to develop interbank bond markets, OMOs have become a key instrument for implementing monetary policy in China. In market-based economies, OMOs are commonly used to influence money market rates and, consequently, overall interest rates. However, in China, where bank deposit and lending rates are set by the government, these rates are not entirely affected by OMOs. Instead, OMOs play a crucial role in managing liquidity and influencing the monetary base. The PBOC utilizes repurchase agreements (repos) involving government bonds to reduce the monetary base and reverse repurchase agreements (reverse repos) to inject liquidity. Additionally, the PBOC issues central bank bills to achieve similar objectives.

In 1994, the People's Bank of China expanded the use of promissory notes in coal and other industries and developed the note-discounting market, reaching a rediscount amount of 32.6 billion RMB by the end of 1995. Extensive preparations for open market operations included altering Treasury bond issuance and approving financial institutions to handle bonds. Formal open market operations began in April 1996, and by 1998, Treasury bond issues hit a record high. The interbank market, established in 1997, flourished, enabling significant transactions. By 1998, the Bank conducted 36 transactions, financing 176.13 billion RMB and supplying 70.15 billion RMB of base money. After three years of macroeconomic rectification, China achieved a soft landing in 1996, balancing supply and demand. Inflation dropped from 21% in 1994 to 0.8% in 1997. Money supply control tightened, reducing M2 growth from 34.6% in 1994 to 17.3% in 1997, and M1 growth from 26.8% to 16.5%.

Source: Barth et al. (2013) and Tinggui (2001).

16.2.3.2 Policy Interest Rates

Central banks adjust short-term interest rates to influence the economy by impacting borrowing costs for banks and consumers. When central banks increase interest rates, borrowing becomes more expensive. This discourages both banks and consumers from taking out loans, leading to reduced spending and investment. The goal is to cool down an overheating economy and control inflation. Conversely, when central banks lower interest rates, borrowing costs decrease. This encourages banks and consumers to take out loans, boosting spending and investment. The aim is to stimulate economic activity, particularly during periods of economic slowdown or recession. For example, this tool became successful for the US Federal Reserve in reducing inflation during stagflation (see Box 16.2).

Box 16.2: U.S. Federal Reserve Use Policy Interest Rate Under Paul Volcker

In the late 1970s, the United States faced stagflation, marked by high inflation and unemployment rates. Upon assuming the chairmanship of the Federal Reserve in 1979, Paul Volcker made reducing inflation a top priority, despite anticipating short-term economic hardships. The Fed, exer-

(continued)

Box 16.2: (continued)

cising its independence, aggressively raised interest rates, peaking at approximately 20% by 1981. This move aimed to make borrowing costlier, thereby dampening spending and investment to combat rampant inflation. The resultant recession in the early 1980s brought about high unemployment and economic contraction, prompting significant political opposition and pressure to reverse course. Despite the immediate economic challenges and political resistance, Volcker's policies significantly lowered inflation. By 1983, inflation rates dropped to approximately 3–4%, stabilising the economy. Volcker's steadfast approach underscored the Federal Reserve's commitment to long-term economic stability over short-term political expediency. This stance bolstered the Fed's credibility, earning the trust of investors, businesses, and the public alike. Investors gained confidence in the U.S. economy's resilience under the Fed's stewardship. Businesses benefited from clearer economic expectations and were assured that inflation would remain controlled. The public's confidence in the Fed strengthened, recognising its focus on safeguarding long-term economic health.

Source: Bordo and Siklos (2017).

16.2.3.3 Reserve Requirements

Central banks regulate the money supply by imposing reserve requirements on commercial banks, mandating they hold a portion of deposits as reserves with the central bank. Adjusting these requirements directly impacts banks' lending capacity. Lowering reserves frees up funds for lending, stimulating economic activity by increasing credit availability. Conversely, raising requirements compels banks to hold more reserves, reducing lending to control inflation. Reserve requirements influence banking liquidity and practices, strategically adjusted based on economic conditions. They provide liquidity during downturns to spur growth and restrict lending during inflationary periods for financial stability. Overall, reserve requirements are vital for central banks to manage the money supply, influence interest rates and foster economic stability through prudent banking regulation. For example, BoG used reserve requirements to ensure financial stability and boost investors' confidence in the banking sector (see Box 16.3).

Box 16.3: Bank of Ghana Used Reserve Requirement to Clean the Banking Sector in 2018

In 2018, the Bank of Ghana (BoG) initiated comprehensive reforms to address critical issues within the banking sector, such as high non-performing loans and weak capitalization. The Minimum Capital Directive issued in September 2017 mandated universal banks to increase their minimum paid-up capital to GHC400 million by December 2018. Banks could achieve compliance through fresh capital injections, income surplus capitalization, or a combination of both, aiming to enhance sector resilience and support Ghana's economic growth These measures aimed to bolster banks' liquidity positions, enabling them to manage risks more effectively and maintain stability amidst financial pressures. By December 2019, these reforms had yielded significant results. This proactive approach not only enhanced the sector's stability but also promoted a healthier banking environment overall. For instance, the banking sector's total assets grew to GH¢129.06 billion, marking a robust 22.8% year-on-year increase, compared to 12.3% growth in the previous year. This growth was driven by substantial gains in both domestic and foreign assets. Domestic assets surged by 23.1% to GH¢118.69 billion, while foreign assets grew by 19.8% to GH¢10.38 billion. This shift increased the share of domestic assets in total assets to 92.2% from 90.3% in December 2018, indicating a stronger local banking sector. The stricter reserve requirements and heightened capital thresholds incentivised banks to strengthen their risk management practices and adhere more closely to regulatory standards. Concurrently, the increased reserve ratios facilitated sector consolidation, prompting weaker banks to merge with stronger counterparts or exit the market entirely. This consolidation streamlined operations, reducing systemic risks and enhancing the resilience of Ghana's financial industry to future economic challenges. Despite this, there was a lot of job losses which increased poverty and inequality. Job losses in Ghana's banking sector led to profound psychosocial challenges, including identity confusion, shame, and harassment from customers, impacting both immediate and extended families emotionally and economically.

Source: Bank of Ghana (2020) and Agyemang et al. (2024).

16.2.3.4 Unconventional Tools

Apart from the traditional tools discussed, there are emerging tools central banks can use for monetary policy. For instance, forward guidance which is central banks can communicate their future monetary policy intentions to influence market expectations. This guidance can signal future changes in interest rates or other policy measures, affecting long-term interest rates and economic decisions. Additionally, an important emerging tool, QE, has been utilised by a lot of central banks including the Federal Reserve in the USA for monetary policy. QE is the purchase of long-term securities and other financial assets to increase the money supply and lower long-term interest rates. During periods of economic crisis or when traditional policy tools are ineffective, central banks may implement QE. To boost economic growth during the 2008 financial crisis, the US Federal Reserve initiated a QE policy, buying $4.5 trillion in assets by 2014 (Li, 2016). QE encouraged private bank lending and economic recovery, though concerns about large-scale money supply increases inducing inflation remain.

16.3 Central Bank Independence and Monetary Policy

The effectiveness of monetary policy is more often dependent on the level of Central Bank Independence (CBI) (Abor et al., 2022; Frenkel, 1999; Garriga, 2016). This is because independent central banks are better positioned to control inflation. Without political pressure to finance government spending or stimulate the economy before elections, they can maintain price stability more effectively. According to some literature (Abor et al., 2022; Frenkel, 1999; Garriga, 2016), CBI can best be described as follows:

(a) The independence of central bank members restricts politicians from influencing the composition of the central bank's board or its decisions.
(b) The central bank's autonomy in managing its operations without being subject to government decisions (i.e. financial independence).
(c) The central bank's authority to formulate and implement its monetary policy (i.e. policy independence).

CBI can either be instrument independence or goal independence. Instrument independence allows the central bank to select the tools it uses for monetary policy without political interference. Goal independence on the other hand, enables the central bank to determine its monetary policy objectives, ensuring that its decisions are made based on economic considerations rather than political pressures.

The following shows how CBI enhances monetary policies

1. Inflation Control: Independent central banks are better positioned to control inflation. Without political pressure to finance government spending or stimulate the economy before elections, they can maintain price stability more effectively. For example, as shown

in Boxes 16.1 and 16.2, the selected central banks were able to control inflation to the desired target.

2. Credibility and Trust: An independent central bank can make decisions that prioritise long-term economic health over short-term political gains. This independence allows the central bank to implement policies aimed at maintaining economic stability, controlling inflation and fostering sustainable growth without political interference. By focusing on these long-term goals, the central bank can build credibility and trust among investors, businesses and the public. This trust is crucial, as it ensures that monetary policies are seen as reliable and effective, which can lead to more stable financial markets, increased investment and a healthier overall economy. Additionally, the perceived independence of the central bank can enhance its ability to respond to economic crises with appropriate measures, further reinforcing confidence in its commitment to economic stability.
3. Policy Consistency: In the Eurozone, the ECB exemplifies the benefits of an independent central bank. During the Eurozone crisis, the ECB implemented measures such as low interest rates and asset purchases to stabilise the economy.[8] Despite political pressure from various member states to prioritise national interests, the ECB focused on long-term stability and inflation control. Conversely, in countries with politically influenced central banks, such as Argentina, monetary policy often becomes erratic, leading to high inflation and economic instability. This contrast highlights how independence fosters consistent and predictable policies, while political interference can destabilise the economy.
4. Focus on Economic Goals: Independent central banks can prioritise key economic objectives such as controlling inflation, managing unemployment and ensuring financial stability, without being swayed by political agendas. For example, in the UK, the Bank of England's independence allows it to prioritise key economic objectives such as controlling inflation, managing unemployment and ensuring financial stability.[9] For instance, during the 2008 financial crisis, the Bank of England implemented QE and adjusted interest rates to stabilise the economy, focusing on long-term recovery rather than short-term political pressures. This independence ensures that monetary policy decisions are based on economic data and objectives, rather than political agendas, enabling more effective management of the economy and fostering confidence among investors and the public.
5. Reduced Risk of Hyperinflation: Political interference often leads to excessive money printing to finance deficits, causing hyperinflation. An independent central bank can resist these pressures, focusing instead on maintaining a stable monetary environment by controlling inflation and ensuring economic stability, which builds confidence among investors and the public.
6. Long-Term Planning: Independence allows central banks to focus on long-term economic health, implementing policies that might be unpopular in the short term but beneficial in the long run. Therefore, if there is political interference, long-term economic health may be affected. For example, some central banks in India, Canada, the UK and others encountered political interference which affected their monetary policy (see Box 16.4).

Box 16.4: Some Central Banks Experienced Political Interference

On December 10, 2018, central banks in Canada, the United Kingdom, and the United States likely took note of Reserve Bank of India (RBI) Governor Urjit Patel's resignation, reflecting on the broader issue of political interference. Patel's abrupt departure

[8] https://www.ecb.europa.eu/press/key/date/2024/html/ecb.sp240528~a4f151497d.en.html

[9] https://committees.parliament.uk/writtenevidence/120469/pdf/

marked the end of a power struggle with the Indian government over banking regulations, indicating a loss of RBI's operational independence shortly after adopting a formal inflation target. Patel's predecessor, Raghuram Rajan, had also been pressured out by Prime Minister Narendra Modi's administration. This situation resonates with other major central banks, as their leaders faced similar political turbulence in 2018. Bank of Canada Governor Stephen Poloz, appointed by former Conservative Prime Minister Stephen Harper, clashed with a Conservative senator in April over his supportive remarks about the current Liberal government's gender equality efforts. Mark Carney, head of the Bank of England, was dismissed as a "second-tier Canadian politician" by a Conservative party member displeased with the bank's Brexit forecasts.

In the United States, Federal Reserve Chairman Jerome Powell faced repeated criticisms from President Donald Trump, who appointed him over the incumbent, Janet Yellen. Trump publicly expressed dissatisfaction with Powell, criticizing the Fed's policies. In 2018, 17 central banks experiencing unusual political interference. Besides the RBI, this includes the Central Bank of Turkey, which dealt with a currency crisis exacerbated by President Recep Tayyip Erdogan's pressure to keep borrowing costs low, and the Reserve Bank of New Zealand, which was directed by Prime Minister Jacinda Ardern's government to add employment goals to its inflation target.

Source: Carmichael (2018).

16.4 Challenges Central Bank Face When Implementing Monetary Policy

There are some challenges that may limit the efficiency of the monetary policy. These are but not limited to the following.

16.4.1 Political Pressures

Central banks must remain independent from political influence to focus on long-term economic stability, making objective decisions based on economic fundamentals (Carmichael, 2018). Political pressure could lead to short-term, popular measures, resulting in suboptimal outcomes like inflation or asset bubbles. In democratic economies, political pressures arise because presidents are judged by indicators like inflation, employment and economic growth, which affect their re-election chances (Bivens, 2024). When monetary policies aimed at controlling inflation led to higher unemployment, citizens may criticise the president, threatening re-election. This can prompt presidents to interfere in monetary policy decisions to avoid unpopular outcomes and maintain public support, potentially compromising long-term economic stability. The opposition party or president may criticise the central bank's implementation of monetary policy, particularly if the policies lead to adverse economic outcomes or are perceived as ineffective. This critique often reflects differing economic philosophies or political strategies aimed at gaining public support or challenging the ruling party's economic stewardship.

16.4.2 Technological Changes

Financial innovations and digital currencies present significant challenges to traditional monetary policy frameworks by altering the mechanisms through which money is created, circulated and controlled (Poloz, 2021). Innovations like decentralised finance (DeFi), blockchain technology and cryptocurrencies introduce new forms of transactions and assets that operate outside conventional banking systems, potentially reducing the effectiveness of central banks' tools such as interest rate adjustments and open market operations. These changes demand that central banks adapt by developing new regulatory measures, understanding the implications of digital assets on monetary supply and demand and exploring the integration of digital currencies into the broader financial system to ensure economic stability and policy efficacy.

16.4.3 Difficulty in Attaining Inflation Target

A key challenge of monetary policy is achieving and maintaining a stable inflation rate without triggering deflation or excessive inflation, which is critical for economic stability and growth (Bordo & Siklos, 2017). Central banks must carefully calibrate their tools, such as interest rate adjustments and open market operations, to influence inflation while avoiding the extremes. Excessive inflation erodes purchasing power and can diminish consumer and business confidence, leading to reduced spending and investment. Conversely, deflation can cause consumers to delay purchases in anticipation of lower prices, reducing demand and potentially sparking a downward economic spiral.[10] Balancing these dynamics is complex, requiring precise and responsive policy measures to foster a predictable economic environment conducive to sustainable growth, investment and employment.

16.4.4 Global Interdependencies

One of the major challenges of monetary policy is managing the spillover effects from the policies of large economies on global markets, which can influence exchange rates, capital flows and worldwide financial stability (Georgiadis & Zhu, 2019). When central banks in major economies modify their interest rates or employ unconventional monetary measures, it can cause investors to shift their capital across borders in pursuit of higher returns. This can lead to fluctuating exchange rates, impacting trade balances and economic stability in other nations. Furthermore, sudden changes in capital flows can destabilise financial markets, particularly in emerging economies, making it difficult for these countries to sustain their own economic stability and growth. International coordination of monetary policies and effective management of these spillover effects are essential but challenging, requiring a careful balance to minimise adverse impacts on the global financial system.

16.5 Inclusive Growth: An Overview

Despite its increasing presence in policy discussions, the definition of inclusive growth remains contested. Some scholars align it closely with pro-poor growth, while others emphasise broader societal benefits and the equitable sharing of growth opportunities. For example, initially used by Kakwani and Pernia (2000), inclusive growth focuses on the active participation of the poor in economic activity and the equitable distribution of growth benefits across various social groups. According to Ali and Son (2007), inclusive growth is achieved through two elements such as average opportunities available to the population and how opportunities are shared among the population. Habito (2009) also understands inclusive growth as gross domestic product (GDP) growth that leads to significant poverty reduction. Klasen (2010) also viewed this concept as disadvantage-reducing growth. From the various definitions, it can be deduced that inclusive growth is the equal distribution of countries' resource to all populaces. These resources can be income growth, access to basic social amenities and access to infrastructure.

However, this concept has been enhanced over time due to some crises and the implementation of policies. For instance, in the post-World War II, the dominant belief was that rapid industrial growth was essential for improving the living conditions of the poor (Ranieri & Almeida Ramos, 2013). This belief was rooted in Kuznets (1955) hypothesis, which suggested that economic growth would initially increase income inequality but eventually lead to broader improvements in living standards as nations reached higher levels of per capita income. Development

[10] https://www.europarl.europa.eu/RegData/etudes/BRIE/2015/559492/EPRS_BRI(2015)559492_EN.pdf. Accessed July 14 2024

theories, such as Rostow (1959, 1960) stages of growth model, reflected this linear view, positing that all countries would naturally progress through stages of economic growth, leading to higher average incomes. However, this expectation did not hold true for many developing countries (Berg & Ostry, 2011a, b). While advanced economies like the United States and the United Kingdom followed this trajectory, many developing nations experienced erratic growth patterns. In some cases, rapid growth was accompanied by worsening inequality, and the anticipated reduction in inequality did not occur (Berg & Ostry, 2011a, b). This defied the assumption that growth would automatically benefit the poorer segments of society, leading to a reevaluation of the relationship between growth and inequality.

The experiences of the East Asian 'tigers' (Hong Kong, Singapore, South Korea and Taiwan) further challenged the perceived trade-off between growth and equity (Ranieri & Almeida Ramos, 2013). These countries achieved rapid economic growth while maintaining relatively low levels of inequality, demonstrating that growth and equity could coexist. Their success underscored the importance of strategic state policies in fostering inclusive growth. By the turn of the century, the belief that growth and equity could be compatible gained traction. Studies found that growth accompanied by progressive redistribution was more effective in reducing poverty than growth alone. High initial inequality was identified as a barrier to poverty reduction and long-term growth sustainability. This shift in thinking led to a consensus that inequality matters not just for social justice but also for sustaining economic growth over the long term.

As a result, the development literature began to focus on ensuring that economic growth benefits the poor, challenging the traditional 'trickle-down' theory (Ranieri & Almeida Ramos, 2013). It became clear that growth alone was not sufficient for poverty reduction; it needed to be deliberately structured to achieve pro-poor outcomes. This marked a departure from mid-twentieth-century theories that assumed growth would naturally reduce poverty. The concept of 'pro-poor growth' gained attention, although its definition remains debated. One approach, 'weak absolute pro-poor growth', defines it as any growth that reduces poverty, regardless of income distribution. However, this approach can lead to greater inequality if the wealthy's income grows faster. A second approach, 'relative pro-poor growth', argues that growth is pro-poor when the income of the poor increases at a faster rate than that of the wealthy, aiming to reduce inequality. This includes variations where the poor's share of incremental income exceeds their current share or even their share of the population.

Recent discussions have expanded pro-poor growth to include non-income dimensions such as education, health and welfare, acknowledging the multidimensional nature of poverty. However, the debate on pro-poor growth remains unresolved, with differing views on whether targeted pro-poor policies are necessary or if general growth-maximising policies suffice. Alongside pro-poor growth, the concept of inclusive growth emerged, emphasising the importance of broad-based participation and benefits from economic growth. Unlike pro-poor growth, which traditionally focused on income outcomes, inclusive growth encompasses broader aspects, including education and health, and stresses the need to reduce disadvantages among marginalised groups.

16.6 The Role of Monetary Policy in Inclusive Growth

The impact of an independent central bank's monetary policy on the inclusiveness of economic growth has sparked debates within monetary policy-economic growth literature (see Ahiadorme, 2022; Chang & Jaffar, 2014; Goshit, 2015; Pérez-Moreno et al., 2022; Yien et al., 2017). Inclusive growth involves ensuring that economic opportunities created by growth are accessible to and beneficial for all, particularly the economically disadvantaged (Iddrisu et al., 2023a, b, c, d). This concept has gained prominence due to growing awareness that economic growth often fails to equitably distribute its benefits (Goshit, 2015). Growth becomes inclusive when it not only generates economic opportunities but also ensures fair and equal access to these opportunities. This section discusses these debates by identifying how monetary policy influences inclusive growth.

First, monetary policy fosters sustainable economic activity and benefits all segments of society by ensuring stable prices and controlling inflation. Lower and stable inflation rates safeguard the purchasing power of lower income groups and retirees reliant on fixed incomes (Iddrisu et al., 2022). This prompts greater savings and reduces poverty as disposable incomes decline, ensuring that even marginalised groups can set aside sufficient income for savings (Demirguc-Kunt et al., 2018; Iddrisu, 2024). This conforms with an empirical study by Ahiadorme (2022), who found that monetary reduces poverty and enhances inclusive growth. On the other hand, it has also been identified that ensuring stable prices and controlling inflation can hamper inclusive growth. Excessively tight control over prices can hinder economic stimulus efforts like interest rate adjustments, critical for fostering growth and job creation, especially among marginalised communities. This strict approach may also curtail access to credit, making it challenging for small businesses and individuals from lower income backgrounds to secure essential loans for investment and entrepreneurship. Moreover, overly stable or controlled prices could dampen consumer spending, potentially stalling demand for goods and services crucial for economic expansion and employment. Additionally, rigid price controls may inadvertently worsen income inequality by limiting wage growth and discouraging investments that could benefit marginalised groups while also stifling innovation and new business opportunities essential for inclusive growth. The study of Ahiadorme (2022) also showed that disinflation harms equity by exacerbating unemployment and compromising inclusive growth in advanced economies with significantly lower inflation rates.

Second, monetary policy plays a crucial role in influencing employment levels through adjustments in interest rates (IMF, 2023). Lowering interest rates can stimulate economic activity by making borrowing cheaper for businesses, which in turn can lead to increased investment, expansion and hiring (Abor et al., 2022). This stimulation is especially beneficial for marginalised groups who often face higher unemployment rates during economic downturns. By strategically adjusting interest rates, central banks aim to foster a more inclusive economy where job opportunities are more readily available, supporting overall economic stability and prosperity. Despite enhancing employment, some countries faced unemployment when monetary policy was implemented, especially with reserve requirements. For example, in Ghana, the higher reserve requirements implemented in 2018 led to significant job losses as several banks exited the market or merged with others (see Box 16.3).

Lastly, access to credit is crucial for the growth and sustainability SMEs and entrepreneurs (Ayyagari et al., 2014; Iddrisu et al., 2023b). Central banks can facilitate this access through monetary policy measures, such as reducing interest rates or injecting liquidity into the banking system. Lower interest rates make borrowing cheaper for businesses, enabling SMEs to expand operations, invest in technology and create jobs. Liquidity injections ensure that banks have sufficient funds to lend, especially during economic downturns when credit availability may be constrained. By supporting SMEs and entrepreneurship in this way, central banks contribute to economic dynamism, innovation and job creation, fostering a more robust and inclusive economy. For example, in Africa, many countries have reduced interest rates and allocated funds to support SMEs during COVID-19.[11] In countries with limited microfinance and development banks that are specifically dedicated to financing SMEs, there is a risk of cherry-picking, where the focus may shift to large firms or institutions with thorough documentation or to less risky clients.[12]

[11] https://www.csis.org/analysis/supporting-small-and-medium-enterprises-sub-saharan-africa-through-blended-finance. Accessed July 13 2024

[12] https://www.worldbank.org/en/topic/smefinance. Accessed July 13 2024

16.6.1 Mitigating Economic Uncertainty on Inclusive Growth: The Role of Monetary Policy

Economic uncertainty, driven by global crises, geopolitical risks and domestic disruptions, poses significant challenges to achieving inclusive growth. In this context, monetary policy plays a crucial role in stabilising the economy and promoting equitable development. Here's an in-depth discussion on how monetary policy can address economic uncertainty while fostering inclusive growth.

16.6.1.1 Enhancing Economic Activity

Monetary policy influences the economy through adjustments in interest rates. In times of economic uncertainty, central banks can lower interest rates to stimulate borrowing and investment. By making credit more affordable, these policies can spur business expansion, job creation and consumer spending, all of which contribute to inclusive growth. Conversely, in an overheated economy, raising interest rates can prevent inflationary pressures that might otherwise widen income inequality which disproportionately affect lower income groups. For instance, during post-COVID-19, the ECB used monetary policy tools such as increasing interest rates, which led to an increase in employment and wages, although some other economic factors such as GDP were negatively affected (see Box 16.5).

Box 16.5: European Central Bank Used Interest Rate to Reduce COVID-19 Impact on Inclusive Growth

The European Central Bank (ECB) raised its main refinancing rate from 0% in July 2022 to 4.5% by September 2023. This policy aimed to address rising inflation, which had peaked at 10.6% in October 2022 but decreased to around 4.3% by mid-2023, reflecting the impact of the ECB's tightening measures. However, the higher interest rates contributed to a slowdown in economic growth, with the Eurozone economy expanding by only 1.0% in 2023, down from 3.5% in 2021. Despite these challenges, the unemployment rate remained relatively stable at approximately 6.6% in 2023, showing resilience in the labor market. Wages in the Eurozone increased by about 3.8% in 2023, which helped partially offset the effects of inflation.

Source: ECB (2023).

16.6.1.2 Protecting Vulnerable Institutions

In periods of severe economic downturns or financial crises, traditional monetary policy tools may be insufficient. In such cases, central banks can employ unconventional measures like QE, which involves purchasing government securities and other financial assets to inject liquidity into the economy. This increase in money supply can lower long-term interest rates, boost asset prices and encourage lending. Ensuring that credit flows to SMEs and underserved communities can enhance the inclusivity of growth by supporting sectors that provide employment and essential services to a broad base of the population. An empirical study has shown that in Japan, monetary policy tool like QE increases the access to credit for all the populace (see Box 16.6). This helps to minimise the poverty rate and expansion of businesses.

Box 16.6: Quantitative Easing on Credit Standards and Systemic Risk: The Japanese Experience

This empirical paper by Vu (2020) investigates the impact of quantitative easing (QE) on financial intermediaries in Japan, focusing on bank credit standards (CSs) and systemic risk during QE periods (03/2001–03/2006 and from 10/2010,

(continued)

Box 16.6: (continued)

when comprehensive monetary easing was reactivated, with quantitative and qualitative easing (QQE) starting in 04/2013). Empirical evidence suggests that QE, by increasing liquidity and lowering interest rates, encourages banks to relax their lending standards, potentially leading to riskier lending practices. This environment can heighten systemic risk through the risk-taking channel, where banks, seeking higher returns, invest in riskier assets. The study finds that QE has significantly relaxed CSs, particularly for small firms and households, while low interest rates have had a mixed impact—tightening standards for large firms but generally increasing systemic risk. The paper also highlights that while QE can reduce default risk by improving macroeconomic conditions, the low-interest rate environment may offset these benefits by raising systemic risk. The findings support the competition-fragility hypothesis, indicating that increased competition among banks can exacerbate systemic risk. Overall, the study contributes to understanding the complex relationship between QE, bank behavior, and financial stability, providing insights into the effectiveness of QE in Japan's prolonged low-interest environment.

Source: Vu (2020).

16.6.1.3 Financial Inclusion

Monetary policy can also support inclusive growth by promoting financial inclusion in times of economic uncertainty. Financial inclusion has been identified as a key factor to reduce poverty (Ayyagari et al., 2017; Demirgüç-Kunt & Singer, 2017; Iddrisu et al., 2022). Therefore, central banks can implement policies that encourage the development of digital financial services and the extension of credit to underserved populations. For example, reducing reserve requirements for banks that lend to SMEs or offering incentives for banks to expand services to rural areas can help bring more people into the formal financial system. Increased access to financial services enables individuals and businesses to invest in education, health and entrepreneurial activities, which are critical drivers of inclusive growth.

16.6.1.4 Managing Exchange Rate Volatility

Exchange rate stability is another critical aspect of monetary policy that impacts inclusive growth. Volatile exchange rates can create uncertainty in trade and investment, particularly in emerging markets that rely heavily on exports. By managing exchange rate fluctuations through interventions in the foreign exchange market or by adopting a managed float system, central banks can reduce the risks associated with currency instability. A stable exchange rate environment supports trade, attracts foreign investment and helps maintain the affordability of essential imports, benefiting a broad spectrum of the population.

16.7 Challenges and Opportunities of Monetary Policy on Inclusive Growth in Times of Economic Uncertainty

16.7.1 Challenges

While monetary policy can be used to enhance inclusive growth even in times of economic uncertainties caused by global crises and geopolitical risk, there are some challenges associated with such policies. This subsection discusses some of the major challenges such as the issue of balancing inflation and other macroeconomic performance, transmission mechanism variability and policy expectations.

16.7.1.1 Balancing Inflation and Other Macroeconomic Performance

Monetary policy faces the challenge of balancing inflation control with other macroeconomic performance such as enhancing growth and reducing poverty, a task complicated by current global and geopolitical risks. These risks, such as geopolitical tensions, supply chain disruptions and shifting trade dynamics, exacerbate the conflict between these objectives. For example, the conflict in Ukraine has driven up commodity prices, especially in energy and food, fuelling inflation. In response, central banks may tighten monetary policy by raising interest rates or reducing asset purchases. However, such actions can slow economic activity, potentially leading to higher unemployment and disproportionately affecting vulnerable groups like low-income households and small businesses. These groups often lack the financial resilience to cope with rising costs and reduced access to credit, which can hinder their ability to manage expenses and sustain operations. This can undermine inclusive growth and exacerbate economic inequality. For instance, the quest to minimise inflation as a result of COVID-19 caused the ECB to rise interest rates, which indeed reduced inflation and created employment but slowed economic growth and increased poverty (see Box 16.5). Additionally, geopolitical risks contribute to global capital flow volatility and exchange rate instability, further complicating monetary policy.

16.7.1.2 Transmission Mechanism Variability

The effects of monetary policy often differ across various segments of the economy, amplifying inequality if not carefully managed. For example, interest rate hikes typically benefit savers by increasing returns on deposits but can simultaneously burden borrowers, particularly in sectors with high leverage, such as real estate or small businesses. This creates a differential impact, where wealthier individuals and large corporations, who are more likely to have savings and financial assets, may benefit from higher rates, while those reliant on borrowing—such as low-income households or heavily indebted firms—suffer increased costs. In the context of developing regions, like parts of Africa, this disparity is compounded by political instability, which can exacerbate economic volatility. Governments in politically unstable countries may struggle to implement consistent monetary policies, leading to higher borrowing costs, reduced access to credit and diminished investor confidence. The political uncertainty may also deter foreign investment, further restricting credit availability and deepening the economic divide. Without targeted interventions to address these imbalances, such as social protection programs or targeted credit support, monetary policy actions could inadvertently worsen economic inequality, leaving vulnerable populations more exposed to both inflationary pressures and the consequences of political and economic instability.

16.7.1.3 Policy Expectations

If public expectations are poorly managed, it can lead to increased economic uncertainty, which negatively impacts growth. When businesses and consumers are uncertain about the future direction of monetary policy—whether it's the trajectory of interest rates, inflation targets or regulatory changes—they may become cautious, delaying spending and investment decisions. This hesitation can lead to a slowdown in economic activity, as businesses hold off on expanding operations or hiring, and consumers reduce their spending on big-ticket items like homes or cars. In today's context, where global and geopolitical risks such as climate change are becoming increasingly pronounced, the stakes are even higher. Climate change introduces additional layers of uncertainty, such as unpredictable weather patterns, regulatory shifts toward sustainability and potential disruptions in global supply chains. If monetary policy does not convincingly address these risks—either by failing to integrate climate considerations into policy frameworks or by sending mixed signals about future actions—businesses may be even more reluctant to invest in long-term projects, particularly those involving significant environmental impact or requiring large capital outlays. This could further slow economic growth and exacerbate vulnerabilities, especially in sectors heavily exposed to climate risks, like agriculture, insurance and energy.

For instance, investors in Japan shifted their investment into the US due to policy uncertainty (see Box 16.7).

Box 16.7: The Japanese Investors Shifted Their Investment to USA

After the COVID-19 pandemic, investors in Japan experienced significant uncertainty regarding the country's monetary policy, leading many to shift their investments to the U.S. market. This uncertainty stemmed from concerns about Japan's economic recovery, the effectiveness of its monetary policy response, and the potential for prolonged low interest rates and deflationary pressures. In contrast, the U.S. offered a more attractive investment climate, bolstered by aggressive monetary policy measures from the Federal Reserve, including substantial interest rate cuts and quantitative easing. These measures provided a more promising environment for growth and returns, attracting Japanese investors seeking higher yields and stability. Additionally, the U.S. economic recovery, driven by robust fiscal stimulus and a relatively faster vaccination rollout, enhanced investor confidence in American markets. This shift reflects how global economic uncertainties and divergent monetary policies can influence investment flows, with investors moving capital to regions perceived as more stable and potentially profitable amid uncertain conditions.

Source: EAF editors (2024).

16.7.2 Opportunities

Although the central bank aims at implementing monetary policy to promote inclusive growth in the era of economic uncertainty, there is also an opportunity the central bank can enjoy. Some of these opportunities are enhanced policy tools, focus on financial inclusion, innovation in financial systems and enhanced collaboration and partnership.

16.7.2.1 Enhanced Policy Tools

Economic uncertainty often leads central banks to use a wider range of policy tools beyond traditional methods. For example, during the 2008 global financial crisis, the Federal Reserve implemented QE to counteract the limitations of conventional monetary policy, which had reached the zero lower bound on interest rates. QE involves purchasing long-term securities to lower long-term interest rates, boost asset prices and stimulate investment (Eckhold et al., 2024). Similarly, forward guidance, which involves central banks communicating future policy intentions, was used during the COVID-19 pandemic to maintain low interest rates and manage market expectations. These unconventional tools helped stabilise financial markets, encourage spending and support economic growth. By using such measures, central banks can address immediate challenges and promote a more inclusive recovery by creating favourable conditions for broad-based economic activity.

16.7.2.2 Focus on Financial Inclusion

During periods of economic instability, central banks have a crucial opportunity to prioritise financial inclusion as a means to support broader economic stability and growth. For instance, during the COVID-19 pandemic, many central banks, including the RBI, took significant steps to enhance financial inclusion. The RBI introduced measures to expand access to digital financial services, such as mobile banking and digital payments, to reach underserved and rural populations who had limited access to traditional banking services. By implementing initiatives like financial literacy programs and increasing the availability of low-cost banking services, the RBI aimed to include more individuals in the financial system. This approach not only helped those directly affected by the economic downturn but also supported the overall economy by broadening access to credit, which in turn facilitated increased consumer spending and investment. Expanding financial inclusion in such times ensures that the benefits of monetary policy reach a wider segment of the population, fostering more equitable economic participation and contributing to a more resilient recovery.

16.7.2.3 Innovation in Financial Systems

Economic uncertainty often drives innovation within financial systems as institutions seek novel solutions to emerging challenges. Central banks have a key role in advancing financial technologies that improve accessibility, efficiency and inclusivity. For instance, during the COVID-19 pandemic, the People's Bank of China accelerated its digital currency electronic payment (DCEP) initiative to meet the rising demand for contactless transactions and enhance financial inclusion. Similarly, central banks have explored fintech innovations like blockchain technology to strengthen payment systems and security. In the UK, the Bank of England has supported financial technology through initiatives such as the Fintech Hub, promoting innovation in financial services. These advancements address immediate needs and contribute to building a more inclusive and resilient financial system for the future.

16.7.2.4 Enhanced Collaborations and Partnerships

Economic uncertainty often requires closer collaboration among central banks, governments and other institutions to create more coordinated and effective policies. For instance, during the Eurozone debt crisis, the ECB, national governments and the European Commission collaborated to address the crisis comprehensively. The ECB implemented the Outright Monetary Transactions (OMT) program to stabilise bond markets by buying government bonds from distressed nations. At the same time, national governments enacted fiscal consolidation and structural reforms to stabilise their economies. The European Stability Mechanism (ESM) provided financial aid and supported reforms. This coordinated approach ensured that monetary, fiscal and structural policies were aligned, enhancing their effectiveness in managing the crisis and promoting recovery. Such collaboration helped create a more robust and inclusive framework for economic recovery, addressing both immediate challenges and long-term structural issues.

16.7.3 Future Prospect of Monetary Policy

The role of monetary policy remains dynamic and indispensable, regardless of evolving factors. Its significance in economic decision-making is enduring and essential. Key opportunities for enhancing its impact include the expansion of central banking, the core mandates that guide central banks, alignment with global policies, ongoing partnerships and collaborations, and strategic coordination with fiscal policy. These factors ensure that monetary policy continues to play a crucial role in promoting inclusive growth.

16.7.3.1 Increase in Central Banking

Unlike the 1980s, when central banks were limited to a few countries, almost every country now has a central bank. This widespread presence provides a significant opportunity to use monetary policy to promote inclusive growth, even in times of economic uncertainty. Central banks play a key role in shaping economic conditions by implementing policies such as adjusting interest rates, managing inflation and employing unconventional measures like quantitative easing. These tools help stabilise economies during recessions and financial crises by ensuring liquidity, reducing borrowing costs and encouraging investment. Additionally, central banks can target policies to support SMEs, enhance financial inclusion and stimulate investment in underserved areas. This approach helps distribute the benefits of growth more widely, reducing inequality. The presence of central banks also allows for better coordination with fiscal policies, creating a more cohesive strategy for promoting inclusive and sustainable development.

16.7.3.2 The Mandate or Objective of Central Banking

The primary mandate of central banking is to stabilise the economy by controlling inflation, managing employment and ensuring financial stability. This mandate provides a significant opportunity for using monetary policy to promote inclusive growth. Central banks influence economic conditions through tools such as regulating interest rates and controlling the money supply, which can make borrowing more affordable for businesses and consumers, thereby stimulating investment and supporting underdeveloped sectors. Central banks can also target monetary policies to support vital sectors like SMEs and underserved regions, reducing economic disparities. Additionally, by enhancing financial inclusion through accessible banking services and digital financial solutions, central banks can enable broader economic participation. Supporting sustainable development by directing resources toward green technologies further contributes to inclusive and long-term economic growth. Overall, central banks play a crucial role in fostering a balanced and equitable economic environment.

16.7.3.3 Global Policies

Global policies and agreements such as the Sustainable Development Goals (SDGs), the Paris Agreement and the Kyoto Protocol guide central banks and financial regulators in promoting inclusive growth, even during economic uncertainty. The SDGs aim for a more equitable and sustainable world by 2030, with targets like poverty eradication and quality education. Central banks and regulators align their policies with these goals to support broad-based development and reduce inequalities. Climate change frameworks stress the need for investments in green technologies and sustainable practices, which central banks can support through environmental, social and governance (ESG) criteria in monetary policy and financial regulation. Financial sustainability policies ensure long-term stability and resilience in the financial sector. By integrating these global mandates, central banks and regulators can stabilise the economy, support vulnerable sectors and enhance financial inclusion, creating a more resilient and equitable economic environment.

16.7.3.4 Continued Partnerships and Collaborations

Development partners, including Multilateral Development Banks (MDBs) and other development agencies, are essential in collaborating with central banks and financial regulators to implement monetary policies that foster inclusive growth, especially during economic uncertainty. Institutions like the World Bank, IMF and regional entities such as the African Development Bank (AfDB) provide crucial support through financial assistance, technical expertise and policy advice. Their involvement enhances the effectiveness of monetary policies and helps countries address complex economic challenges. For instance, during the 2008–2009 global financial crisis, the IMF offered financial support and guidance to stabilise economies, while the World Bank funded social safety nets and infrastructure projects to support vulnerable populations. Similarly, during the COVID-19 pandemic, the AfDB worked with the Central Bank of Kenya to manage economic impacts by providing emergency funding and technical assistance, helping address liquidity issues and supporting SMEs.[13] Other agencies like the United Nations Development Programme (UNDP) and the International Finance Corporation (IFC) also contribute by promoting financial inclusion, digital finance and sustainable investments. Their expertise aids central banks in creating policies that are inclusive and responsive to societal needs.

16.7.3.5 Fiscal Policy Prefer to Coordinate with Monetary Policy

The coordination between fiscal and monetary policies is crucial for implementing effective measures, especially during economic uncertainty. This alignment ensures that monetary and fiscal actions complement each other, enhancing

[13] https://www.afdb.org/en/news-and-events/press-releases/kenya-eu188m-african-development-bank-loan-boost-covid-19-response-35735. Accessed August 22 2024

their overall impact and contributing to economic stability and inclusive growth. During the COVID-19 pandemic, countries exemplified this coordination by integrating fiscal and monetary strategies to mitigate economic impacts. For instance, the European Union demonstrated effective policy coordination by combining expansive monetary measures from the ECB with significant fiscal interventions. The ECB implemented interest rate cuts and asset purchases to provide liquidity and stimulate economic activity, while the EU and member states introduced substantial fiscal measures, including the EU Recovery Fund and national stimulus packages.[14] This approach supported businesses, protected jobs and spurred economic activity during the crisis, showcasing how synchronised fiscal and monetary policies can effectively address economic challenges and facilitate recovery. By working together, fiscal and monetary authorities can amplify the effectiveness of their measures and foster a more resilient and inclusive economic environment.

16.8 Conclusion

Monetary policy is a fundamental instrument in shaping economic performance across diverse economies. This chapter explores the interplay between monetary policy and inclusive growth during periods of economic uncertainty. Initially, it defines monetary policy, outlines the tools employed and examines the role of CBI in policy implementation. It then delves into the challenges faced by central banks in executing monetary policy and assesses its impact on fostering inclusive growth amidst economic instability. The chapter highlights the evolution of monetary policy from its traditional focus on currency stability to broader objectives, including inflation control, unemployment management and the promotion of inclusive growth. Central banks have increasingly adopted a range of monetary tools, including unconventional measures such as quantitative easing and forward guidance, to address economic challenges and stabilise markets in the aftermath of global crises. The effectiveness of monetary policy in fostering inclusive growth is closely tied to central bank independence, which insulates policy decisions from short-term political influences and bolsters credibility in financial markets. The chapter identifies several key challenges associated with using monetary policy to enhance inclusive growth during economic uncertainty. These challenges include balancing inflation control with other macroeconomic objectives, managing policy uncertainty and addressing variability in the transmission mechanisms of monetary policy. Despite these obstacles, the chapter also highlights several opportunities presented by monetary policy, such as the development of advanced monetary tools, a heightened focus on financial inclusion, innovation in financial systems and improved collaboration and partnerships.

From a policy perspective, several implications emerge: Firstly, to effectively manage economic uncertainty and promote inclusive growth, central banks should maintain a high degree of independence from political pressures. This ensures that monetary policy decisions are made based on long-term economic stability rather than short-term political considerations. Also, policymakers should recognise and manage the impact of global economic interdependencies on domestic monetary policy. This includes coordinating with international monetary policies and adapting strategies to global financial trends. Additionally, the integration of technological innovations in financial systems can enhance the effectiveness of monetary policy. Embracing digital currencies and fintech solutions can improve access to credit and financial inclusion. Furthermore, effective monetary policy should be complemented by coordinated fiscal measures. Collaboration between monetary and fiscal authorities can help address economic uncertainties and achieve broader economic objectives. Lastly, central banks should focus on policies that enhance financial inclusion, particularly during times of economic instability. This includes supporting vulnerable firms and expanding access to financial services.

[14] https://www.ecb.europa.eu/pub/pdf/scpops/ecb.op273~fae24ce432.en.pdf. Accessed August 22 2024

References

Abor, J. Y., Agoba, A. M., Mumuni, Z., & Yawson, A. (2022). Monetary policy, central banks' independence, and financial development in Africa (pp. 227–267). https://doi.org/10.1007/978-3-031-04162-4_7

Abor, J. Y., Subramanian, L., Iddrisu, K., & Nsor-Ambala, R. (2024). Multilateral development banks: Contributions and challenges. In *Perspectives on development banks in Africa* (pp. 277–298). Springer Nature. https://doi.org/10.1007/978-3-031-59511-0_12

Agyemang, C. B., Agyei, F., Hansen-Garshong, R., Parimah, F., & Kekesi, E. (2024). Psychosocial consequences of job loss after the financial sector clean-up in Ghana. *Journal of Workplace Behavioral Health, 1–20*. https://doi.org/10.1080/15555240.2024.2316233

Ahiadorme, J. W. (2022). Monetary policy in search of macroeconomic stability and inclusive growth. *Research in Economics, 76*(4), 308–324. https://doi.org/10.1016/j.rie.2022.08.002

Ajayi, S. (1999). *Evolution and functions of central banks*. Economic and Financial Review

Ali, I., & Son, H. H. (2007). Measuring inclusive growth. *Asian Development Review, 24*(1), 11–31.

Ayyagari, M., Demirguc-Kunt, A., & Maksimovic, V. (2014). Who creates jobs in developing countries? *Small Business Economics, 43*(1), 75–99. https://doi.org/10.1007/s11187-014-9549-5

Ayyagari, M., Demirgüç-Kunt, A., & Maksimovic, V. (2017). *SME finance*. (8241; Policy Research Working Paper).

Bagehot, W. (1873). *Lombard street*. King.

Bank of Ghana. (2020). Banking Sector Report.

Barth, J. R., Li, L., Li, T., & Song, F. (2013). Policy issues of China's financial globalization. In *The evidence and impact of financial globalization* (pp. 355–365). Elsevier. https://doi.org/10.1016/B978-0-12-397874-5.00006-3

Berg, A., & Ostry, J. (2011a). Equality and efficiency: Is there a trade-off between the two or do they go hand in hand? *Finance and Development, 48*(3), 12–15.

Berg, A., & Ostry, J. (2011b). *Inequality and unsustainable growth: Two sides of the same coin?* (SDN/11/08; Staff Discussion Notes).

Bivens, J. (2024). Economic performance is stronger when democrats hold the White House.

Bordo, M. D. (2008). Monetary policy, history of. In *The new Palgrave dictionary of economics* (pp. 1–9). Palgrave Macmillan. https://doi.org/10.1057/978-1-349-95121-5_2743-1

Bordo, M., & Siklos, P. L. (2017). *Central Banks: Evolution and Innovation in Historical Perspective*. https://doi.org/10.3386/w23847

Carmichael, K. (2018, December 20). *Central Banks and the year of political interference*. Centre for International Governance Innovation. https://www.cigionline.org/articles/central-banks-and-year-political-interference/

Chang, D., & Jaffar, J. (2014). Monetary policy towards inclusive growth: The case of Korea (5/2014).

Cobham, D. (2023). *Monetary policy frameworks since Bretton Woods, across the world and its regions*. Review of World Economics. https://doi.org/10.1007/s10290-023-00517-1

Demirgüç-Kunt, A., & Singer, D. (2017). *Financial inclusion and inclusive growth: A review of recent empirical evidence*. (8040; World Bank Policy Research Working Paper).

Demirguc-Kunt, A., Klapper, L., Singer, D., & Ansar, S. (2018). *The Global Findex Database 2017: Measuring financial inclusion and the fintech revolution*. World Bank Publications.

EAF editors. (2024). *Japanese monetary policy under the economy's new normal*. https://doi.org/10.59425/eabc.1713168000.

ECB. (2023). *Monetary policy decisions*.

Eckhold, K., Faltermeier, J., King, D., Mak, I., & Petrov, D. (2024). *EMDE Central Bank Interventions during COVID-19 to support market functioning*. (WP/24/101; IMF Working Paper).

Frenkel, J. A. (1999). Central-Bank Independence and Monetary Policy. In *Central banking, monetary policies, and the implications for transition economies* (pp. 11–30). Springer. https://doi.org/10.1007/978-1-4615-5193-5_2

Garriga, A. C. (2016). Central Bank independence in the world: A new data set. *International Interactions, 42*(5), 849–868. https://doi.org/10.1080/03050629.2016.1188813

Georgiadis, G., & Zhu, F. (2019). *Monetary policy spillovers, capital controls and exchange rate flexibility, and the financial channel of exchange rates* (2267/April 2019).

Goodhart, C. (1988). *The evolution of central banks*. MIT Press.

Goodhart, C. A. E. (2011). The changing role of central banks. *Financial History Review, 18*(2), 135–154. https://doi.org/10.1017/S0968565011000096

Goshit, G. G. (2015). Monetary policy and inclusive growth in Nigeria: Theoretical issues, challenges and prospects. *Journal of Economics and Sustainable Development, 6*(20).

Habito, C. F. (2009). *Patterns of inclusive growth in developing Asia: Insights from an enhanced growth-poverty elasticity analysis* (issue 145). Asian Development Bank Institute (ADBI). https://hdl.handle.net/10419/53607

Iddrisu, K. (2024). Foreign bank presence and income inequality in Africa: What role does economic freedom play? *Future Business Journal*. https://doi.org/10.1186/s43093-024-00357-x

Iddrisu, K., Abor, J. Y., & Banyen, K. T. (2022). Fintech, foreign bank presence and inclusive finance in Africa: Using a quantile regression approach. *Cogent Economics & Finance, 10*(1). https://doi.org/10.1080/23322039.2022.2157120

Iddrisu, K., Doku, J. N., Abor, J. Y., & Dziwornu, R. (2023a). Financial inclusion and inclusive growth in Africa: What is the moderation role of financial stability? *Cogent Economics & Finance, 11*(2). https://doi.org/10.1080/23322039.2023.2267857

Iddrisu, K., Ofoeda, I., & Abor, J. Y. (2023b). Inward foreign direct investment and inclusiveness of growth: Will renewable energy consumption make a difference? *International Economics and Economic Policy, 20*(3), 367–388. https://doi.org/10.1007/s10368-023-00562-z

Iddrisu, K., Abor, J. Y., & Banyen, K. T. (2023c). Foreign bank presence and inclusive growth in Africa: The moderating role of financial development. *African Journal of Economic and Management Studies*. https://doi.org/10.1108/AJEMS-11-2022-0444

Iddrisu, K., Abor, J. Y., & Banyen, K. T. (2023d). The effects of foreign bank presence on financial development in Africa: The role of institutional quality. *Review of Financial Economics*. https://doi.org/10.1002/rfe.1183

IMF. (2023, January). *Monetary policy and central banking*. https://www.imf.org/en/About/Factsheets/Sheets/2023/monetary-policy-and-central-banking

Kakwani, N., & Pernia, E. M. (2000). What is pro-poor growth? *Asian Development Review, 18*(01), 1–16. https://doi.org/10.1142/S0116110500000014

Khera, P. (2023). India's financial system. In A. Schipke, J. Turunen, N. Choueiri, & A. M. Gulde (Eds.), *India's financial system building the foundation for strong and sustainable growth*. International Monetary Fund. https://doi.org/10.5089/9798400223525.071

Klasen, S. (2010). *Measuring and monitoring inclusive growth: Multiple definitions, open questions, and some constructive proposals*. (ADB Sustainable Development Working Paper Series).

Kuznets, S. (1955). Economic growth and income inequality. *The American Economic Review, 45*(1), 1–28.

Li, S. (2016). The currency exchange market in East Asia. In *East Asian business in the New World* (pp. 95–102). Elsevier. https://doi.org/10.1016/B978-0-08-101283-3.00007-5

Meltzer, A. H. (2011). Politics and the fed. *Journal of Monetary Economics, 58*(1), 39–48. https://doi.org/10.1016/j.jmoneco.2010.09.009

Ocampo, J. A. (2017). A brief history of the international monetary system since Bretton woods. In *Resetting the international monetary (non)system* (pp. 1–44). Oxford University PressOxford. https://doi.org/10.1093/oso/9780198718116.003.0001

Pérez-Moreno, S., Martín-Fuentes, N., & Albert, J.-F. (2022). Rethinking monetary policy in the framework of inclusive and sustainable growth. In *Economic policies for sustainability and resilience* (pp. 319–364). Springer International Publishing. https://doi.org/10.1007/978-3-030-84288-8_8

Poloz, S. S. (2021). Technological progress and monetary policy: Managing the fourth industrial revolution. *Journal of International Money and Finance, 114*, 102373. https://doi.org/10.1016/j.jimonfin.2021.102373

Ranieri, R., & Almeida Ramos, R. (2013). Inclusive growth: Building up a concept (104).

Rocheteau, G., Wright, R., & Xiaolin Xiao, S. (2018). Open market operations. *Journal of Monetary Economics, 98*, 114–128. https://doi.org/10.1016/j.jmoneco.2018.04.012

Rostow, W. W. (1959). The stages of economic growth. *The Economic History Review, 12*(1), 1–16.

Rostow, W. W. (1960). *The five stages of growth—A summary*. Cambridge Press.

Thornton, H. (1802). *An enquiry into the nature and effects of the paper credit of Great Britain*. George Allen and Unwin.

Tinggui, C. (2001). The People's Bank of China and its monetary policy (14).

Vu, A. N. (2020). On the impact of quantitative easing on credit standards and systemic risk: The Japanese experience. *Economics Letters, 186*, 108520. https://doi.org/10.1016/j.econlet.2019.07.005

Yien, L. C., Abdullah, H., & Azam, M. (2017). Monetary policy inclusive growth: Empirical evidence from Malaysia. *International Journal of Academic Research in Accounting, Finance and Management Sciences, 7*(1). https://doi.org/10.6007/IJARAFMS/v7-i1/2634

17 The Effect of Monetary Policy and Global Economic Policy Uncertainty on Bank Credit Risk

Mubarik Salifu, Mohammed Gbanja Abdulai, William Angko, Paul Bata Domanban, Ibrahim Nandom Yakubu, and Saani Mohammed Ridwan

17.1 Introduction

Over the years, there has been significant interest in understanding banks' motivations for engaging in risky behavior due to their vital role in the overall stability of the financial system. Past studies have attributed the risk profiles of banks to various factors, such as manager compensation (DeYoung et al., 2013), government bailout policies (Gropp et al., 2011), and the presence of deposit insurance (Hovakimian & Kane, 2000). More recently, scholars have increasingly focused on monetary policy and economic policy uncertainty (Berger et al., 2018; Bordo et al., 2016) as potential catalysts for banks' inclination to assume risk. Building on this literature, our study investigates whether monetary policy and global economic policy uncertainty influence bank credit risk in Ghana.

In the wake of the 2007/08 financial crisis, considerable attention has been drawn to the potential impact of monetary policy on bank risk-taking incentives. It has been claimed that accommodative federal monetary policies led to increased risk-taking among financial institutions (Borio & Zhu, 2008; Diamond & Rajan, 2009). Earlier studies on the relationship between interest rates and the risk-taking behavior of financial institutions argue that interest rates can increase banks' appetite for risk in several ways. Rajan (2006) suggested that in a "search for yield," financial institutions might make riskier investments to achieve substantial nominal rates of return during periods of low-interest rates. Adrian and Shin (2009) noted that interest rates affect the valuation of financial institutions, impacting their risk behavior. For instance, low-interest rates can inflate collateral and asset values, reducing perceived risk and encouraging more aggressive investment strategies. Dell'Ariccia and Marquez (2013) suggested that if adverse selections are reduced due to low-interest rates, banks' incentives to screen loan applicants diminish, leading to extending credit to riskier borrowers.

M. Salifu (✉) · S. M. Ridwan
Department of Economics, School of Economics, University for Development Studies, Tamale, Ghana
e-mail: smubarik@uds.edu.gh; msaani@uds.edu.gh

M. G. Abdulai
Department of Economic Studies, School of Economics, University of Cape Coast, Cape Coast, Ghana

W. Angko
Department of Economics, SD Dombo University of Business and Integrated Development Studies, Wa, Ghana
e-mail: wangko@ubids.edu.gh

P. B. Domanban
Department of Development Studies, SD Dombo University of Business and Integrated Development Studies, Wa, Ghana
e-mail: pdomanban@ubids.edu.gh

I. N. Yakubu
School of Business, University for Development Studies, Tamale, Ghana
e-mail: ibrahim.yakubu@uds.edu.gh

I. N. Yakubu (ed.), *Strategic Approaches to Banking Business and Sustainable Development Goals*, Sustainable Development Goals Series, https://doi.org/10.1007/978-3-031-80744-2_17

These insights highlight how monetary policy can shape banks' risk-taking behaviors, which is crucial for understanding the dynamics of financial stability.

Alongside monetary policy, economic policies—including trade, government expenditure, tax regulations, and fiscal strategies—also significantly influence financial institutions, especially in environments characterized by high uncertainty (Danisman et al., 2020). The lack of clarity, ambiguity, and unpredictability of these policies often creates disturbances (Chi & Li, 2017; Louhichi & Boujelbene, 2017). Alessandri and Bottero (2020) argue that during periods of uncertainty, banks prefer to delay investments to avoid sensitivity to interest rate fluctuations, which likely reduces credit volumes (Hu & Gong, 2019). Additionally, uncertainty prompts households to reduce expenditures due to unpredictable future cash flows and financial situations, while firms may cancel or defer investments, waiting for more accurate information. This behavior can exacerbate economic downturns, leading to reduced lending and increased credit risk for banks. Understanding these dynamics is essential for policymakers and financial institutions aiming to navigate uncertain economic environments effectively.

Moreover, inclusive growth has emerged as a key consideration in the broader context of economic development and financial stability. Inclusive growth is critical for sustainable economic development, aiming to ensure that the benefits of growth are equitably distributed across society. It involves improving access to quality education, healthcare, financial services, and employment opportunities for all population segments, especially the marginalized (Rauniyar & Kanbur, 2010). In developing economies, inclusive growth can reduce income inequality, enhance social cohesion, and foster a more stable economic environment. For banks, inclusive growth means expanding access to credit and financial services to underserved communities, contributing to overall economic stability and reducing systemic risk (Demirgüç-Kunt & Klapper, 2013). By supporting inclusive growth, banks can play a crucial role in fostering a more resilient financial system that benefits a broader segment of society. For instance, targeted financial inclusion initiatives can help integrate marginalized groups into the formal economy, enhancing their economic opportunities and reducing poverty (Beck et al., 2007). This perspective underscores the importance of examining the broader economic implications of financial policies and practices.

Following Ghana's implementation of inflation targeting, the central bank has primarily employed monetary policy to control the money supply and stabilize prices. Through the interest rate channel, the central bank sets the monetary policy rate, influencing the interest rates that commercial banks charge for loans. Reserve requirements are used to determine the economy's liquidity level (Pérez Juanazo & Ruíz Martínez, 2016). Changes in interest rates impact the valuations of banks' assets and liabilities and are significant factors in determining banks' profitability (Sun et al., 2017). Therefore, it is anticipated that any modifications to the monetary policy stance will affect banks' propensity for taking on risk (Borio & Zhu, 2012; Maddaloni & Peydró, 2012; Ioannidou et al., 2015). Understanding these mechanisms is vital for comprehensively analyzing the factors influencing bank risk-taking behavior.

Despite the importance of these issues, there has been limited research in Ghana analyzing the empirical relationship between monetary policy, international economic policy, and bank credit risk. Most research has focused on macroeconomic and bank-specific factors contributing to non-performing loans (Asare-Bekoe, 2010; Kwashie, 2013; Adusei et al., 2014; Asamoah & Takyiwaa, 2015; Odonkor et al., 2016; Amuakwa-Mensah & Boakye-Adjei, 2015; Amuakwa-Mensah & Marbuah, 2017). Only a few studies (Asiama & Amoah, 2019) examine the dynamics of monetary policy and non-performing loans in Ghana. However, the impact of uncertain global economic policies on credit risk has not received sufficient attention. Addressing this gap is critical for developing a more nuanced understanding of the factors influencing bank risk behavior in developing economies.

Our study makes several key contributions to the literature. First, we enhance the understanding of the relationship between monetary policy and bank risk-taking using recent data from Ghana. Second, we provide evidence of the connection between bank risk-taking and the uncertainty surrounding global economic policy. Although this link has been studied in developed economies, there is limited evidence of the relationship in developing countries. By investigating whether monetary policy and global economic policy uncertainty influence bank credit risk in Ghana, this paper contributes to the literature on bank risk behavior. Policymakers and bank regulators will benefit from the study's findings, gaining insights into the factors influencing banks' risk-taking practices and helping to develop more effective policies to mitigate the risks associated with bank failures. By understanding these dynamics, policymakers can craft strategies that balance financial stability with economic growth, ensuring a more robust banking sector.

This chapter is organized as follows: Section 17.2 presents recent empirical evidence on the relationship between monetary policy, global economic policy uncertainty, and bank credit risk. Section 17.3 outlines the methods employed in the analysis. Sections 17.4 and 17.5 discuss the results and provide recommendations, respectively.

17.2 Review of Literature

The increase in attention toward the effect of monetary policy on financial risk after the 2007/08 crisis stands in stark contrast to the almost complete lack of any mention of risk in the earlier body of literature on the transmission of monetary policy. Also, recently, attention has shifted to how external shocks including uncertainties could affect financial stability. The economic policies implemented by the government frequently bewilder individuals engaged in business operations or compel governments to adopt an opposing stance to the original objective of a policy due to their lack of transparency, ambiguity, and unpredictability. Consequently, this gives rise to the creation of shocks of uncertainty. Relating to the risk-taking channel, it has been suggested that monetary policy is transmitted through a risk channel (De Nicolò et al., 2010; Adrian & Shin, 2009; Borio & Zhu, 2012; Morris & Shin, 2014; Afanasyeva & Güntner, 2020; González-Aguado & Suarez, 2015; Cesa-Bianchi & Rebucci, 2017). It has been identified that these channels include the search for yield (Dell'Ariccia et al., 2011), risk shifting (Bruno & Shin, 2015), and distortion of incentives otherwise known as the agency problem (Acharya & Naqvi, 2012).

Demirgüç-Kunt and Klapper (2013) found that financial inclusion efforts, such as the proliferation of microfinance institutions, can significantly reduce the credit risk of banks. By diversifying their loan portfolios to include a wider range of borrowers, banks can reduce their exposure to systemic risks associated with lending to a narrow segment of the population. This diversification can enhance the resilience of banks to economic shocks, thereby lowering overall credit risk. Beck et al. (2007) argue that financial inclusion, a key component of inclusive growth, can lower poverty levels and mitigate income inequality. By integrating more individuals and businesses into the formal financial system, inclusive growth can create a broader base of economically active participants, which in turn can lead to more stable and diversified sources of income for banks. Inclusive growth also involves promoting financial literacy, which can have a direct impact on bank credit risk. Xu and Zia (2012) emphasize that financially literate individuals are better equipped to make informed borrowing decisions, manage their finances effectively, and avoid over-indebtedness. By improving financial literacy, inclusive growth can reduce the incidence of default and delinquency, thereby lowering credit risk for banks. Empirical studies in developing economies provide mixed evidence on the effect of inclusive growth on bank credit risk. For example, in India, Sahay et al. (2015) found that regions with higher financial inclusion experienced lower non-performing loan (NPL) ratios, suggesting a positive impact of inclusive growth on credit risk. However, in some

African countries, rapid financial inclusion efforts have led to concerns about rising credit risk due to inadequate borrower screening and financial literacy (Cull et al., 2014).

There have been numerous noteworthy empirical additions to the corpus of work on the risk-taking channel, with the majority of these contributions providing substantiated evidence in support of the presence of this channel (Dell'Aricia et al., 2017; Morais et al., 2015; Paligorova & Santos, 2017). Generally, these studies indicate an inverse relation between the risk and the monetary policy rates' level associated with the loan portfolio of the banking institutions. However, the data pertaining to the United States seem to be somewhat weaker. In the medium run, low policy rates considerably increase the risk associated with banks' loan portfolios, while in the near term, they minimize it (Delis et al., 2017). On the other hand, Buch et al. (2014) find no evidence in favor of the banking system as a whole, but they do observe some notable differences across different kinds of banks.

Bonfim and Soares (2018) also examined the risk-taking channel using various methods. Their results give interesting insights into the risk-taking channel. Overall, their study's results suggest that a short-term drop in policy interest rates lowers the total credit risk that the banking industry faces. This is primarily due to the positive effect it has on borrowers' ability to repay their outstanding loans. However, over the medium term, the increased willingness to take on higher levels of risk may eventually result in a decline in the quality of banks' assets. This is particularly evident in situations where a severe contraction in monetary policy or an economic downturn occurs after a period of low policy interest rates. When the interest rates of monetary policy are decreased, there is an increased probability of banks providing loans to borrowers who possess higher levels of ex-ante risk. Nevertheless, the overall level of exposure to current borrowers with existing risk does not observe any rise. When examining the performance of loans over time, it becomes evident that loans that were granted during periods of low monetary policy rates are not generally more prone to default in the ex-post period. Nevertheless, a notable exception emerges when interest rates rise after a prolonged period of being low. In such instances, loans that were granted during low-interest rate periods are markedly more inclined to default when monetary policy tightens due to abrupt shifts in expectations. About the diversity of banks, it is discovered that banks with less capital exhibit stronger tendencies toward risk-taking behaviors. Research indicates that banks that are closer to the minimal regulatory ratios are more inclined to take on risk because they do not fully understand the possible repercussions of doing so.

Angeloni et al. (2015) also examined the monetary policy and the risk-taking channel by developing a model that combines two sources of risk (assets and liability). On the liability side, banks assume the role of relationship lenders, working on behalf of external financiers and acquiring funds through short-term liabilities, which are bound by a service constraint, as well as capital. As outlined in the contractual agreement between the bank and the external investors, instances of short-term liability run occur endogenously as a means of maintaining discipline. However, when the risk-free rate decreases because of the expansionary monetary policy, banks tend to expand their short-term liabilities (which become more affordable), subsequently increasing the likelihood of a run on said liabilities occurring. On the asset side, loans are extended to entrepreneurs to initiate investment projects, with financing provided partially through internal funds, partially through the intermediary, and partially through corporate loans. The return on these loans is susceptible to idiosyncratic shocks, which are exclusively observable by the entrepreneurs. Ex-post, the bank can observe the impact of these shocks upon payment of a monitoring cost. Their findings suggest that risks on the asset side and the liability side of banks tend to exhibit a mutual movement and mutually strengthen one another: a decrease in the policy rate stimulates a higher number of companies to necessitate bank financing; the augmented scope of investment ventures, which necessitate funds, along with the diminished expense of short-term obligations, compels banks to amplify their leverage, thereby augmenting the overall probability of a bank run.

Rajan (2006) revealed how the incentive structures within the financial system may prompt asset allocation managers to assume greater levels of risk in the presence of persistently low-interest rates. Within an environment characterized by low-interest rates, portfolio managers who receive compensation based on nominal returns are motivated to seek out higher yields by engaging in riskier investments. The accumulation of risk during periods of monetary accommodation transforms into instability when policy measures are tightened, thereby giving rise to crises of confidence and potentially leading to sudden halts in credit. There are two notable implications for central banks as a result of these findings: first, the monetary policy ought to avoid extended periods of excessively low-interest rates. Second, in cases where high levels of risk are already deeply entrenched within the financial sector, sudden and significant policy tightening can have highly contractionary effects or even cause destabilization.

According to Borio and Zhu (2012), banks may be required to keep a greater proportion of riskier assets to achieve the fixed or targeted level of returns if expansionary monetary policy lowers the return on safe assets. In the meantime, the return above the banks' minimal aim may potentially have an impact on bank managers' income. The "search for yield" theory is consistently supported by empirical data by Buch et al. (2014), especially in small banks during expansionary monetary shocks. Research by Valencia (2014), Dell'Ariccia et al. (2014), and Angeloni et al. (2015) shows that when interest rates drop, bank risk and leverage rise. According to the theory put forth by De Nicolò et al. (2010), lower interest rates might increase the value of a bank's franchise by boosting earnings, which would reduce moral hazard and encourage less risk-taking. According to Drakos et al. (2016), foreign banks—that is, banks with their headquarters abroad—do not take more risks as a result of monetary policy changes. The economic policies implemented by the government frequently bewilder individuals involved in commercial endeavors or prompt governments to adopt a stance contrary to the policy's initial objective due to their lack of transparency, ambiguity, and unpredictability, thereby creating disturbances of uncertainty. This uncertainty is referred to as economic policy uncertainty (EPU).

As shown by Bordo et al. (2016), economic policy uncertainty (EPU) has a negative effect on credit growth, especially when it comes to the rise of commercial and industrial loans in the United States. Using Italian loan application data, Alessandri and Bottero (2020) investigate how uncertainty affects credit availability. According to their findings, an increase in EPU lowers the acceptance rate of credit applications and makes it more difficult for money to enter the market on time. According to Chi and Li (2017), EPU causes an increase in loan concentrations, non-performing loan ratios, and the typical loan migration rate. As a result, loan amounts decrease and credit risks increase. According to Berger et al. (2018), during uncertain times, banks tend to hoard liquidity as a protective mechanism, which results in a bigger cash reserve and a restricted availability of credit to both enterprises and consumers. Hu and Gong (2019) examine closely how EPU affects the expansion of bank lending in 19 significant economies. According to their analysis, EPU causes bank credit growth to decline; however, the degree of this effect differs throughout institutions. Larger and riskier banks bear the brunt of the negative effects, while institutions with greater liquidity and diversity fare better.

A decline in the availability of business loans is the result of an exogenously generated increase in EPU, as demonstrated by Barraza and Civelli (2020). This decrease is ascribed to a decreased demand for funds, indicating that EPU shocks impact the economy by causing the supply of money to the business sector to contract. Nguyen et al. (2020) apply feasible generalized least squares and panel-corrected standard errors to show that a higher level of domestic EPU has a dampening effect on bank credit growth. Interestingly, nevertheless, variations in EPU volatility have a positive impact on the expansion of bank lending. Larger, better capitalized, and/or riskier banking systems see a reduction in the negative effect of EPU on credit expansion, but this effect is more noticeable in highly liquid and/or lucrative banking systems. Moreover, it is seen that worldwide EPU has adverse effects on the expansion of credit. The negative impacts of EPU are more severe in emerging economies than in developed ones. Using national-level data from eighteen countries, Caglayan and Xu (2019) investigate the effects of EPU on credits, loan-loss provisions, and non-performing loans. Their data indicates that credit levels and uncertainty are negatively correlated. In particular, uncertainty raises the ratio of loan-loss provisions and gross loans to non-performing loans.

The literature has focused on the relationship between the uncertainty of global economic policy and monetary policy with much of this centered in developed countries and little on the effect of inclusive growth. Severally, the risk-taking channel discussed in the literature has been tested also in developed countries. However, little is known of the risk-taking channel in developing countries and the effect of global economic policy uncertainty on bank risk-taking especially in Ghana. Thus, this study looks at how Ghanaian bank credit risk is affected by monetary policy and uncertainties surrounding global economic policy. We set the following hypothesis.

H_1: Monetary policy affects bank credit risk both in the long run and short run.

H_2: Global Economic policy uncertainty affects bank credit risk both in the long run and short run.

17.3 Methodology

17.3.1 Model Specification and Data Sources

Our sample includes quarterly aggregated data of banks collected from Q1 2010 to Q4 2022. To understand the short-run and long-run dynamics in the relationship between monetary policy, global economic policy uncertainty, and bank credit risk, employing quarterly data is ideal. The variable on global economic policy uncertainty is sourced from the index created by Ahir et al. (2022), following Baker et al. (2016), and hosted on the policy uncertainty website. Other macroeconomic variables used in the study, such as GDP growth, inflation, real exchange rate, and the monetary policy rate, are all sourced from the Bank of Ghana's time series database.

17.3.2 Measurement of Variables

Credit risk: This is the dependent variable measured using the proxy non-performing loan ratio. Credit risk may arise from a debtor failing to make prompt repayments of interest on loans and the principal at agreed periods. This measure indicates that banks have ex-ante taken a higher lending risk and therefore in the process have accumulated a higher amount of bad loans on their balance sheet. To measure the credit risk, we use the ratio of non-performing loans to total loans issued by the bank. It is expected that the risk level as measured by NPLR will be affected by monetary policy stances and bank-specific factors as indicated in the works of Klein (2013).

Monetary Policy Rate Inflation targeting is the framework under which Ghana's central bank conducts monetary policy, with the monetary policy rate (MPR) serving as the primary policy tool for enforcing monetary policy and anchoring inflation expectations in the economy. In this research, the major instrument of interest is the monetary policy rate. It is the best option for this study since it indicates the economy's interest rates (Fiador & Biekpe, 2015). The monetary policy rate is expected to have a negative impact on the amount of risk that banks take on. A loosening of monetary policy is projected to lower commercial banks' borrowing costs. This will induce them to increase their level of loan extension and thereby increasing their credit risk.

Global Economic Policy Uncertainty We measure global economic policy uncertainty using the index developed by Baker et al. (2016). The index is constructed as a weighted average of three different components. The primary and foremost factor is determined through a count of print media articles encompassing crucial phrases associated with the state of uncertainty in policies. The second component analyzes the uncertainty surrounding future changes in tax regulations by evaluating the financial impact of expiring tax clauses. The final component employs the distribution of economic forecasts of the Consumer Price Index and government expenditures as a substitute measure to assess the level of uncertainty regarding financial and monetary policies (Gulen & Ion, 2016). It is expected that global economic policy uncertainty will make lending institutions and commercial banks cautious and thus tightening their lending criteria. It is thus expected that GEPU will negatively influence credit risk.

Capital Adequacy Ratio This is measured based on bank risk and this is maintained according to the regulations set by the central regulatory body of commercial banks. The risk-based capital ratio, which is defined as the ratio of capital to risk-weighted assets, is used to calculate this capital. The study employs the capital adequacy ratio as a measure of regulations due to its connection with the Basel I, II, and III accords. This is computed as:

$$\mathrm{CAR} = \frac{(\mathrm{Tier1capital} + \mathrm{Tier2capital})}{\mathrm{Risk-weighted\,assets}} \quad (17.1)$$

Ordinary shares capital, equity capital, intangible assets, and audited revenue reserves make up Tier 1 capital, often known as core capital. This is capital that banks may quickly use to protect themselves against losses without having to halt operations. Unaudited retained profits, unaudited reserves, and general loss reserves make up Tier 2 capital. This capital is necessary to sustain losses if a bank goes bankrupt.

Inflation This is the quarterly change in the Consumer Price Index (CPI). This variable captures the rate at which the general level of prices for goods and services is rising, affecting the purchasing power of money.

Return on Equity (ROE) The performance of banks is proxied using the return on equity (ROE). It is expected that higher ROE will be associated with higher risk, explained by the return-risk trade-off. Smithson and Simkins (2005) suggest that higher risks are linked to higher profitability, based on their use of both ROA and ROE as proxies for profitability.

Economic Growth GDP per capita growth is an important indicator of a country's investment activity. In this study, the rate of GDP per capita growth per quarter is used. The Gross Domestic Product (GDP) provides a reliable measure of economic activity, income, and demand within a country. It is expected that a higher GDP per capita growth rate will positively affect bank credit risk.

17.3.3 Model Specification

Assuming that bank credit risk is a function of monetary policy, global economic policy uncertainty, and other control variables, we can specify the empirical model as:

$$Y_t = \alpha 0 + \beta_1 \mathrm{MP}_t + \beta_2 \mathrm{GEPU}_t + \beta_3 \mathrm{INF}_t + \beta_4 \mathrm{CAR}_t + \beta_5 \mathrm{ROE}_t + \beta_6 \mathrm{GROWTH}_t + \beta_7 \mathrm{REER}_t + \varepsilon_t \quad (17.2)$$

Where Y_t represents credit risk (non-performing loan ratio). MP, GEPU, INF, CAR, ROE, GROWTH, and REER represent the Monetary Policy Rate, Global Economic Policy Uncertainty, Inflation Rate, Risk-Adjusted Capital Adequacy Ratio, Return on Equity, GDP Growth Rate, and Real Exchange Rate, respectively. t is the time period and ε_t represents the error term.

17.3.4 Model Estimation

To determine the relationship between monetary policy, global economic policy uncertainty, and bank credit risk, we use the ARDL bounds testing approach to cointegration. The ARDL methodology employs a single reduced-form equation to investigate the cointegration of the variables, in contrast to the conventional Johansen test, which utilizes a system of equations. Moreover, an additional advantage is its robustness with small sample sizes. Other cointegration approaches often have restrictive assumptions concerning the order of cointegration (Solarin et al., 2013). Another benefit of employing the ARDL method for cointegration is its independence from the necessity for the underlying variables to possess a similar degree of integration, such as being integrated of order zero I(0), integrated of order one I(1), or even fractionally. Additionally, it provides unbiased estimations of the long-term model, even in scenarios where certain variables may be endogenous (Odhiambo, 2009). Following Dingela and Khobai (2017), we specify our ARDL model as follows:

$$\begin{aligned}\Delta Y_t &= \gamma_0 + \sum_{i=1}^{n}\gamma_{1i}\Delta Y_{t-1} + \sum_{i=1}^{n}\gamma_{2i}\Delta \text{MP}_{t-1} + \sum_{i=1}^{n}\gamma_{3i}\Delta \text{GEPU}_{t-1} + \\ &\sum_{i=1}^{n}\gamma_{4i}\Delta \text{INF}_{t-1} + \sum_{i=1}^{n}\gamma_{5i}\Delta \text{CAR}_{t-1} + \sum_{i=1}^{n}\gamma_{6}\Delta \text{ROE}_{t-1} + \sum_{i=1}^{n}\gamma_{7i}\Delta \text{GROWTH}_{t-1} + \sum_{i=1}^{n}\gamma_{8i}\Delta \text{REER}_{t-1} + \lambda_1 Y_{t-1} + \\ &\lambda_2 \text{MP}_{t-1} + \lambda_3 \text{GEPU}_{t-1} + \lambda_4 \text{INF}_{t-1} + \lambda_5 \text{CAR}_{t-1} + \lambda_6 \text{ROE}_{t-1} + \lambda_7 \text{GROWTH}_{t-1} + \lambda_8 \text{REER}_{t-1} + \varepsilon_{1t}\end{aligned} \tag{17.3}$$

To determine the long-run cointegrating relationship, we test the null hypothesis of no cointegration. We compare the computed F-statistic against the critical values provided by Pesaran et al. (2001). The null hypothesis is stated as follows:

$$H_0 : \lambda_1 = \lambda_2 = \lambda_3 = \lambda_4 = \lambda_5 = \lambda_6 = \lambda_7 = \lambda_8 \tag{17.4}$$

We reject the null hypothesis of no cointegration when the computed F-statistic is greater than the critical value of the upper bounds and fail to reject the null hypothesis if it is otherwise.

$$H_0 : \lambda_1 \neq \lambda_2 \neq \lambda_3 \neq \lambda_4 \neq \lambda_5 \neq \lambda_6 \neq \lambda_7 \neq \lambda_8 \tag{17.5}$$

The short-run error correction model is then specified as:

$$\begin{aligned}\Delta Y_t &= \gamma_0 + \sum_{i=1}^{n}\gamma_{1i}\Delta Y_{t-1} + \sum_{i=1}^{n}\gamma_{2i}\Delta \text{MP}_{t-1} + \sum_{i=1}^{n}\gamma_{3i}\Delta \text{GEPU}_{t-1} + \\ &\sum_{i=1}^{n}\gamma_{4i}\Delta \text{INF}_{t-1} + \sum_{i=1}^{n}\gamma_{5i}\Delta \text{CAR}_{t-1} + \sum_{i=1}^{n}\gamma_{6}\Delta \text{ROE}_{t-1} + \sum_{i=1}^{n}\gamma_{7i}\Delta \text{GROWTH}_{t-1} \\ &+ \sum_{i=1}^{n}\gamma_{8i}\Delta \text{REER}_{t-1} + \phi_1 \text{ECM}_{t-1} + \varepsilon_{1t}\end{aligned} \tag{17.6}$$

The expectation is that ϕ_1 is negative and statistically significant, indicating that short-run disequilibria are easily corrected to equilibrium in the long run.

17.4 Results and Discussion

17.4.1 Descriptive Statistics

Table 17.1 shows the descriptive statistics of the variables used in the study. The non-performing loan ratio averaged 16.068 percent per quarter between 2010 and 2022. Monetary policy during the same period averaged 17.71 percent per quarter, with a maximum of about 25.167 percent. Global economic policy uncertainty averaged 183.67, while the GDP growth rate averaged 10.508 percent per quarter. The banks' return on equity during the period was quite high, averaging 22.451 percent per quarter, with the capital adequacy ratio also averaging about 18.36 percent. The average quarterly inflation during the period was 13.282 percent, reaching a maximum of about 48.267 percent.

17.4.2 Unit Root Test

Table 17.2 presents the results of the Dickey-Fuller unit root tests. We find that the variables are either significant at $I(0)$ or $I(1)$. Since the variables have different degrees of integration, using the ARDL bounds test for cointegration is justified for variables that are $I(1)$.

17.4.3 Coefficient Symmetry Test

To justify the use of a linear ARDL estimation approach, we estimate a non-linear ARDL and test for the presence of an asymmetric relationship in monetary policy in the long run, short run, and a joint asymmetric relationship. We test this against the null hypothesis of a symmetric relationship existing between non-performing loans and the monetary policy rate. The results of the coefficient symmetry test in Table 17.3 indicate that there is a symmetrical relationship between non-performing loans and monetary policy both in the long run and in the joint test. Therefore, we are unable to use a non-linear ARDL approach.

Table 17.1 Descriptive statistics

Variable	Mean	Std dev	Minimum	Maximum
NPL	16.068	3.051	11.31	22.927
Policy rate	17.71	4.18	12.5	26.167
CAR	18.36	1.676	15.063	21.483
ROE	22.451	4.776	14.757	32.623
GROWTH	10.508	14.311	−5.523	62.513
INFLATION	13.282	6.962	7.800	48.267
GEPU	183.67	68.287	97.028	370.154
REER	76.656	9.971	58.623	99.91

Source: Authors' Computation

Table 17.2 Unit root test using ADF

Variables	Levels	First difference
NPL	−3.819**	−4.913***
Policy rate	−2.272	−3.556**
CAR	−1.903	−5.999***
ROE	−2.066	−8.247***
GROWTH	−3.874**	−5.069***
INFLATION	−0.751	−4.925***
GEPU	−1.478	−7.482***
REER	−1.365	−5.635***

Source: Authors' Computation

*** and ** are statistical significance at 1% and 5% respectively

Table 17.3 NARDL Coefficient Symmetric test

Period	Test	Test statistic	*P*-value
Long run	*F*-statistic	0.988	0.3369
	Chi-square	0.988	0.3201
Short run	*F*-statistic	3.469	0.0836
	Chi-square	3.469	0.0625
Joint test	*F*-statistic	1.973	0.1758
	Chi-square	3.946	0.1390

Source: Authors' Computation

Table 17.4 Bounds test cointegration test

F-statistic	Significance level	*I*(0)	*I*(1)
6.1530	10%	1.92	2.80
	5%	2.17	3.21
	1%	2.73	3.90

I(0) and *I*(1) are the stationary and non-stationary bounds, respectively.

Source: Authors' Computation

17.4.4 Bounds Test Cointegration Test

The study used the ARDL bounds test approach to determine the presence of cointegration among the variables. The results in Table 17.4 indicate that we reject the null hypothesis of no cointegration at the 1% significance level. The *F*-statistic value of 6.1530 exceeds the critical value of 3.90 at the 1% significance level. Given the presence of a non-degenerate cointegrating relation, it is possible to identify the normalized long-term coefficients corresponding to each of the distributed lag variables within the cointegrating space. This suggests that there is a long-run linear relationship between the explanatory variables and non-performing loans.

17.4.5 Long-Run Effect of Monetary Policy Rate, Inclusive Growth, and Global Economic Policy Uncertainty on Bank Credit Risk

Results from Table 17.5 indicate the cointegrating relationship between monetary policy, global economic policy uncertainty, and bank credit risk. As indicated in Table 17.5, only the coefficient of the real exchange rate is insignificant in the long run; all other variables are significant. The results suggest that the monetary policy rate is a significant determinant of bank credit risk, as there is a positive and significant long-run relationship between the variables. In the long run, a 1 percentage point increase in the monetary policy rate raises bank credit risk by approximately 0.829 percent. Contractionary monetary policy increases the interest rate on both new and existing loans. This rise in the interest rate results in higher credit costs, which can lead to payment defaults on existing loans and significantly contribute to an increase in the non-performing loan ratio. The study's outcome aligns with the conclusions of Mahrous et al. (2020), who find that monetary policy positively influences credit risk.

Global economic policy uncertainty also exhibits a significant negative long-run relationship with credit risk. The result indicates that a 1-unit increase in the index for global economic policy uncertainty should reduce the non-performing loan ratio by about 0.019 percent in the long run. Uncertainty about global policy could lead financial and lending institutions to be more cautious about the profiles of borrowers when extending credit. Financial institutions may become stricter in borrower screening, which could, in the long run, reduce the ratio of non-performing loans to their total loan portfolio.

Table 17.5 Estimates of cointegration relations between Variables

Variable	Coefficient	Standard error
Policy rate	0.829***	0.103
CAR	0.813***	0.161
ROE	−0.525***	0.042
GROWTH	0.046***	0.016
INFLATION	−0.298***	0.097
GEPU	−0.019***	0.003
REER	−0.008	0.019
C	5.259	5.151
CE = NPL(−1) − (0.828773*POLICYRATE(−1) + 0.813440*CAR(−1) − 0.525137*ROE(−1) + 0.045685*GROWTH(−1) − 0.298056*INFLATION(−1) − 0.018953*GEPU(−1) − 0.008167*REER(−1) + 5.258620)		

Source: Authors' Computation
Note: $*p < 0.10$, $**p < 0.05$, $***p < 0.010$

Bordo et al. (2016) argue that banks, in response to economic policy uncertainty, may reduce their credit growth to lower their risk exposure. These findings are also consistent with Hu and Gong (2019).

The results also show that growth has a significant positive relationship with bank credit risk in the long run. Inclusive growth can lead to structural changes in the economy, such as increased participation of small and medium-sized enterprises (SMEs) and individuals in economic activities. While this boosts economic growth, it can also mean a higher proportion of loans are extended to sectors or individuals with inherently higher credit risk profiles. Additionally, as the banking sector broadens its customer base, the overall systemic risk can increase if there is insufficient financial education and risk management. This broader base of borrowers might still be more vulnerable to economic shocks, leading to sustained higher credit risk for banks, as argued by Beck et al. (2014). These results also corroborate the findings of Loukoianova et al. (2018).

The results further indicate that the capital adequacy ratio has a positive long-run relationship with the non-performing loan ratio, while the return on equity and the rate of inflation (0.298) have significant negative long-run relationships with the non-performing loan ratio.

17.4.6 Short-Run Effect of Monetary Policy Rate and Global Economic Policy Uncertainty on Bank Credit Risk

Table 17.6 presents the short-run impacts of monetary policy and global economic policy uncertainty on the non-performing loan (NPL) ratio. The analysis indicates a negative and significant cointegrating term, confirming the existence of a long-run cointegrating relationship among the variables. This suggests that any short-run deviations from equilibrium will be corrected over time, reinforcing the stability of the long-term relationship. In the short run, an increase in the policy rate is associated with a reduction in the NPL ratio. Specifically, a 1% increase in the monetary policy rate is estimated to decrease the NPL ratio by 1.314%. This finding can be attributed to the fact that higher policy rates lead banks to adopt stricter credit eligibility criteria, thereby extending credit only to more creditworthy borrowers. Consequently, this reduces the proportion of non-performing loans in the short term.

Additionally, the results reveal a significant negative relationship between global economic policy uncertainty and the NPL ratio in the short run. An increase in global policy uncertainty is estimated to reduce the NPL ratio by approximately 0.017%. This counterintuitive result may suggest that during periods of high global uncer-

Table 17.6 ARDL short-run results

Variable	Coefficient	Standard error
Δpolicy rate	0.006	0.117
Δpolicy rate (−1)	−1.314***	0.173
ΔCAR	0.372***	0.091
ΔCAR (−1)	−0.233**	0.099
Δroe	−0.156***	0.031
ΔROE (−1)	0.305***	0.059
Δgrowth	0.017**	0.008
ΔGROWTH (−1)	−0.014	0.009
ΔINFLATION	0.040	0.048
ΔINFLATION (−1)	0.537***	0.079
ΔGEPU	−0.017***	0.003
ΔGEPU (−1)	0.001	0.002
ΔREER	0.007	0.022
ΔREER (−1)	−0.088***	0.023
COINTEG*	−1.129***	0.125
R-squared	0.913	
F-statistics	12.004***	
AIC	1.493	
Selected model	ARDL (1,3,3,4,2,3,3,4)	

Source: Authors' Computation
Note: *$p < 0.10$, **$p < 0.05$, ***$p < 0.010$

tainty, banks become more cautious and stringent in their lending practices, leading to a reduction in non-performing loans. Other variables, including return on equity, capital adequacy ratio, and real exchange rate, have a significant negative relationship with the NPL ratio in the short run. In contrast, the GDP growth rate and inflation have a significant positive relationship with the NPL ratio in the short run.

17.4.7 Shock Evolution of Non-performing Loans

Next, we estimate the shock evolution of the non-performing loan ratio in response to changes in the monetary policy rate and global economic policy uncertainty. Shock evolution paths measure the responses of a dependent variable to cumulative unitary positive shocks in distributed lag variables. As indicated by Shin et al. (2014), shock evolution paths can trace out the adjustment patterns as they evolve to converge to their cointegrating equilibrium states in the long run. Figure 17.1 illustrates the aforementioned discussion. Figure 17.1a shows that a 1% positive shock to the monetary policy rate increases the non-performing loan ratio after the first year (4th quarter), with some considerable short-run boosts in the first 9 months. The same trend is observed in Fig. 17.1b, which traces the evolution pattern of the non-performing loan ratio in response to a 1% positive shock in global economic policy uncertainty.

17.4.8 Diagnostics Tests

In Table 17.7, we perform regression diagnostics to test for model specification using the RAMSEY RESET test, normality using the Jarque-Bera statistic, heteroskedasticity using the Breusch-Pagan-Godfrey test, and serial correlation. The results indicate that our model does not suffer from model misspecification, normality issues, or serial correlation. However, errors have been found to be heteroscedastic.

We further assess the stability of our short-run and long-run models by computing the

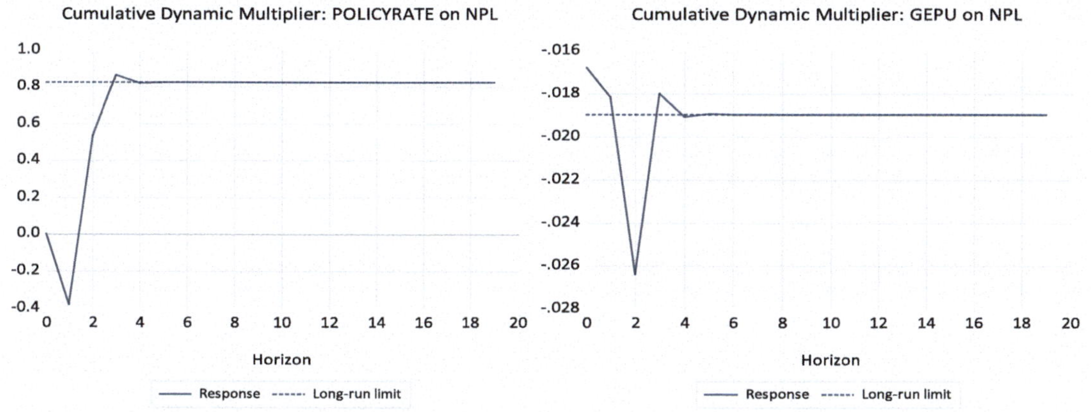

Fig. 17.1 Shock evolution of monetary policy and global economic policy uncertainty on non-performing loan. (*Source*: Authors' Computation)

Table 17.7 ARDL Test Diagnostics

Test	Statistic	Decision
RESET	0.141	Model is correctly specified
Jarque-Bera	1.534	Residuals are normally distributed
Breusch-pagan-Godfrey	1.896*	Heteroskedastic errors
Serial correlation	1.582	No serial correlation

Source: Authors' Computation

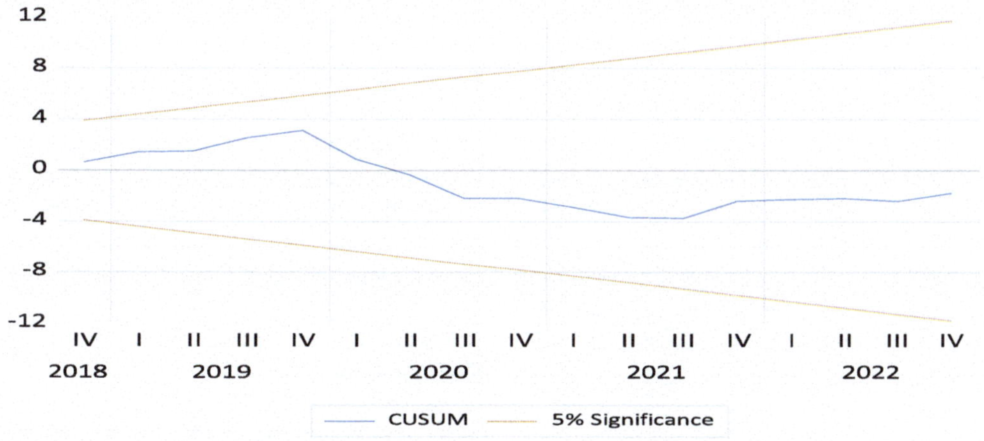

Fig. 17.2 CUSUM. (*Source*: Authors' Computation)

cumulative sum of squares (CUSUM) and the CUSUM of squares, as shown in Figs. 17.2 and 17.3, respectively. By comparing the CUSUM statistics to the 5% significance level, we demonstrate that the model does not suffer from stability issues in either the short run or the long run.

17.5 Conclusions and Recommendations

In this study, we examine the effect of monetary policy and global policy uncertainty on credit risk in Ghana using an ARDL bounds testing approach to cointegration. We use quarterly

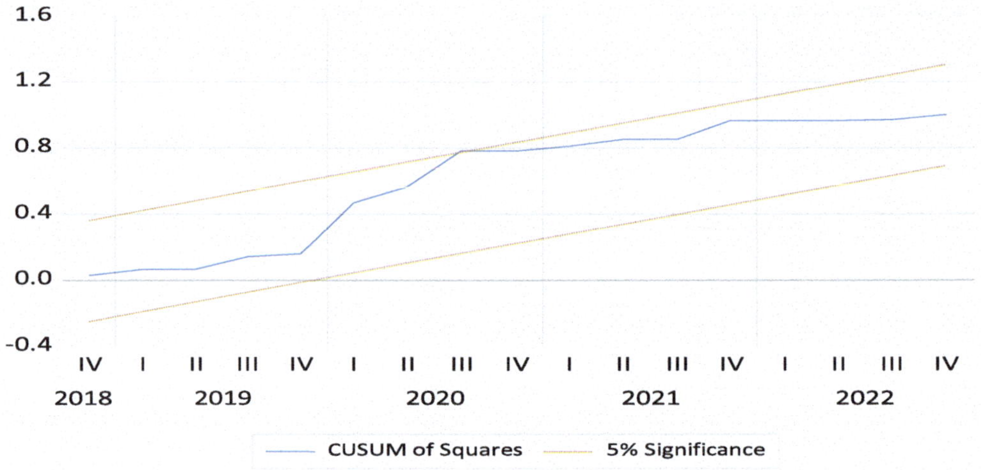

Fig. 17.3 CUSUM of squares. (*Source*: Authors' Computation)

aggregated data for the period from 2010Q1 to 2022Q4. Applying the ARDL bounds testing approach, we find evidence of a cointegrating relationship between monetary policy, global economic policy uncertainty, and credit risk, as measured by the non-performing loan ratio. However, we are unable to establish an asymmetric relationship between monetary policy and the non-performing loan ratio. The results indicate a positive long-run relationship between the monetary policy rate, growth, and the non-performing loan ratio, and a negative short-run relationship. We also find that global economic policy uncertainty negatively influences credit risk both in the long run and the short run. Other variables that significantly affect the non-performing loan ratio in both the long and short runs include the GDP growth rate, inflation rate, capital adequacy ratio, and bank performance (as measured by the return on equity).

Based on the findings of the study, the researchers propose the following: Given the observed positive long-run relationship between the monetary policy rate and the non-performing loan (NPL) ratio, monetary authorities should consider improving their communication strategies. Providing clear and unambiguous information can help banking institutions and borrowers anticipate and adjust to changes in monetary policy, potentially reducing credit risk. Banks should strengthen their risk management methodologies, specifically in evaluating credit risk. Meticulous assessments of borrowers, continuous surveillance, and efficient strategies to mitigate risk can aid in reducing non-performing loans. Continued diligence in monitoring banks' capacity to meet capital requirements and their performance is crucial. It may be imperative to implement more stringent regulatory initiatives to ensure that banks maintain sufficient capital reserves to withstand economic disruptions, thereby reducing the likelihood of credit deterioration. Further research into the specific mechanisms through which global economic policy uncertainty affects credit risk in Ghana could provide deeper insights and potentially lead to more targeted policy interventions.

References

Acharya, V., & Naqvi, H. (2012). Bank liquidity and bubbles: Why central banks should lean against liquidity. *Evanoff et al.*

Adrian, T., & Shin, H. S. (2009). Money, liquidity, and monetary policy. *American Economic Review, 99*(2), 600–605.

Adusei, M., Akomea, Y. S., & Nyadu-Addo, R. (2014). Predicting bank credit risk: Does board structure matter? *International Journal of Business & Finance Research (IJBFR), 8*(5).

Afanasyeva, E., & Güntner, J. (2020). Bank market power and the risk channel of monetary policy. *Journal of Monetary Economics, 111*, 118–134.

Ahir, H., Bloom, N., & Furceri, D. (2022). *The world uncertainty index* (No. w29763). National Bureau of Economic Research.

Alessandri, P., & Bottero, M. (2020). Bank lending in uncertain times. *European Economic Review, 128*, 103503.

Amuakwa-Mensah, F., & Boakye-Adjei, G. (2015). Determinants of non-performing loans in Ghana banking industry. *International Journal of Computational Economics and Econometrics, 5*(1), 35–54.

Amuakwa-Mensah, F., & Marbuah, D. (2017). Re-examining the determinants of non-performing loans in Ghana's banking industry: The role of 2007-2009 financial crisis. *Journal of African Business, 18*, 357–379. https://doi.org/10.1080/15228916.2017.1308199

Angeloni, I., Faia, E., & Duca, M. L. (2015). Monetary policy and risk taking. *Journal of Economic Dynamics and Control, 52*, 285–307.

Asamoah, LA., & Takyiwaa, D. A. (2015). *Determinants of credit risk of commercial banks in Ghana*. Research Gate. Retrieved February, 18, 2017 from https://www.researchgate.net/publication/293476975.

Asare-Bekoe, K.M. (2010). *A risk-based assessment of Ecobank Ghana Limited (EGH)* (p. 113).

Asiama, R. K., & Amoah, A. (2019). Non-performing loans and monetary policy dynamics in Ghana. *African Journal of Economic and Management Studies, 10*(2), 169–184.

Baker, S. R., Bloom, N., & Davis, S. J. (2016). Measuring economic policy uncertainty. *The Quarterly Journal of Economics, 131*(4), 1593–1636.

Barraza, S., & Civelli, A. (2020). Economic policy uncertainty and the supply of business loans. *Journal of Banking & Finance, 121*, 105983.

Beck, T., Demirgüç-Kunt, A., & Levine, R. (2007). Finance, inequality and the poor. *Journal of Economic Growth, 12*, 27–49.

Beck, T., Lin, C., & Ma, Y. (2014). Why do firms evade taxes? The role of information sharing and financial sector outreach. *The Journal of Finance, 69*(2), 763–817.

Berger, A. N., Guedhami, O., Kim, H. H., & Li, X. (2018). *Economic policy uncertainty and bank liquidity creation*. SSRN Scholarly Paper. Social Science Research Network.

Bonfim, D., & Soares, C. (2018). The risk-taking channel of monetary policy: Exploring all avenues. *Journal of Money, Credit and Banking, 50*(7), 1507–1541.

Bordo, M. D., Duca, J. V., & Koch, C. (2016). Economic policy uncertainty and the credit channel: Aggregate and bank level US evidence over several decades. *Journal of Financial Stability, 26*, 90–106.

Borio, C., & Zhu, H. (2008). Capital regulation, risk-taking and monetary policy: A missing link in the transmission mechanism? *BIS Working Papers, 8*, 236–251.

Borio, C., & Zhu, H. (2012). Capital regulation, risk-taking and monetary policy: A missing link in the transmission mechanism? *Journal of Financial Stability, 8*(4), 236–251.

Bruno, V., & Shin, H. S. (2015). Capital flows and the risk-taking channel of monetary policy. *Journal of Monetary Economics, 71*, 119–132.

Buch, C. M., Eickmeier, S., & Prieto, E. (2014). In search for yield? Survey-based evidence on bank risk taking. *Journal of Economic Dynamics and Control, 43*, 12–30.

Caglayan, M., & Xu, B. (2019). Economic policy uncertainty effects on credit and stability of financial institutions. *Bulletin of Economic Research, 71*(3), 342–347.

Cesa-Bianchi, A., & Rebucci, A. (2017). Does easing monetary policy increase financial instability? *Journal of Financial Stability, 30*, 111–125.

Chi, Q., & Li, W. (2017). Economic policy uncertainty, credit risks and banks' lending decisions: Evidence from Chinese commercial banks. *China Journal of Accounting Research, 10*(1), 33–50.

Cull, R., Demirgüç-Kunt, A., & Morduch, J. (2014). Banks and microbanks. *Journal of Financial Services Research, 46*, 1–53.

Danisman, G. O., Ersan, O., & Demir, E. (2020). Economic policy uncertainty and bank credit growth: Evidence from European banks. *Journal of Multinational Financial Management, 57*, 100653.

De Nicolò, G., Dell'Ariccia, G., Laeven, L., & Valencia, F. (2010). Monetary policy and bank risk taking. Available at SSRN 1654582.

Delis, M. D., Hasan, I., & Mylonidis, N. (2017). The risk-taking channel of monetary policy in the US: Evidence from corporate loan data. *Journal of Money, Credit and Banking, 49*(1), 187–213.

Dell'Ariccia, G., & Marquez, R. (2013). Interest rates and the bank risk-taking channel. *Annual Review of Financial Economics, 5*(1), 123–141.

Dell'Ariccia, G., Laeven, L., & Marquez, R. (2011, November). Financial frictions, foreign currency borrowing, and systemic risk. In *12th Jacques Polak Annual Research Conference, Hosted by the International Monetary Fund*, .

Dell'Ariccia, G., Igan, D., Laeven, L., & Tong, H. (2014). Policies for macrofinancial stability: Dealing with credit booms and busts. In *Financial crises: Causes, consequences, and policy responses*, pp. 325–364.

Demirgüç-Kunt, A., & Klapper, L. (2013). Measuring financial inclusion: Explaining variation in use of financial services across and within countries. *Brookings Papers on Economic Activity, 1*, 279–340.

DeYoung, R., Peng, E. Y., & Yan, M. (2013). Executive compensation and business policy choices at U.S. commercial banks. *Journal of Financial and Quantitative Analysis, 48*, 165–196. https://doi.org/10.1017/S0022109012000646

Diamond, D. W., & Rajan, R. G. (2009). The credit crisis: Conjectures about causes and remedies. *American Economic Review, 99*(2), 606–610.

Dingela, S., & Khobai, H. (2017). Dynamic impact of money supply on economic growth in South Africa. An ARDL approach.

Drakos, A. A., Kouretas, G. P., & Tsoumas, C. (2016). Ownership, interest rates and bank risk-taking in

Central and Eastern European countries. *International Review of Financial Analysis, 45*, 308–319.
Fiador, V. O., & Biekpe, N. (2015). Monetary policy and exchange market pressure–Evidence from sub-Saharan Africa. *Applied Economics, 47*(37), 3921–3937.
González-Aguado, C., & Suarez, J. (2015). Interest rates and credit risk. *Journal of Money, Credit and Banking, 47*(2–3), 445–480.
Gropp, R., Hakenes, H., & Schnabel, I. (2011). Competition, risk-shifting, and public bail-out policies. *The Review of Financial Studies, 24*(6), 2084–2120.
Gulen, H., & Ion, M. (2016). Policy uncertainty and corporate investment. *The Review of Financial Studies, 29*(3), 523–564.
Hovakimian, A., & Kane, E. J. (2000). Effectiveness of capital regulation at US commercial banks, 1985 to 1994. *The Journal of Finance, 55*(1), 451–468.
Hu, S., & Gong, D. (2019). Economic policy uncertainty, prudential regulation and bank lending. *Finance Research Letters, 29*, 373–378.
Ioannidou, V., Ongena, S., & Peydró, J. L. (2015). Monetary policy, risk-taking, and pricing: Evidence from a quasi-natural experiment. *Review of Finance, 19*(1), 95–144.
Klein, N. (2013). Nonperforming loan in CESEE: Determinants and impact on macroeconomic performance. *IMF working Papers, 13*(72).
Kwashie, D. G. (2013). Determinants of credit risk in the banking industry of Ghana. *Developing Country Studies, 3*(11), 64–77.
Louhichi, A., & Boujelbene, Y. (2017). Bank capital, lending and financing behaviour of dual banking systems. *Journal of Multinational Financial Management, 41*, 61–79.
Loukoianova, M. E., Yang, Y., Guo, M. S., Hunter, M. L., Jahan, M. S., Jamaludin, M. F., ... & Wu, M. Y. (2018). *Financial Inclusion in Asia-Pacific*. International Monetary Fund.
Maddaloni, A., & Peydró, J. (2012). The low monetary rates paradox, banking stability and credit: Evidence from the euro area. In *12th Jacques Polak Annual Research Conference* (pp. 1–39).
Mahrous, S. N., Samak, N., & Abdelsalam, M. A. M. (2020). The effect of monetary policy on credit risk: Evidence from the MENA region countries. *Review of Economics and Political Science, 5*(4), 289–304.
Morais, B., Peydró, J. L., & Ruiz, C. (2015). The international bank lending channel of monetary policy rates and quantitative easing.
Morris, S., & Shin, H. S. (2014). Risk-taking channel of monetary policy: A global game approach. *Unpublished working paper, Princeton University, 24*(8), 316–340.
Nguyen, C. P., Le, T. H., & Su, T. D. (2020). Economic policy uncertainty and credit growth: Evidence from a global sample. *Research in International Business and Finance, 51*, 101118.
Odhiambo, N. M. (2009). Electricity consumption and economic growth in South Africa: A trivariate causality test. *Energy Economics, 31*(5), 635–640.
Odonkor, T. A., Osei, B. A., & Sjö, B. (2016). Risk-taking, ownership and excess reserves in the Ghanaian Banking system. *Journal of Emerging Market Finance, 15*(2), 147–168.
Paligorova, T., & Santos, J. A. (2017). Monetary policy and bank risk-taking: Evidence from the corporate loan market. *Journal of Financial Intermediation, 30*, 35–49.
Pérez Juanazo, G., & Ruíz Martínez, M. (2016). Deposit insurance and moral Hazard in Ecuadorian credit unions.
Pesaran, M. H., Shin, Y., & Smith, R. J. (2001). Bounds testing approaches to the analysis of level relationships. *Journal of Applied Econometrics, 16*(3), 289–326.
Rajan, R. G. (2006). Has finance made the world riskier? *European Financial Management, 12*(4), 499–533.
Rauniyar, G., & Kanbur, R. (2010). Inclusive growth and inclusive development: A review and synthesis of asian development bank literature. *Journal of the Asia Pacific Economy, 15*(4), 455–469.
Sahay, M. R., Cihak, M., N'Diaye, M. P., Barajas, M. A., Pena, M. D. A., Bi, R., ... & Yousefi, M. R. (2015). *Rethinking financial deepening: Stability and growth in emerging markets*. International Monetary Fund.
Shin, Y., Yu, B., & Greenwood-Nimmo, M. (2014). Modelling asymmetric cointegration and dynamic multipliers in a nonlinear ARDL framework. *Festschrift in honor of Peter Schmidt: Econometric methods and applications*, 281–314.
Smithson, C., & Simkins, B. J. (2005). Does risk management add value? A survey of the evidence. *Journal of Applied Corporate Finance, 17*(3), 8–17.
Solarin, S. A., Shahbaz, M., Mahmood, H., & Arouri, M. (2013). Does financial development reduce CO_2 emissions in Malaysian economy? A time series analysis. *Economic Modelling, 35*, 145–152.
Sun, P. H., Mohamad, S., & Ariff, M. (2017). Determinants driving bank performance: A comparison of two types of banks in the OIC. *Pacific-Basin Finance Journal, 42*, 193–203.
Valencia, F. (2014). Monetary policy, bank leverage, and financial stability. *Journal of Economic Dynamics and Control, 47*, 20–38.
Xu, L., & Zia, B. (2012). Financial literacy around the world: an overview of the evidence with practical suggestions for the way forward. *World Bank Policy Research Working Paper*, (6107), World Bank.

Index

I. N. Yakubu (ed.), *Strategic Approaches to Banking Business and Sustainable Development Goals*, Sustainable Development Goals Series, https://doi.org/10.1007/978-3-031-80744-2

The manufacturer's authorised representative in the EU is Springer Nature Customer Service Centre GmbH, Europaplatz 3, 69115 Heidelberg, Germany. If you have any concerns regarding our products, please contact ProductSafety@springernature.com

Printed and bound by CPI Group (UK) Ltd, Croydon, CR0 4YY
01/10/2026
02225958-0020